Being Modern

Being Modern

*The Cultural Impact of Science
in the Early Twentieth Century*

Edited by

Robert Bud
Paul Greenhalgh
Frank James
Morag Shiach

First published in 2018 by
UCL Press
University College London
Gower Street
London WC1E 6BT
Available to download free: www.ucl.ac.uk/ucl-press

A CIP catalogue record for this book is available from The British Library.

ISBN: 978-1-78735-395-4 (Hbk.)
ISBN: 978-1-78735-394-7 (Pbk.)
ISBN: 978-1-78735-393-0 (PDF)
ISBN: 978-1-78735-396-1 (epub)
ISBN: 978-1-78735-397-8 (mobi)
ISBN: 978-1-78735-398-5 (html)
DOI: https://doi.org/10.14324/111.9781787353930

Foreword

History of science lacks organising narratives for the twentieth century. This is especially true when we widen the lens to the discipline's more-inclusive coterie: science, technology, engineering, mathematics and medicine. Mostly, we've chosen war as a narrative structure. Add imperialism. Add globalisation, though that seems simply to be imperialism by another name. We seek narratives that either describe or explain science's growing presence, resonance and (dare we suggest) hegemony across a plenitude of landscapes. Try as we might, these continue to prove elusive.

One viable choice engages the century's endlessly nuanced encounter with Modernity. Whatever Modernity is, or was, we seem certain science is somehow intimately associated. At once science seems causal for and caused by this thing, this philosophy, this miasma. Our quest to delineate precisely what and how has led us scholars towards ever more refined species of its genus. We seem to be getting somewhere, though the going is slow and the way is sometimes lost.

Being Modern shifts our perspective from observer to participant. The aim is to capture Modernity at work within mentalities, within cultural and biographical aesthetics, within the collisions between scientific and other things occurring in the lived experience of the people we study and from within their perspective. This anthology is a collective study of potency, infection and resistance.

The result is a refreshing alternative to scholastic delineations of movements seen from abstracting distances. This collection of original papers delivers richly researched, critical and thought-filled case studies of Modernity as an actor's category, observed in situ. It ranges across familiar and new settings. It certainly will help us as we build a better conceptualisation of the Modern both as project and product.

Joe Cain
Professor of History and Philosophy of Biology
Head of Department of Science and Technology Studies
UCL

Acknowledgements

The editors would like to thank the Arts and Humanities Research Council, which made possible the 2015 conference also entitled 'Being Modern' which underpins this collection, through Grant no. AH/L014815/1. The authors and editors are grateful to participants at this conference for their comments. We are also most appreciative to the Science and Society Picture Library of the Science Museum Group and to the Sainsbury Gallery for their offer of images without charge. The editors wish to express their gratitude to Dr Ellen Catherine Jones whose invaluable editorial assistance helped convert a group of separate papers into a coherent volume.

Contents

List of illustrations

Chapter 1

Contributors

Mitchell G. Ash is Professor Emeritus of Modern History and Speaker of the PhD programme 'The Sciences in Historical, Philosophical and Cultural Contexts' at the University of Vienna, Austria. He holds a PhD from Harvard University, and is a member of the Berlin-Brandenburg Academy of Sciences and Humanities and the European Academy of Arts and Sciences. He is author or editor of 16 books and more than 150 articles and chapters focussing on the sciences in political, social and cultural contexts in the nineteenth and twentieth centuries, the history of human–animal relations and the history of the human sciences.

Tim Benton is Professor of Art History (Emeritus) at the Open University, England. His recent books include *The Rhetoric of Modernism: Le Corbusier as Lecturer* (Basel, 2009), *Lc Foto: Le Corbusier: Secret Photographer* (Zürich, 2013) and a new edition of his book *The Villas of Le Corbusier and Pierre Jeanneret* (Basel, 2007). More recently he has been working with the Association Cap Moderne on the restoration of the villa E-1027, publishing a book *Le Corbusier peintre à Cap Martin* (Paris 2015) which was awarded the Prix du Livre de la Méditérrannée.

Tim Boon is Head of Research and Public History for the Science Museum Group and a historian and curator of the public culture of science. He is currently (part-time) Visiting Cheney Fellow at the University of Leeds. His published research (the books *Films of Fact* (2008) and *Material Culture and Electronic Sound* (co-edited with Frode Weium, 2013) and more than 30 papers) is mainly concerned with the history of science in documentary films, television, museums and, latterly, music. He has contributed to the exhibitions *Health Matters* (1994), *Making the Modern World* (2000) and *Oramics to Electronica* (2011).

Kevin Brazil is a Lecturer in Twentieth and Twenty-First Century Literature at the University of Southampton. He is the author of *Art, History, and Postwar Fiction*, forthcoming from Oxford University Press, and the co-editor of *Doris Lessing and the Forming of History* (2015).

Robert Bud is Research Keeper at the Science Museum in London. He has served at the museum as Head of Research (Collections) and as Keeper of Science and Medicine. He is a Fellow of the Royal Historical Society, past winner of the Bunge Prize awarded by the German Chemical and Physical Societies, and holder of the Sarton Medal and Sarton Professorship at the University of Ghent. He has published extensively on the histories of applied science, chemistry and biotechnology, including his books *The Uses of Life: A History of Biotechnology* (1994) and *Penicillin: Triumph and Tragedy* (2007). His current research is on the history of the concept of 'applied science' in the public sphere, over two centuries.

Nina Engelhardt joined the English Department at the University of Cologne as lecturer in English and American Literature in 2016. She received her PhD from the University of Edinburgh and has held research and teaching positions at the University of Edinburgh, the Institute for Advanced Studies in the Humanities, Edinburgh, and the Research Group 'Transformations' at the University of Cologne. She is author of the monograph *Modernism, Fiction and Mathematics* (Edinburgh University Press, 2018), and co-editor, together with Julia Hoydis, of the special issue 'Doing Science', *Interdisciplinary Science Reviews* 42, no. 3 (2017).

Craig Gordon's research and teaching focusses on the relationship between literature and science in early twentieth-century British culture. He is currently completing a monograph exploring the ways in which different forms of organicist thought constitute a crucial point of mutually determining interaction between the period's biological science, philosophical discourse and literary culture.

Paul Greenhalgh trained originally as a painter before organically drifting into being a writer/historian. He has worked in a number of countries, and is Director and Professor of Art History of the Sainsbury Centre for the Visual Arts at the University of East Anglia, UK. Previous roles include Director and President of the Corcoran Gallery of Art, Washington, DC President of NSCAD University, Halifax, Canada; Head of Research at the Victoria and Albert Museum, London. He has published many books and articles and is currently working on books about skill in the modern period, and the history of ceramics.

Michael Guida is a Research Associate and Tutor in the Department of Media & Cultural Studies at the University of Sussex. His research is concerned with the place of nature, experienced and imagined, in modern British culture. He has written about early twentieth-century ideas linking birdsong and emotion in *The Routledge Companion to Animal-Human History* (2018). He is completing a book about the cultures of listening to nature during and after the Great War.

Jeff Hughes is Honorary Senior Lecturer at the Centre for the History of Science, Technology and Medicine at the University of Manchester. His research focusses on the history of science in twentieth-century Britain, particularly how and why science developed as it did – institutionally and intellectually – over the last hundred years. He is currently researching the history of radioactivity and nuclear physics, including the work of Ernest Rutherford at Manchester and Cambridge, and the history of the Royal Society of London in the post-war and Cold War periods. His publications include *The Manhattan Project: Big Science and the Atom Bomb* (2002).

Frank James is Professor of the History of Science at UCL and at the Royal Institution. He has written widely on science and technology from the late eighteenth to the mid-twentieth century and how they relate to other areas of society and culture, for example technology, art, religion and the military. He edited in six volumes *The Correspondence of Michael Faraday* (IET) and wrote *Michael Faraday: A Very Short Introduction* (OUP) and is currently studying Humphry Davy's practical work. He has been President of both the Newcomen Society for the History of Engineering and Technology and the British Society for the History of Science and is now Chair of the Society for the History of Alchemy and Chemistry.

Esther Leslie is Professor of Political Aesthetics at Birkbeck, University of London. She has written extensively on Walter Benjamin. She published *Hollywood Flatlands* in 2002, on the intersections of critical theory and cartooning. Extension of these ideas into questions of representation and European history and politics appeared as *Derelicts: Thought Worms from the Wreckage* (Unkant, 2014). Latterly, interest in the proximities of poetics and science led to a study of the implications of the synthetic dye industry in Germany, *Synthetic Worlds* and *Liquid Crystals: The Science and Art of a Fluid Form* (both Reaktion Books). She runs a website: www.militantesthetix.co.uk.

Judi Loach trained in Architecture at the Architectural Association, London, then worked freelance in architectural journalism, exhibitions and commercial research before undertaking a PhD in architectural history at Cambridge University with a thesis on seventeenth-century Lyons. While researching that she became involved in the campaign for Le Corbusier's work at Firminy-Vert and worked on the Hayward Gallery exhibition *Le Corbusier: architect of the century* (1987) and has published widely on Le Corbusier. She has held academic posts as an architectural historian at Oxford Brookes and Cardiff Universities, and chairs in that but also in Early Modern and Modern European Cultural History at Cardiff, where she was also Director of the Research and Graduate School in Humanities.

Ruth Oldenziel (History PhD Yale 1992) is Professor of American-European history and innovation at the Technical University of Eindhoven. She has published and lectured widely in history at the intersection of American technology and gender studies. Currently, she is focussing on the history of cycling in a global context.

Annebella Pollen is Principal Lecturer in the History of Art and Design at the University of Brighton, where she researches craft, design, dress and photography across a range of periods and case studies. Her recent publications include *Mass Photography: Collective Histories of Everyday Life* (I.B. Tauris, 2015), *The Kindred of the Kibbo Kift: Intellectual Barbarians* (Donlon Books, 2015), the co-edited collections *Dress History: New Directions in Theory and Practice* (Bloomsbury, 2015) and *Photography Reframed: New Visions in Contemporary Photographic Culture* (I.B. Tauris, 2018).

Lewis Pyenson is Professor of History at Western Michigan University. Before returning to the faculty, he was Graduate Dean there and at the University of Louisiana at Lafayette, serving for ten years under four presidents and six provosts. He is completing a book about art and science in Modernity.

Morag Shiach is Professor of Cultural History in the School of English and Drama at Queen Mary University of London. Her publications include: *Cultural Policy, Innovation and the Creative Economy: Creative Collaborations in Arts and Humanities Research* (2017), *The Cambridge Companion to the Modernist Novel* (2007), *Modernism, Labour and Selfhood in British Literature and Culture, 1890–1930* (2004) and *Discourse on Popular Culture: Class, Gender and History in Cultural Analysis 1730 to the Present* (1989). Her current research is on language reform in the early twentieth century, immaterial labour, knowledge exchange with the creative economy and marginal modernisms.

Charlotte Sleigh is Professor of Science Humanities at the University of Kent and current editor of the *British Journal for the History of Science*. She has published widely on science and culture; her co-edited volume on twentieth-century science (with Don Leggett) is *Scientific Governance in Britain, 1914–79* (Manchester University Press, 2016).

Being Modern: Introduction

Robert Bud and Morag Shiach

At the beginning of the year 1901, *The Scotsman* newspaper peered into the future and, under the title 'Poetry of the Twentieth Century' reported:

> Two things we seem to see, that it will be an era of empire, or the struggle for it; an era perforce of larger national aggregations, and an era of scientific discovery, progressing in an accelerated ratio.[1]

Expectations of 'the era of scientific discovery' would become a commonplace in the new century. At a time when new findings in electricity and chemistry seemed to be revolutionising life, invocations of 'science' entailed much more than reference to esoteric knowledge and methodology. Cars, aeroplanes, telephones, 'talkies' and poison gas could all be included within its ambit at a time when use of the word 'technology' had not yet been standardised.[2] A 1941 study by Mass Observation reported that people 'look upon progress in the familiar and by now traditional light, as an amalgam of scientific invention, social improvement and increased knowledge and educational opportunity'.[3]

In line with contemporary ways of doing things and even of knowing, during the early part of the century the concept of 'being modern' captured a reinterpretation of aesthetics and even ethics. Thanks to radically new techniques in fields from dentistry to road transport, ways of life seemed to a new generation fundamentally different from those of the past. Repeatedly, the experience of the First World War would be drawn upon to emphasise the caesura. It is now some years since Robert Hughes argued that in art, literature, theatre, music, cinema, dance, architecture, design and beyond, there was a clear sense of a change of aesthetic which, although it had its beginnings long before the war, was able to flourish following the disintegration of long-standing social structures during and after the conflict.[4]

The arts drew upon the ideas, metaphors, symbolic meaning and practical potential of science. In widespread and rich discourses about and between sciences and other cultural areas, science was used variously as a coherent hegemonic authority and as an epistemologically and methodologically diverse category. It was also used to generate an important series of metaphors and images that supported experimental cultural practices.

The nature of this inter-relationship is the subject of this book. Contributors engage with a broad range of disciplines and a variety of contexts in which the notion of 'being modern' was deployed, and allied with conceptions of science. Respecting the 'modern' and 'science' as categories deployed by our historical actors, the disciplines discussed include psychiatry, physics, literary studies, mathematics and theology, while the contexts range from popular movements to science journalism, and from religion to electrical engineering. The volume is distinctive in its own interdisciplinarity, drawing on literature studies, the study of art and design, and the history of science and technology. It seeks to enable cross-disciplinary insights and understandings, drawing in contributors from a diverse set of disciplinary perspectives to interrogate a shared set of questions that offer insight into the experience of modernity and the relations of this to scientific discourse and practice. It also explores the engagement with and interpretation of science and technology by an unusually wide-ranging set of diverse actors ranging from elite writers and artists to Kibbo Kift, local politicians, radio hams, science fiction enthusiasts, historians of science and architects.

The modern

This volume interrogates the category of 'the modern' as an actors' category, teasing out the richness, the complexity, and the shared concerns of a range of articulations of the idea of 'being modern', particularly as it informed thinking about science and culture in the early years of the twentieth century, rather than proposing any higher-order analytic definition of 'modernity'. Such a methodological approach is commonly used within a range of disciplines, including sociology, history of science, and science and technology studies, but has been less common in interdisciplinary work on literature, culture and science. There are, of course, both advantages and disadvantages to such an approach. As Charles Rosenberg argued in 1988, 'An actor-oriented approach seeks to appropriate the individual in the service of transcending the individual and thus the idiosyncratic; it seeks to use an individual's experience as a sampling device for gaining an understanding of the structural and normative'.[5] This approach is designed to avoid anachronism, and to build a framework on which it is possible to move from the particular to the more general. But Rosenberg also highlighted the importance of generating a productive tension between actors' categories and more analytic and retrospective categories to the generation of historical understanding. Similarly, in 'Actors' and Analysts' Categories in the Social Analysis of Science', Harry Collins proposed that 'it be accepted that sociological explanation must begin with the perspective of the actor. The causes that give rise to anything that can be seen as consistent actions among actors turn on regularities as perceived by the actors first.'[6]

The twenty-first century digitisation of literature and the press enables us to gauge the rapid ascent of 'modern' as such an actors' category. Google Books Ngram Viewer offers some insight into the usage of words and phrases in a large corpus of books in English across decades or indeed across centuries. The word 'modern' saw a sharp increase in frequency between 1870 and 1930, with its annual usage doubling over these years. Its most common usage was in phrases that sought to capture the specific character of the contemporary world, with 'modern times' and 'modern world' being the most frequent uses between 1870 and 1950, while 'modern life' was overtaken in frequency by 'modern science' only in 1940. The term 'modern science' itself had seen a certain currency as a phrase in the late sixteenth century, but then more or less disappeared until the early nineteenth century. It then grew steadily in usage throughout the nineteenth century, reaching peak frequency of usage in 1930. A cognate term such as 'modern physics' saw its usage increase significantly between 1880 and 1920, and grow five-fold between 1920 and 1960; while 'modern mathematics' increased in usage significantly between the 1890s and the 1960s, but declined rapidly after this date. The phrase 'modern poetry' tripled in frequency of usage between 1910 and 1960, while 'modern industry' was most frequently deployed between 1915 and 1950. Usage of 'modern' in all these combinations declined significantly in the second half of the twentieth century, suggesting that it became less useful in capturing what felt to people like the most significant aspects of culture, of economy and of science in that later period. The British Library sponsored 'British newspaper archive online', which covers a large number of regional newspapers, provides a comparable insight. 'Modernity', a term known but hardly used in the 1870s, was deployed over two thousand times, across this corpus, in the first decade of the twentieth century.

The association between science, modernity, modernism and progress was envied, drawn upon and exploited.[7] The material presented here builds on the many studies of the interactions between science and other aspects of culture. Modernity too has been the subject of a rich literature, and the distinctive American context has already been the subject of many studies.[8] Scholars have, for instance, explored how this nexus was expressed in the World Fairs of the period.[9] We have also drawn upon the scholarship detailing the interest of modernist writers in the relations between literary writing (and creativity more broadly) and science.

Scholarly thinking and writing about the relations between literature and science have been driven by different epistemological, historical and aesthetic impulses since the early years of the twentieth century. The institutionalisation of thinking about the relations between science and literature increased over the century, with a steady growth in the number of learned societies, research centres, and specialist degree programmes in these fields. Critical and theoretical work on the relations between literature and science became important and influential in Britain in the 1980s, with studies such as Gillian Beer's *Darwin's Plots* (1983) and Sally Shuttleworth's *George Eliot and Nineteenth-Century Science* (1984) providing new methodologies for thinking about the ways in which scientific inquiry shaped forms of narrative.[10] Scientific writing and practice offered rich metaphorical possibilities; and scientific thought offered new understandings of historical time for nineteenth-century writers, particularly novelists. This kind of exploration was then further developed in the late-twentieth and early twenty-first centuries in a number of scholarly works that focussed on early twentieth-century scientific innovations, addressing in particular new work in Physics. Examples of this are Daniel Albright, *Quantum Poetics: Yeats, Pound Eliot, and the Science of Modernism* (1997), Michael Whitworth, *Einstein's Wake: Relativity, Metaphor and Modernist Literature* (2002) and Katie Price, *Loving Faster than Light: Romance and Readers in Einstein's Universe* (2012).[11]

Such work had its theoretical and critical roots in literary studies, albeit very significantly influenced by the critical theory that had become so important for that discipline by the 1980s, as well as by the method-ologies of cultural history and cultural studies. This literary grounding remains very significant for the majority of work addressing science and literature today. Looking at the exciting range of research societies, centres and journals focussed on this field, such as the *Journal of Literature and Science*, the British Society for Literature and Science, or Literature and Science, Oxford, it is notable that while they express a clear and important commitment and enthusiasm for promoting 'interdisciplinary research into the relationships of science and literature in all periods',[12] the majority of the editors of, contributors to, and participants in their activities come from the discipline of literary, and specifically English, studies. They are however moving into fields which could profitably engage other disciplines.

The breadth of approach now called upon is illustrated by recent research on modernism and science which has focussed particularly on the social, cultural and artistic impacts of new communication technologies. For example, Friedrich Kittler, a Professor of Aesthetics and Media Studies, has produced an important and influential body of work that draws on discourse analysis, media theory, psychoanalysis and history to consider how modernity is shaped by technology. In a volume such as *Gramophone, Film, Typewriter* he examines the ways in which subjectivity, artistic practices and social forms of organisation are shaped by the developing technologies of communication. Thus, for example, he writes that:

> *Mechanization Takes Command* – Siegfried Giedion could not have come up with a better title for a book that retraces the path from Marey's chronophotographic gun via modern art to military-industrial ergonomics. The automatized weapons of world wars yet to come demanded similarly automatized, average people as 'apparatuses' whose motions – in terms of both precision and speed – could only be controlled by filmic slow motion.[13]

Kittler draws tight, and compelling, links between technologies, cultural forms and individual subjectivity that create a powerful but constraining sense of the nature of modernity. He later writes that, 'films are more real than reality, and that their so-called reproductions are, in reality, productions. A psychiatry beefed up by media technologies, a psychiatry loaded with scientific presumptions, flips over into an entertainment industry'.[14] Drawing on the theoretical insights of Heidegger and Lacan, and the technological innovations of Edison and Turing, however, Kittler offers a model of history that, despite its subtlety and sophistication, can be seen as amounting to a form of technological determinism. Contributors in this volume challenge such determinism in a variety of ways, while acknowledging the psychic and social significance of science and technology for early-twentieth-century thinkers. They do this by focussing on the negotiations that take place between science, technology, and culture, which generate never fully resolved tensions and productive sites of resistance and of imagination.

A more recent and important study of the interconnections between modernity and technology can be found in *Moving Modernisms: Motion, Technology and Modernity* (2016). This collection of essays focusses on the idea of 'movement' as expressed in relation to space, place, energy, mathematics, cinema, cycling and urban transport. Movement, the editors argue, 'becomes definitional of modernity', and contributors to the volume also argue that 'new technologies of transportation, communication, and representation in the urban context' are central to understanding this relationship.[15] Again, the contributors to *Moving Modernisms* all write broadly from within the space of literary studies, but bring to that field a strong sense of the ideas, metaphors and images that were being generated in other disciplines in the modern period.

As the reference to the work of Giedion makes clear, historians of design have long been concerned with the meaning of science and technology.[16] The *Journal of Design History* was established in 1988 and its second article dealt with the theory of machine design in the second industrial age.[17] The idea of 'the modern', 'modernity' and modernism run strongly through this journal too. In 1998, a special issue devoted to 'Craft, Modernism and Modernity' discussed the tension between the modern movement, mass production and handwork. In the introduction, Tanya Harrod provides a context from the history of the visual arts; however, she is dealing also with the tension between the attractions and threats of modern life and modern means of production.[18] Such arguments can be linked to discussions elsewhere within the broader space of 'history', and need locating within common cultural frameworks.

This book

Within Europe, as Daunton and Rieger explained in the first lines of their 2001 *Meanings of Modernity*, 'The study of British modernity is in its infancy when compared with the prominence of this field in other national historiographies.'[19] This volume will go some way to building understanding of the British case in an international context. It is therefore important that it does also include several chapters dealing with German, French and Austrian topics. Far from emphasising a British exceptionalism, this volume will be conducive to interpreting the British experience within an understanding of broader European conversations.

In the past, the disciplinary diversity that has always underpinned approaches to thinking about the relations 'between science and the arts', as well as investigations into the history of science, has led to separation. Now there are new opportunities, and we have sought to benefit from a confluence of scholars from disciplines that do not always talk to each other in depth or productively, drawing upon diverse disciplinary approaches and interdisciplinary methods.

Contributors do seek to develop insights into both science and culture in the early twentieth century by beginning with the ways in which writers, artists, designers, scientists, journalists and other relevant actors sought to understand 'being modern'. But they also seek to develop an understanding of the significance of 'consistent actions' and 'regularities' among actors, that are necessary to a broader historical understanding of the relations between science, culture and 'being modern'.

In their different ways, the chapters in this book deal with a problematised view of science. Tensions between the modern indus-trialised science and a postulated purer or more innocent past science are explored. The alternative attractions of mathematics and physics on the one hand, and biology and living shapes, on the other, for writers and designers provide the stimulus for authors. For the contributors have shared a common interest in understanding how the centrality of science in twentieth-century culture generated a range of ideas, images and practices that shaped what it meant to be 'modern'.

As Oldenziel argues in Chapter 12, 'what being modern meant was contested and unstable' and 'how-to-be-modern could be declassed, reclassed, and recycled'. She argues that only by paying close attention to local discursive and social contexts can the semantic and social complexity of 'being modern' be grasped: 'focussing on the bicycle let us view what being modern has meant, how it was contested, and who controlled its narrative'. Jeff Hughes points out in Chapter 4 that 'the modern' does not designate an 'epistemologically uncontentious or ontologically fixed' category.

Yet our understanding of the flexibility of the term need not undermine our recognition of its potency. In Chapter 1, Mitchell G. Ash deals with Vienna, perhaps the single site most associated with the shock of 'modernity'. From visual arts to psychiatry, engaging with the 'modern' was the ambition, and thinking about science was a central part of the means. Ash argues that there were affinities and even linkages between 'modern ways of thinking about science' and the radical development of the visual arts. The significance of technological modernism, he argues following Mehrtens, 'presupposes a concept of knowledge based less on self-referential abstraction than on what can be done with, or to, nature as well as other human beings'. The plurality of the 'modern' extended even into the sciences themselves, because what counted as 'modern' was different, often fundamentally so, in different disciplines. In Chapter 9, Pyenson also builds upon Mehrtens in his study of the hugely influential mathematician Felix Klein. Amongst his many legacies, as well as the Göttingen Institute of Applied Physics, was the building and use of mathematical models for teaching (he may be best remembered today as the conceiver of the four-dimensional Klein bottle). Pyenson explores the role of the work of the Italian sculptor Umberto Boccioni as a mediator between the models of Klein and the cubism of Picasso. In Chapter 8, Nina Engelhardt, dealing with Musil and Zamyatin, looks at how these writers explore implications of new mathematics for literary fiction. Concepts of Life, as deployed by the German writer Alexander Döblin, are explored by Esther Leslie in Chapter 16.

Conceptions of what science is like would, repeatedly, serve as a reference. Charlotte Sleigh, in Chapter 7, explores the science fiction writing of engineers. She points out, 'Science, for the fans, was a generalised toolkit that allowed one to have one's say, to apply general technical skills to any area of culture'. Yet science too was contested. The breadth of the meanings of 'science' for writers in the early twentieth century is a key concern for contributors to this volume. They show how the term could denote a general sense of abstract and structured thinking, or suggest a particular scientific discipline that was seen as demonstrating the characteristics of modernity with significant force, or indeed invoke popular and applied forms of scientific knowledge as central parts of their sense of modernity.

Thus, for example, in his 1920 essay on Dante, T.S. Eliot refers to 'the greater specialization of the modern world' and connects this to his cultural and historical argument about the 'dissociation of sensibility'.[20] Here the experience of modern culture and modern science is one of fragmentation and specialisation, which act as inhibitors rather than producers of significant understanding. Strength and solidity are also ascribed to modern science by Hugo Münsterberg, the organiser of a Congress of Arts and Sciences that was part of the St Louis World Fair held in 1904, as Kevin Brazil observes in Chapter 4. Münsterberg described his aim in organising the Congress as being to offer a synthesis of the 'specialization which makes our modern science and scholarship solid and strong'. As Brazil points out here, the sense that specialisation was what made both science and culture 'modern' can also be found in various writings by Eliot, for whom this was understood broadly as a negative element of 'being modern', which he associated with a loss of the scientific and cultural sensibility of earlier periods. As Virginia Woolf polemicises about modern fiction as a necessary counter to the weight and solidity of naturalist novels, or Eliot develops his arguments about modern tendencies in poetry (writing for example in his essay 'Tradition and the Individual Talent' that 'it is in depersonalization that art may be said to approach the condition of science'), so we can, as this volume so clearly demonstrates, map what is at stake for them in the question of 'being modern'.[21]

By contrast with Eliot's sense of fragmentation across culture and the common thread of specialisation, two articles here deal with Le Corbusier, the best-known architectural exponent of modernism. Judy Loach in Chapter 10 examines his concept of 'purism', his distillation in a single word of the 'modern spirit'. Loach here is dealing with the young Corbusier enamoured of geometry and lines that underlay both science and art. In the final chapter, 17, Tim Benton looks at how Corbusier later sought to reconcile with this geometrical Nature, a more messy nature, which Benton distinguishes with a lower-case 'n'.

This was an era in which the scope of science was expanding. The doyen of British science between the wars, Lord Rutherford, was however famous for distinguishing between what he saw as the only two sorts of science, physics and stamp collecting.[22] This frequently cited aphorism can be seen as the arrogant self-affirmation of the Director of the Cavendish Laboratory, but also as the characteristic defence of the qualities of a private club at the time of its expansion to a rich and diverse movement. As Michael Guida shows in Chapter 13, the naturalist's recording of birdsong won a wide public following. Eliot famously argued in 1919 that 'poetry *is* a science', and that a mature poet 'works like the chemist,'[23] while Ezra Pound insisted on the significance for the modern creative artist of the cultural and imaginative changes he associated with his contemporary scientific world:

> The mind, even the musician's mind, is conditioned by contemporary things, our minimum, in a time when the old atom is 'bombarded' by electricity, when chemical atoms and elements are more strictly considered, is no longer the minimum of the sixteenth century pre-chemists.[24]

There has been significant discussion of the meaning of Eliot's claim, and Kevin Brazil argues here that for T.S. Eliot reference to 'science' might invoke 'not the natural sciences, but the disciplines which have come to be called in the English-speaking world the social sciences'. In his 1926 *Science and Poetry*, I.A. Richards asked how our estimate of poetry is going to be affected by science, and argued that 'if only something could be done in psychology remotely comparable to what has been achieved in physics, practical consequences might be expected even more remarkable than any engineer can contrive'.[25] He further argued that 'the first positive steps in the science of mind have been slow in coming, but already they are beginning to change man's whole outlook'. Richards is privileging psychology in his representation of 'science' and its impacts on poetry, while for other literary figures disciplines such as physics, mathematics or, on the other hand, the studies of biological organisms and of life, were key to their conceptions of the new aesthetic of modernism.

From the end of the nineteenth century, issues raised by the phenomena of 'life' raised increasing interest. Organicism emphasised the emergent properties of the whole as greater than merely the interaction of the parts and the significance of biological form. The work of Donna Haraway on crystals, of Oliver A.I. Botar and Isabel Wünsche on biocentrism, Smith on organicism and Huxley, and, more recently of such scholars as Juler and Esposito, has explored the stimulus given to both science and the arts.[26] The approach of the naturalist Julian Huxley, known for his contribution to the integration of Mendelian Genetics and Darwinian natural selection was underpinned by his love of nature, and he has been described as fundamentally a 'vitalist'. His greatest book was the study of a bird, the crested grebe.[27] Rather than privileging any particular interpretation of science, this book is organised around the competing attractions of the physical and the organicist sciences.

From a distance, it might have seemed obvious that science and modernity should go hand in hand in a spirit of mutual support. This has been suggested by the constant contemporary refrains of politicians, scientists and engineers that we should look to the future, and that our continuing prosperity demands a scientific knowledge base. But the success of such rhetoric depends to a large extent on endowing the term and concept of modernity with a positive progressive aura and this has not always been the consensus. Mass Observation introduced its 1941 report on everyday attitudes to science with its dominant impression that people felt 'science has got out of control'.[28]

Interpreters of science could cope by redefining the reference of the term from the urgent, contemporary and scary to the beautiful, enduring and old. In Chapter 6, Frank James cites the views of Rupert Hall, who was appointed to a lectureship in the History of Science at the University of Cambridge in 1948, as found in his influential study, *The Scientific Revolution 1500–1800: The Formation of the Modern Scientific Attitude*:

'[science] is the one product of the West that has had decisive, probably permanent, impact upon other contemporary civilizations. Compared with modern science, capitalism, the nation-state, art and literature, Christianity and democracy, seem regional idiosyncrasies, whose past is full of vicissitudes and whose future is full of dark uncertainty'. In an age that had just seen the use of atomic weapons and of gas in the concentration camps, Hall had sought to rescue the reputation of science by disconnecting it from modernity.[29] Bud too, in Chapter 5, deals with the competing images of science as urgently contemporary, expressed in the design of the Hornsey Town Hall complex and the appeal to deep time in the Museum of History of Science in Oxford.

The chapters[30]

The book is structured in four sections. The first section deals with 'science, modernity and culture'. It includes four chapters. The first by Ash looks at the relationships between scientific and artistic modernism in Vienna, the heartland of both movements. Two theses are advanced: first, that there are certain affinities, and in some cases actual linkages, between the breakthrough to modern ways of thinking in the natural sciences and mathematics and the radical changes in the arts that occurred at the same time; and second, that modernism, and hence 'cultural modernity' in these fields was nonetheless fundamentally plural and not the "totalizing project" that postmodernist thinkers have claimed that it was. Modernism in the sciences, as in the arts, often involved a break with pictorial representation of nature and a turn toward giving free play to abstraction and theoretical imagination. Alongside this, Ash points to technological modernism, which presupposed a concept of knowledge based on what can be done with nature as well as human beings. As he argues, these two styles of modernism in science were not fundamentally opposed, but in certain respects deeply related to one another.' Boon then, in Chapter 2, looks at an exemplary case of convergence between scientific and artistic modernism. He deals with a different and completely modern art form: the sound track of a film. The cinema itself was a new and apparently science-based art form and synchronised sound only became routinely possible around 1930. His chapter argues that filmmakers in the first decade of sound on film used sound to represent industrial modernity. 'Taking the contrasting examples of Paul Rotha's 1935 British Documentary *The Face of Britain* and René Clair's 1931 French feature, *À Nous la Liberté*, it listens to their respective soundtracks and situates the sonic practice of the directors and those responsible for the soundtracks within the debates of the period.' Shiach follows in Chapter 3 by exploring in more detail the ways in which three modernist writers engaged with scientific ideas and deployed explicitly scientific metaphors in the year 1919. She offers 'new insights into the extent to which, at this particular historical moment, the theorisation and the creation of what was understood as "modern" writing happened in the interstices between science and literature. The writers discussed are T.S. Eliot, Virginia Woolf and Dorothy Richardson, and the analysis of their texts engages with the metaphor of "the atom", and "the catalyst" and the idea of "waves of light".' In Chapter 4 Brazil looks in more detail at the work of Eliot. He focuses on a different meaning of 'science', central to Eliot's graduate studies in philosophy – science in the sense of an academic discipline.

> Tracing Eliot's engagement with issues of disciplinarity in relation to fin-de-siècle debates about the epistemological foundations of the social sciences, and situating them in relation to broader theories of disciplinarity in modernity, this essay argues that it was science's disciplinary status as a social form that made it an ambiguous model of emulation for modernist poetry.

The second section looks at tensions over science. Strikingly, in rather different chapters, a few key figures appear, including William Inge, the Dean of St. Pauls, and the writer C.S. Lewis. Although none would commonly feature in a history of science, as agents and as targets both were important. Bud deals in Chapter 5 with the circulation of the applied science and its use as a rich and contested concept in a range of discourses. He looks at the role in 'dissemination of a few national politicians such as R.B. Haldane and such well-known writers as H.G. Wells, but also of local figures such as Alderman Moritz, responsible for the selection of the pioneering modernist design of the Hornsey town hall complex.' The term came alive too through the arguments of the 'other side' concerned with the impact of applied science on the development of inhuman weapons, of which gas was the archetype, on culture and unemployment. In Chapter 6 James looks at the same disagreements to argue that

> history of science in Britain began as a response to issues raised in the inter-war years by the potential role of science in the future development of society and culture. Authors with wildly different views and agendas, for instance Dean Inge and Aldous Huxley,

predicted far future utopias and dystopias that contemporaries found both disturbing and damaging to the contemporary status of science. To circumvent such issues scientists and historians with similar outlooks began studying seventeenth-century science in the belief that this illustrated that science was an independent epistemic entity and its cognitive content therefore had no relation to other areas of culture and society.

Sleigh's contribution in Chapter 7 shifts attention to

> the work of young science fiction fans in 1930s Britain and their creation of home-made magazines as a means of participating in the mediatised era of knowledge and culture in the interwar period. It asserts that science was a flexible cultural resource from which they attempted to assemble their identity, and most especially that this assemblage needs to be understood as an exercise in literary criticism.

The third and fourth sections deal with two competing scientific modes used as metaphors and agency in the arts. They deal, respectively, with mathematics and physics on the one hand, and, on the other, with biology and the natural form. The third section, 'mathematics and physics' contains five chapters. Engelhardt and Pyenson (Chapters 8 and 9) deal with mathematics as a metaphor. Engelhardt treats 'Austrian author Robert Musil and Russian writer Yevgeny Zamyatin who draw on the period of redefinition and the modern(ist) notion of mathematics in their attempts to develop new literary forms with which to respond to the pressures of modernity'. Both authors set out their visions for a mathematically inspired literature in their essayistic work as well as in fictional texts. Engelhardt's essay 'examines both in order to explore the role of mathematics in their ideas of being modern and writing modern literature'. Pyenson interprets Picasso's 1914 sculptures called *Glass of Absinthe* in

> the light of mathematician Felix Klein's promotion of plaster models for abstract mathematical surfaces, notably the non-orientable surface known as a Klein bottle, which received attention as a popular, mathematical curiosity. By the first decade of the twentieth century, scores of mathematics faculties and schools had extensive collections of Klein's plaster models, notably the Paris Conservatoire des Arts et Métiers. The Klein bottle, like Picasso's vessels and a related sculpture by Umberto Boccioni, is unable to contain liquid.

In Chapter 10, Loach is concerned with architecture as art, and with the role accorded to science in the aesthetic rather than technological aspect of buildings, in the avant garde around the time of the Great War. Her chapter seeks to explain

> why science, and most overtly mathematics, played a key role in modern architectural theory, and how it did so after being mediated through theories developed in other art forms, most notably painting but also music; and that it did so, at least in part as a result of personal association with artists from other genres, through a wider range of scientific theorising, notably psychological theories of sensual perception and its mental impact. It focuses on the work of the most influential theorist and designer of the period, one working across the fields of fine art and architecture: Le Corbusier. He found, in Parisian avant garde circles at the end of the Great War, a preoccupation with science – most explicitly mathematics – which offered artists a way of vindicating the importance of their contribution to, and thus validating their own position within, a society that saw science as the inevitable means for 'becoming modern'.

The two final chapters of this section are concerned with the techno-logical aspect of science. Focussing on the Cavendish Laboratory, Cambridge, Hughes shows in Chapter 11 how

> the electronics skills of young wireless enthusiasts were channelled and mediated by the Cambridge University Wireless Society, which acted as a space for the sharing of materials and practices and the development of members' wireless technique and ideological values. Wireless and the growth of organised broadcasting in the 1920s transformed the public sphere and many aspects of social, economic and political life, as well as material, spatial and sonic cultures. Embedded in a network created by the University Wireless Society – ranging from the BBC and the General Post Office to Marconi, EMI and other characteristically modern industries, the Cambridge University Officers Training Corps and national military and civil research establishments – these researchers developed skilled wireless techniques that cut across the boundaries between civil and military, university and industry, and 'pure' and 'applied' research. Their practices simultaneously shaped both the development of military field communications and research at

the Cavendish Laboratory, including ionospheric research and Lord Rutherford's experiments in nuclear physics. The chapter thus proposes a radical new framing for interwar nuclear science and its modernity.

Oldenziel is concerned in Chapter 12 with competing modernities between two novel technologies of the early twentieth century: the bicycle and the car.

> By the eve of the First World War, the thriving literary production (cycling travelogues), the explorations of experimental artists (Futurists), and the sponsorship of commercial art (advertising posters) had helped to establish cycling as truly modern. The same market players, civil-society actors, and experts once involved in the bicycle business, shifted to the car industry in the 1910s – and with them the discourse of modern mobility. The debates on modern always implied opposites: 'fast' versus 'slow' traffic; motorized versus non-motorized circulation; flow and disruption; modern versus old fashioned – in short, a distinction between future and past. Mobilizing modernity is as much about one's identity in the present as an aspiration for the future – about the road ahead beyond the horizon. Building that road was not just a metaphor, but also a reality – including the question of who had the right to ride on it.

The final section deals with the interest in 'life, biology and the organicist metaphor'. As good naturalists ourselves, we might note the recurring presence in several of the accounts of Julian Huxley, a leading public intellectual of the interwar years. In Chapter 13, Guida deals with the work of the pioneering natural history broadcaster, the German refugee Ludwig Koch, and his unique British birdsong broadcasts. Though these have been 'little considered for their place and purpose in the crisis of war', Guida argues

> that Koch's broadcasts provided an emotional sustenance that was quite distinct from what could be derived from the BBC staples of talk and music. Koch's work was promoted and supported by several leading naturalists and ornithologists, Julian Huxley and Max Nicholson in particular. They agreed with Koch that birdsong was a universal human pleasure that anyone could experience, without training or knowledge.

Guida considers 'how listening to birdsong on the radio can be conceived of as an act of citizenship through knowing and relishing the sound signatures of the nation's bird heritage'.

Nature was of course widely seen as dynamic, and evolutionary theory provided a model for how change occurred. In Chapter 14, Pollen explores the ways that 'popular scientific ideas about life force, degeneration, cultural evolution and the biogenetic law were disseminated and incorporated into the symbols, philosophies and practices of experimental woodcraft campaign groups in interwar England.' She draws on works that assess the intersection of scientific ideas in application, from Beer and Stephen Gould to Oliver A.I. Botar.

Evolution was one aspect of biological thinking, organicism was another. This has been frequently associated with the authoritarian consequences of neo-Kantianism. In Chapter 15, Gordon argues that the organicism articulated by interanimation of early twentieth-century biological, philosophical and literary discourse provides significant resources for reassessing modernist attempts to theorise the role of aesthetics in mediating different conceptions of individuality and forms of socio-political organisation. He pursues this claim through a detailed exploration of the relationship between Whitehead's *Science and the Modern World*, the organicist biology of Needham and Bertalanffy, and the literary writings of Lewis and Lawrence. In Chapter 16, Leslie takes a look at competition between mechanical and organicist metaphors in chemistry and literature by studying Alfred Döblin's *Our Existence*.

> Its major concern is to categorize life, and to categorize it in relation to questions of crystal formation and the roles of seas in the origins of life. It focuses the debates about life, as mechanical, chemical and physical force, that were extant from the late nineteenth century, more popularly than institutionally received. This essay sets the ideas on life as liquid and crystal in relation to marginalised science of the time.

Finally, Benton addresses some of the fundamental issues of the book in Chapter 17, looking again at Le Corbusier. While in the previous section Loach reflected on the young Corbusier intoxicated by mathematics, Benton looks at a competing strain within his thought.

> From his earliest training at the arts and crafts school at La Chaux-de-Fonds, Switzerland, the young Charles-Édouard Jeanneret was trained to believe that all beauty derived from nature. Through to

the end of his career, nature remained a fundamental issue: its infinite variety and mysterious laws of growth and decay had to be understood. The paper argues that for Le Corbusier there were two forms of nature. Geometry belongs to an order of Nature – let us give this order a capital 'N', from which the so-called Laws of Nature are derived. Absolute, unchanging, universal, the Laws of Nature are the product of human reason and are thought to derive from some higher order. Modern architecture derived from geometry – 'Nature' –, but nature – the wet, messy, sensual and gratifying kind – was also essential for human satisfaction. The paper is centrally concerned with how Le Corbusier sought to reconcile nature and Nature.

This book is distinctive, therefore, in the conjuncture of specialists from a variety of disciplines addressing the interpretation of the role of science in the search for 'modernity' in the arts and culture of the early twentieth century. They show how the meaning of such concepts as Nature, the organic, specialisation, speed, applied science and 'science' itself were explored and deployed to address what seemed to be a new world. Sensitive to the contestation of these categories at the time, the contributions offer rich interrelationships rather than simple generalisations. The volume is offered as evidence of the potential of such studies, and as an inducement to further multidisciplinary historical work on the relationships of science and art.

Notes

1 'The Poetry of the Twentieth Century', *The Scotsman*, 21 January 1901.

2 Before the Second World War the use of the term 'technology' was still unstable. It differed between the UK and the US. For the US, where MIT was influential, see Eric Schatzberg, 'Technik Comes to America: Changing Meanings of Technology before 1930', *Technology and Culture* 47, 3 (2006): 486–512. In Britain references to Imperial College were remarkably prominent. Of 2639 uses of the word 'technology' in *The Times* between 1900 and 1939, more than 20 per cent (582) were in relation to references to 'Imperial College of Science and Technology'.

3 Mass Observation, 'Report on Everyday Feelings about Science' (October 1941),

no. 951. Reproduced with permission of Curtis Brown Group Ltd, London on behalf of The Trustees of the Mass Observation Archive © The Trustees of the Mass Observation Archive.

4 Robert Hughes, *The Shock of the New: Art and the Century of Change* (London: British Broadcasting Corporation, 1980).

5 Charles R. Rosenberg, 'Woods or Trees? Ideas and Actors in the History of Science', *Isis* 79 (1988): 564–70, on 569.

6 Harry Collins, 'Actors' and Analysts' Categories in the Social Analysis of Science', in *Clashes of Knowledge: Knowledge and Space*, eds. Peter Meusburger et al. vol 1. (Dordrecht: Springer, 2008).

7 See for example, Philip Brey, 'Theorizing Modernity and Technology,' in *Modernity*

and Technology, eds. Thomas J. Misa, Philip Brey and Andrew Feenberg (Cambridge, Mass.: London: MIT Press, 2003), 33–72.

8   See for example the dominance of American material and paucity of references to the British context in the chapters relating to the twentieth century in Stephen Kern, The Culture of Time and Space, 1880–1918: With a New Preface (Harvard University Press, 1983).

9   Robert H. Kargon, Karen Fiss, Morris Low and Arthur P. Molella, World's Fairs on the Eve of War: Science Technology & Modernity, 1937–1942 (Pittsburgh, PA: University of Pittsburgh Press, 2015); Paul Greenhalgh, Ephemeral Vistas: The Expositions Universelles, Great Exhibitions and World's Fairs, 1851–1939 (Manchester University Press, Manchester, 1988).

10 Gillian Beer, Darwin's Plots: Evolutionary Narrative in Darwin, George Eliot and 19th-Century Fiction (Cambridge: Cambridge University Press, 1983); Sally Shuttleworth, George Eliot and Nineteenth-Century Science: The Make-Believe of a Beginning (Cambridge: Cambridge University Press, 1984).

11 Daniel Albright, Quantum Poetics: Yeats, Pound Eliot, and the Science of Modernism, (Cambridge: Cambridge University Press, 1997); Michael Whitworth, Einstein's Wake: Relativity, Metaphor and Modernist Literature (Oxford: Oxford University Press, 2002); Katie Price, Loving Faster than Light: Romance and Readers in Einstein's Universe (Chicago, IL: University of Chicago Press, 2012).

12 This is the explicit mission of the British Society for Literature and Science, as announced on their website http://www.bsls.ac.uk.

13 Friedrich A. Kittler, Gramophone, Film, Typewriter, trans. Geoffrey Winthrop-Young and Michael Wutz (Stanford, Ca.: Stanford University Press, 1999), first published 1986, 138.

14 Kittler, Gramophone, Film, Typewriter, 145.

15 Rebecca Roach, Laura Marcus and David Bradshaw (eds.), Moving Modernisms: Motion, Technology and Modernity (Oxford: Oxford University Press, 2016), 2.

16 It would not be possible here to explore the discipline-based literature on the interaction of each art with science and technology. For design see Paul Greenhalgh, ed., Modernism in Design (London: Reaktion Books, 1990); Paul Greenhalgh, The Modern Ideal: The Rise and Collapse of Idealism in the Visual Arts from the Enlightenment to Postmodernism (London: V&A Books, 2005).

17 Tim Putnam, 'The Theory of Machine Design in the Second Industrial Age', Journal of Design History, 1 (1988): 2534.

18 Tanya Harrod, 'Introduction', special issue 'Craft, Modernism and Modernity', Journal of Design History 11, no. 1 (1998): 14.

19 Bernhard Rieger and Martin Daunton, 'Introduction', in Meanings of Modernity: Britain from the Late-Victorian Era to World War II, eds. Bernhard Rieger and Martin Daunton (Oxford: Berg, 2001), 1. This has continued to be true, so the excellent 2007 volume, Legacies of Modernism: Art and Politics in Northern Europe, 1890–2007 does not contain a single article on a British theme among its 17 contributions. The chapter by Sophie Forgan, 'From Modern Babylon to White City: Science, Technology and Urban Change in London, 1870–1914', in Urban Modernity: Cultural Innovation in the Second Industrial Revolution (Cambridge, Mass.: MIT Press, 2010), 75–132, is a rare exception. The monograph by Bernhard Rieger, Technology and the Culture of Modernity in Britain and Germany, 1890–1945 (Cambridge: Cambridge University Press, 2005) is again a noteworthy exception. An approach was taken by the British Journal for the History of Science with a special issue on 'British Nuclear Culture, 45' no. 4, December 2012. For the pre-war era, we still lack studies of the British context analogous to Peter Galison, 'Aufbau/Bauhaus: Logical Positivism and Architectural Modernism', Critical Inquiry 16, no. 4 (1 July 1990): 709–52. However for the post-Second World War era there is an increasingly rich and valuable literature on nuclear culture. See, for example, Catherine Jolivette, ed., British Art in the Nuclear Age (Farnham: Ashgate, 2014); Jonathan Hogg, Nuclear Culture: Official and Unofficial Narratives in the Long 20th Century (London: Bloomsbury, 2016). See also the discussion of the 'Two Cultures Controversy' in the Epilogue of this book.

20 T.S. Eliot, 'Dante' in The Sacred Wood (New York: Alfred A. Knopf, 1921), 144–55, on 145. First published by Methuen, 1920.

21 It is interesting to contrast this vision of poetry as science with that of Coleridge a century earlier. See Trevor Levere, *Poetry Realized in Nature: Samuel Taylor Coleridge and Early Nineteenth-Century Science* (Cambridge: Cambridge University Press, 1981).

22 Kristin Johnson, 'Natural History as Stamp Collecting: A Brief History', *Archives of Natural History* 34, no. 2 (2007): 244–58. For a reflection on the unity and disunity of science, see Peter Galison and David J. Stump, *The Disunity of Science: Boundaries, Contexts and Power* (Stanford, CA: Stanford University Press, 1996).

23 T. S. Eliot made this claim in 'Modern Tendencies in Poetry' (1920), which was written just after his rather better known essay, 'Tradition and the Individual Talent' (1919). 'Modern Tendencies' was initially delivered as a lecture in late 1919, and first published in an Indian Journal, *Shama'a* 1, no. 1 (April 1920), 9–18.

24 Ezra Pound, *Ezra Pound and Music*, ed. R. Murray Schafer (New York: New Directions, 1977), 316.

25 I.A. Richards, *Poetries and Sciences: A Reissue with a Commentary of Science and Poetry* (1926, 1935) (New York: W. W. Norton and Company, 1970), 18, 21. For the subsequent quotation, see 19.

26 Donna Jeanne Haraway, *Crystals, Fabrics and Fields: Metaphors of Organicism in Twentieth-Century Developmental Biology* (New Haven, Conn: Yale University Press, 1976). More recently there has been a huge efflorescence of literature on the impact of organicism on art. See for example Oliver A.I. Botar and Isabel Wünsche, *Biocentrism and Modernism* (Aldershot, Hampshire: Ashgate Publishing, Ltd., 2011); Edward Juler, *Grown but Not Made: Biological Metaphor and Experimental Art in England, 1930–39* (Manchester University, 2008); Maurizio Esposito, *Romantic Biology, 1890–1945* (London: Routledge, 2015); Spyros Papapetros, *On the Animation of the Inorganic: Art, Architecture, and the Extension of Life* (Chicago: University of Chicago Press, 2012); Charissa N. Terranova and Meredith Tromble, *The Routledge Companion to Biology in Art and Architecture* (London: Routledge, 2016).

27 On Huxley's biology, there is a huge literature. See for instance Roger Smith, 'Biology and Values in Interwar Britain: C.S. Sherrington, Julian Huxley and the Vision of Progress', *Past & Present* 178, no. 1 (1 February 2003): 210–42; John C. Greene, 'The Interaction of Science and World View in Sir Julian Huxley's Evolutionary Biology', *Journal of the History of Biology* 23, no. 1 (1990): 39–55.

28 Mass Observation, 'Report on Everyday Feelings about Science' (October 1941), no. 951. Reproduced with permission of Curtis Brown Group Ltd, London on behalf of The Trustees of the Mass Observation Archive © The Trustees of the Mass Observation Archive

29 There is of course a huge literature on the interpretation of science and modernity after the Second World War. See Zygmunt Bauman, *Modernity and the Holocaust* (Cambridge: Polity Press, 1989), and the work of Jacob Bronowski, who wrote and presented the television series, 'The Ascent of Man'. Also see for instance Ralph Desmarais, 'Jacob Bronowski: A Humanist Intellectual for an Atomic Age, 1946–1956', *The British Journal for the History of Science* 45, no. 4 (2012): 573–89; Robert Joseph Emmitt, 'Scientific Humanism and Liberal Education: The Philosophy of Jacob Bronowski' (PhD thesis, University of Southern California, 1982). Lisa Jardine, Bronowski's daughter, was working on a biography of her father dealing with the troubling contradictions of his experience of science at the time of her premature death.

30 Quotations in this section of the chapter are from abstracts prepared by the authors themselves, in the course of the preparation of the book.

Section 1
Science, modernity and culture

1
Multiple modernisms in concert: the sciences, technology and culture in Vienna around 1900

Mitchell G. Ash

Introduction

In his contribution to the volume *Rethinking Vienna 1900* (2001) and in more detail in a Vienna conference paper presented in 2004, Steven Beller pointed to a number of important developments in the sciences other than psychoanalysis that should be considered part of 'Vienna modernism', but have remained understudied by cultural historians.[1] Since then, historians of science have taken significant steps toward addressing the multiple linkages between the sciences and Vienna modernism. To give only a few examples: work on Robert Musil has emphasised the central role of scientific modernism and Musil's own training as a scientist and engineer for his thinking and writing.[2] Veronica Hofer, Cheryl Logan and others have placed the transformation of biology into an experimental science at the 'Vivarium' laboratory, located in the Prater, in the context of Vienna modernism. The work of 'Vivarium' researchers Eugen Steinach (on surgically induced sexual transformations in animals) and of Paul Kammerer (on the supposed inheritance of acquired characteristics in amphibians) attracted considerable media attention – much of it scandalous. The central idea behind that work, the malleability of living things, meshed well with the culture of creation central to modernism.[3]

In her book *Trafficking Materials and Gendered Experimental Practices: Radium Research in Early Twentieth-Century Vienna*, Maria Rentetzi presents a study of radium research in Vienna, the first, and for many years the only, field in the so-called 'hard' natural sciences with significant numbers of female participants.[4] While focussing on cultures of experimentation in the new field, Rentetzi also integrates the architecture of the Radium Institute, founded in 1910, and the multiple interactions of its researchers with the city into her analysis. In this case (like the better-known ones of Lou Andreas-Salomé or Bertha Pappenheim), women's roles in Vienna modernity clearly went beyond those of muse, femme fatale, *Salondame*, or neurotic patient, that have become standard in the literature.

Architectural historian Leslie Topp has published studies of the overall design and building ensemble of the 'Lower Austrian Healing and Care Institution am Steinhof', an asylum for the mentally ill opened in 1907, focussing on but going well beyond the well-known and still highly visible chapel building designed by Otto Wagner.[5] Building in part on Topp's work, Sophie Ledebur has now shown in detail that the asylum itself, the largest such facility in Europe at that time, was presented as an embodiment of 'modern' psychiatry not only because of its size, design and architecture, but also as an independent 'space of knowledge' with a comprehensive approach to the management of mental health knowledge and treatment that could only be realised in a location separate from the university's psychiatric clinic and removed from the city centre.[6] As she writes, citing Topp, the institution's modernity actually lay in the ambiguity of its own goals:

> The established trope of the asylum-as-utopia, gaining its power from its separateness and rejection of the world beyond its walls, and the newer rhetoric of the open, friendly mental hospital existed in unresolved tension in the discourse of early-twentieth-century planners, including the psychiatrists and government officials who planned Steinhof. Otto Wagner's brand of modernism thrived not on tension but on glorious resolution; it filtered out the ambiguity of the asylum planners' goals and focused exclusively on their utopian ambitions.[7]

Vienna's modern asylum, in Ledebur's view, thus embodied a 'heterotopia' in the sense elaborated by Michel Foucault – a spatial 'other', physically and ideally distinct from its illness-inducing urban surroundings, incorporating 'organised freedom' within its ostensibly perfect 'interior order'.[8]

In her recent studies of medicine, culture and politics in nineteenth- and early twentieth-century Vienna, Tatjana Buklijas shows that the

Fig. 1.1   Gustave Klimt, *Danaë* (1907/08), oil on canvas, 77 × 83 cm.
Privately held. Detail.

iconoclastic artists of Vienna 1900 took inspiration from nature and transformed natural motifs into abstract decorative forms. To give only one example: the right corner of Gustav Klimt's famous painting *Danaë*, depicting the mythical moment when Zeus in the shape of golden rain surreptitiously impregnated the princess of Argos and Eurydice, is scattered with the shapes of a blastocyst (Figure 1.1).

This early embryonic structure, first described in humans in 1895, was still a novelty at that time; Klimt must have seen it in embryological literature or, more likely, attending Emil Zuckerkandl's anatomical lectures for artists in the summer of 1903.[9] Here, as in the case of the asylum 'am Steinhof', innovations in medical science and in the arts appear to have gone hand in hand.

In *Vienna in the Age of Uncertainty: Science, Liberalism and Private Life* (2007), Deborah Coen goes still further and challenges Carl Schorske's classical interpretation of *fin-de-siècle* Vienna directly.[10]

Vienna modernism, in her view, was not only a descent into the irrational following the crisis of liberalism. Rather, the probabilistic worldview characteristic of modern physics in Austria was itself an expression of the liberal culture propagated and transmitted by several generations of a single family – the Exners. In characterising the work of Franz Seraphim Exner, professor of physics and physical chemistry and head of the 'Second Physical Institute' at the University of Vienna from 1902 to 1920, Coen draws in part upon work by historian and philosopher of science Michael Stöltzner on what he calls 'Vienna indeterminism'.[11]

All of these studies show that natural scientists clearly deserve to be included, not only as a sideshow to but rather as integral participants in 'Vienna 1900'. At the very least they demonstrate direct personal contact and networking among scientists and artists. This in itself is hardly surprising, given the small number of highly educated people involved. But the point goes beyond 'me too' historiography or anecdotal linkages, and in some cases at least appears to encompass substantive commonalities as well. In this chapter, I present an exploratory effort to advance this discussion – or at least to ask what scientific and artistic modernism may have had in common.

Modernism in the sciences

It is hardly possible to engage in such an effort without at least trying to define modernism. The collective singular noun 'modernism' (*die Moderne* in German) is the established property of the humanities and refers to developments in literature, music, the visual arts and architecture. Yet it is equally established usage to speak also of modern physics, mathematics, chemistry or biology. The term 'modern' can hardly mean the same thing in all of these contexts. As Ruth Oldenziel reminds us in her contribution to this volume, what it meant to be modern has been contested and is unstable in many fields, from the arts to lifestyle choices.[12] As I will show, the term 'modern' also has multiple, contested meanings within the sciences.[13]

My intention here is twofold. I wish to argue, first, that there are certain affinities, and in some cases actual linkages, between the breakthrough to modern ways of thinking in the natural sciences and mathematics and the radical changes in the arts which occurred at the same time. I argue, second, that modernism, and hence cultural modernity, in all of these fields was nonetheless fundamentally plural and not the 'totalising project' that postmodernist thinkers have claimed that it was.

Modernism as an intellectual style in many sciences, as in many of the arts, involved a break with direct, supposedly pictorial represent-ation of nature and a turn towards giving free play to abstraction and theoretical imagination. In order to characterise the affinities involved in at least a preliminary way, I propose to extend a line of interpretation suggested some years ago by historian of mathematics Herbert Mehrtens, and speak here of the breakthrough to 'modern' thinking, at least in part, as an *emancipation of self-referential symbol systems* from representationalism or classical formalistic conventions, giving free play to abstraction and theoretical imagination.[14] Alongside this style of modernistic thinking, Mehrtens places *technological or technocratic modernism*.[15] This presupposes a concept of knowledge based less on self-referential abstraction than on what can be *done* with, or to, nature as well as other human beings, exemplified most clearly in the case of eugenics as a form of biological politics, in the efforts to establish the malleability of living things in the laboratory mentioned above, in the re-conceptualisation of substances in human bodies such as hormones, vitamins and enzymes as effective causal agents (*Wirkstoffe*) that could be synthesised artificially in the laboratory, and in the re-conceptualisation of social science that took society itself as an object that could and should be acted upon by science-based policy, later termed social engineering.[16] As will be shown, these two styles of modernism in science were not fundamentally opposed, but in certain respects deeply related to one another. It should be evident that I am speaking here of ideal types, examples of which may or may not exist in actual science. Moreover, such developments must be seen in the much broader context established by the technological transformations that have been formative of modernity itself. I will return to the latter point in the conclusion.

The plurality of modernism in the sciences and culture of 'Vienna 1900' might best be illustrated with the aid of some examples. In the following I will discuss three cases, two from physics (the work of Ernst Mach and Ludwig Boltzmann) and one from music (Arnold Schoenberg). In each case, I will try to bring out basic features that link the cases in question to one or another of the definitions of modernity as a style of thought just sketched.[17]

The first two cases are from physical science, and focus not on Einstein or relativity,[18] which of course were created elsewhere, but rather on the debate between Ernst Mach and Ludwig Boltzmann on the status of mathematical, and particularly statistical, models in physics, which constituted an important part of the background to the reception of relativity theory and quantum mechanics.

Ernst Mach – modernism as anti-metaphysics

Ernst Mach – a physiologist, psychophysicist and experimental physicist – was professor of 'philosophy, with emphasis on the history and theory of the inductive sciences' in Vienna from 1895 to 1901.[19] Students of Vienna modernism are familiar with the famous, often reproduced image in Mach's *The Analysis of Sensations* showing a reclining man with a (half visible) moustache holding a pencil, which Mach used to show that we cannot entirely perceive our own bodies (Figure 1.2).

From this point Mach went on to make the claim, widely cited at the time, that 'the ego cannot be saved' (in the original: *Das Ich ist unrettbar*), because such an entity 'is not a definite, unalterable, sharply-bounded unity' in our sensations; rather, it is no more than a convenient fiction, 'an ideal mental-economical' abstraction inferred from and then imposed

Fig. 1.2 From Ernst Mach, The *Analysis of Sensations and the Relation of the Physical and the Psychical*, trans. C. M. Williams, rev. and suppl. Sydney Waterlow (La Salle, IL: Open Court 1914 [1886]), Fig. 1, p. 19.

upon sensations.[20] Mach's fundamental claim is that we cannot know the external world directly, but only our sensations. His claim about the ego is an extension of this primary claim to our knowledge of ourselves. On the wider implications and impact of this 'antimetaphysical' claim, cultural historian Philip Blom writes: 'Mach simply threw overboard the constant ego as the midpoint and fulcrum of life experience and presented a conception of personality that could be understood as a pendant to the emptiness at the heart of the Danube monarchy.'[21] The Austrian essayist Hermann Bahr presented Mach's idea to a wider public in his famous text, 'The unsalvageable ego' (*Das unrettbare Ich*) in 1903, declaring it to be no less than 'the philosophy of (literary) impressionism.'[22]

In a major historical work on the foundations of mechanics, Mach presented the development of mechanics as a story of evolutionary progress from the 'practical mechanics' of the great pyramids in Egypt to Newton, Leibniz, and beyond, in which the history of physics and that of human technology intertwined. Specifically, he argued that mechanics developed into a science only when repeated use of handcrafted tools extended the capacity of the senses and eventually drove the intellect to a higher level:

> The doctrines of mechanics have developed out of the collected experiences of handcraft by an intellectual process of refinement … The simple machines – the five mechanical powers – are without question a product of handcraft … The making of rollers (for example) must have gained a great technical importance and have led to the discovery of the lathe. In possession of this, mankind easily discovered the wheel, the wheel and axle, and the pulley.[23]

In a famous essay entitled 'The Economical Nature of Physical Inquiry', Mach argued further that mathematical formulations, too, are no more than convenient tools to summarise sense impressions or laboratory observations.[24] He did not deny the importance of hypotheses or thought-experiments, like those of Galileo and Newton, as aids to discovery, but he opposed going any further beyond direct observation and measurement than necessary, a standpoint characterised at the time as 'phenomenological'. Thus, he argued, for example, that because we experience the external world as a continuum, we should imagine underlying physical reality as consisting of continuities, not particles. Mach opposed physical atomism in part because there was as yet no direct observational evidence of the existence of any such hard, massy, bounded bodies as ultimate components of matter. Finally, Mach argued,

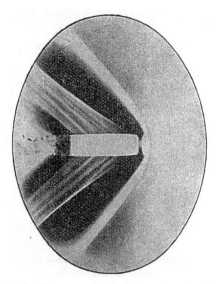

Fig. 1.3 Ernst Mach, Photograph of a blunted projectile, 1887.
Popular Scientific Lectures (1898), p. 324, fig. 53.

using Darwinist imagery, that following simple rules of theoretical self-restraint yields advantages in the struggle for existence, by which he meant competition in science itself.

It is important to note here that for Mach sense impressions were plainly not limited to what the naked eye can see and the unaided ear can hear. Examples of what he did mean are the photographs of projectiles he made with Peter Salcher and first published in 1887 (Figure 1.3).[25]

The avowed aim of this study, according to Mach, was to 'make the process perceivable' – in this case to capture photographically the pressure waves caused by the movement of an object through air at high speed.[26] The similarity of these photos to those in Eadweard Muybridge's widely known movement studies seems obvious, and it was not accidental. Mach later defined the aim of the recording apparatus in general as the 'thickening of immediate perception' (*Verdichtung der Anschauung*).[27] To put it in a nutshell, for Mach the term 'sense impressions' included extensions of ordinary sensation by means of technology; in this respect his work continues a tradition that extends backward at least to Galileo's telescope. The apparent difference between how things appeared to the unaided senses and these technologically produced images was, in Mach's view, merely quantitative, despite the extraordinary difficulties he encountered when he attempted to create such images.

Thus, if Mach was a modernist – and he surely was – then he does not exemplify either of the ideal types sketched above precisely. As shown at the beginning of this section, the central element of Mach's modernism was his antimetaphysics, the effort to overcome traditional, but in his eyes meaningless, concepts. Though he certainly acknowledged that mathematical symbol systems had their uses in physics, he thought that they should be understood only as functional summaries of data from sense impressions, not as self-referential abstractions from these. The photos just described have been cited as examples of what Lorraine Daston and Peter Galison have termed 'mechanical objectivity'.[28] Yet such uses of technology and media as instruments of scientific discovery do not mean that Mach believed that nature or human beings were or ought to be entirely malleable in the sense of Mehrtens' technocratic modernism. Rather, Mach thought that science emerged historically from human technological activity.

Ludwig Boltzmann – 'the independence of mathematical theory'

A primary opponent of Mach on these issues was Ludwig Boltzmann, professor of theoretical physics in Vienna from 1894 to 1900, and again from 1902 to his death in 1906.[29] In contrast to the polymath Mach, Boltzmann stood for the institutionalisation of theoretical physics as an independent field of inquiry. Boltzmann consistently advocated the independent status of mathematical models in physics, asserting that their importance went far beyond the mere summation of measurement data. In an encyclopedia article on mathematical and physical models, apparatus and instruments, he called such models 'instruments' and acknowledged that the 'need to save labor' was an important reason for their use.[30] But he scornfully compared Mach's Darwinist 'economy of thought' with the businessman's need to save time and money. The real purpose of mathematical models, in his view, was 'the need to make results of calculation observable [anschaulich]'; elsewhere in the same text he used the term 'sensorisation' (Versinnlichung).[31]

By terms such as anschaulich and Versinnlichung Boltzmann clearly meant something quite different from Mach's sense impressions. Boltzmann meant to give priority to theory, not observation, and to argue that the former guided the latter, not the other way around. Ever since the time of the great Paris mathematicians, he wrote, 'one has conceived matter as a sum of mathematical points', not as a continuum. Atomism thus became a practical necessity for doing physics at all, quite independent of the observability of atoms:

> One does not believe, however, that one has really created a clear concept of what a continuum is merely by employing the word, continuum, and writing out a differential equation! On closer examination the differential equation is only the expression for the fact that one must begin with a finite number.[32]

The labour involved was not the convenient summary of sense impressions, but the development of 'mental images' (*Gedankenbilder*); indeed, Boltzmann argued that broad collections of facts (*umfassende Tatsachengebiete*) could never be described directly, but could only be depicted by such *Gedankenbilder* – that is, by mathematical equations.[33] Though he spoke of images, representation and *Versinnlichung*, Boltzmann plainly did not mean any analogy to photographic imagery. Rather, theories are symbol systems that stand on their own and are testable only as wholes; they are images of nature only in a very abstract sense. Boltzmann nonetheless calls them 'instruments' and emphasises that they are not entirely self-referential, but rather subject to experimental testing.

Boltzmann's emphasis on the independence of mathematical theory and on abstract, theoretical rather than photographic pictures of nature has come to be regarded as central to the modern standpoint in physics. It is important to note, however, that the synthesis of theoretical physics with high technology in the form of the atomic bomb was hardly dreamed of by anyone involved at the time. Thus, the turn-of-the-century positions I have outlined are two versions of the 'modern' in physics. The two styles of thinking lived alongside one another at first and could even become competitors, but it was also possible to blend them.

Arnold Schoenberg – modernism in music, or: formalism as techno-scientific style

What relevance, if any, has any of this to modernity in music? Though I am not a musicologist, perhaps I may be allowed to note some suggestive references in the writings of Arnold Schoenberg that may provide starting points for discussion. In 'Composition in Twelve Tones', for example, Schoenberg speaks of the 'emancipation of dissonance' and explicitly rejects the designation 'atonal' as a description of his system,

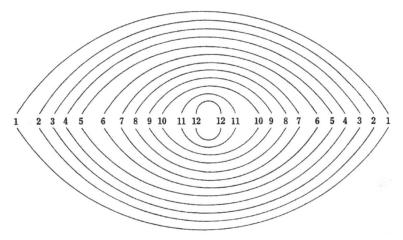

Fig. 1.4 Arnold Schoenberg, 'Figures of Numbers'

Style and Idea (1984), pp. 224–225. Used by permission of Belmont Music Publishers, Los Angeles.

speaking instead of a 'system of components relating only to one another'.[34] Specifically, he described the twelve-tone method in terms of mirrored notes similar to the abstraction of mirrored numbers, as shown in Figure 1.4.

Such remarks appear to place this particularly formalistic version of modern music squarely in the camp of modernism as the emancipation of self-referential symbol systems. Caution is needed here, since one might well say that musical notation itself is a self-referential symbol system. Nonetheless, Schoenberg's willfully austere twelve-tone approach is clearly more abstract than the highly evocative, definitely referential texts Richard Strauss and Gustav Mahler, as well as Schoenberg himself in his early works, attached to their tone poems.

Care is needed for another reason: Schoenberg refers to non-Euclidean geometry only occasionally and vaguely, whereas his references to Mach's phenomenological and evolutionary conception of physics are more frequent and precise, though he does not mention Mach by name. He constantly wrote of the development of musical ideas from lower to higher stages and regarded his own twelve-tone system as a result of gradual evolutionary development. Still more interesting in this regard are Schoenberg's frequent references in the *Theory of Harmony* to the psychological effects of tones and chords.[35]

Though Schoenberg was not a trained scientist, such discussions show a familiarity with the work of the physiologist and physicist Hermann Helmholtz and the philosopher and psychologist Carl Stumpf

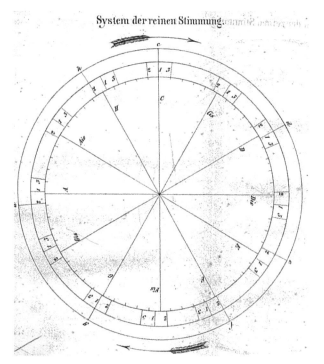

Fig. 1.5 Ernst Mach, 'Pure and Tempered Tuning'.

Einleitung in die Helmholtzsche Musiktheorie (1885), Appendix.

on the psychology of hearing.[36] Mach, too, worked on these subjects. There is no evidence that Schoenberg met Mach in Vienna, but he certainly knew two people who had done so: Guido Adler and David Joseph Bach. Bach, one of the formative influences in Schoenberg's youth by his own account, was a doctoral student of Mach and Friedrich Jodl in Vienna.[37] There is an intriguing similarity between this diagram of Mach's, showing differences between pure and tempered tuning that would remain on a circular keyboard (Figure 1.5) and the following two diagrams by Schoenberg (Figures 1.6a, 1.6b).

Figure 1.6a, a design for Berlin streetcar tickets, shows Schoenberg as a craftsman. Figure 1.6b depicts Schoenberg's 'twelve-tone disc' (*Zwölftondrehscheibe*), a 'tool', as he called it, for generating twelve-tone series.

I do *not* claim that Schoenberg 'borrowed' his design from Mach. Rather, the issue is the way in which they embodied figuratively their shared belief in the evolution of 'higher' creative forms from handcrafted work. Situated within the argument advanced here, both diagrams exemplify syntheses of formalist and technological modernity.

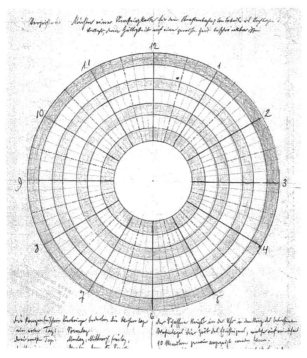

Fig. 1.6a   Arnold Schoenberg, 'Design for Street Car Tickets'.

Courtesy of the Arnold Schoenberg Center, Vienna. Used by permission of Belmont Music Publishers, Los Angeles.

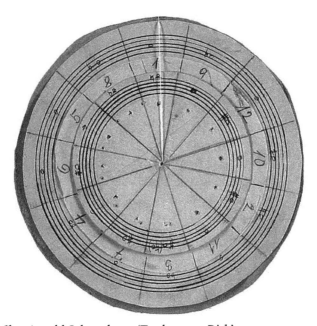

Fig. 1.6b   Arnold Schoenberg, 'Twelve-tone Disk'.

Courtesy of the Arnold Schoenberg Center, Vienna. Used by permission of Belmont Music Publishers, Los Angeles.

Conclusion: technological transformation as the context of modernism

As stated in the introduction, Mehrtens' two styles of modernism in the sciences – abstract, self-referential symbol systems on the one hand and technocratic modernism on the other – were formulated, and are also employed here, as ideal types. The distinction is useful to help us grasp the complexity of scientific modernism, but the three case studies from Vienna in this chapter show that in practice various styles of modernism mingled with one another in interesting ways. Ernst Mach's neo-positivistic effort to overcoming metaphysics in science and philosophy and to reconstruct both on the basis of 'sense impressions' relied in part on using technology and new media to extend the accessibility of physical processes to the senses. In this respect, his work marks a transition from traditional representationalism to modernism. Ludwig Boltzmann's probabilistic physics comes closest to exemplifying the emancipation of abstract symbol systems, yet he regarded mathematics as a tool for constructing symbol systems that he considered to be empirically testable in principle. Finally, Arnold Schoenberg's case represents the use of a technological style in the service of constructing a self-referential symbol system in music.

Let me close by returning to another remark made in the introduction that hints at a broader interpretation for these and other examples of scientific modernism. The common context for the breakthroughs to modern styles of thought in both the sciences and the arts at the turn of the century is the technological transformation of the lived world resulting from the second industrial revolution. Central to that transformation are the abolition of the distinction between day and night by gas and then electric lighting, and the opening of the cities to the countryside and suburbs by means of horse-drawn, then electric street cars and later the automobile.[38] Both developments mark the emergence of the city as an artificial universe of sound and light, seemingly emancipated from any direct dependence on nature and its rhythms – a nature that was itself being transformed by the encroachment of the city. Already here, in these transformations of the lived world, we see again how difficult it is to separate the emancipation of self-referential symbol systems from technocratic modernity. Indeed, we could suggest that the metropolis itself was becoming a large, increasingly self-referential technical system around 1900.[39]

Notes

1   Steven Beller, ed., *Rethinking Vienna 1900* (Oxford; New York: Berghahn, 2001); Beller, 'Vernachlässigte Aufklärung. Mäzenatentum und die Naturwissenschaften in Österreich'. Paper presented at the conference 'Mäzenatentum und naturwissenschaftliche Forschung in Österreich 1865–1938', unpublished.

2   See, e.g., Christoph Hoffmann, *Der Dichter am Apparat. Medientechnik, Experimentalpsychologie und Texte Robert Musils 1899–1942* (Munich: Fink-Verlag, 1997).

3   Veronica Hofer, 'Rudolph Goldscheid, Paul Kammerer und die Biologen des Prater-Vivariums in der Liberalen Volksbildung der Wiener Moderne', in *Wissenschaft, Politik und Öffentlichkeit: von der Wiener Moderne bis zur Gegenwart*, ed. Mitchell G. Ash and Christian Stifter (Vienna: Wiener Universitätsverlag, 2002), 149–84; Cheryl A. Logan, *Hormones, Heredity and Race: Spectacular Failure in Interwar Vienna* (New Brunswick, NJ: Rutgers University Press, 2013); Sonja Walch, *Triebe, Reize und Signale. Eugen Steinachs Physiologie der Sexualhormone. Vom Biologischen Konzept zum Pharmapräparat* (Vienna: Böhlau-Verlag, 2016). On Steinach in broader cultural context see also Heiko Stoff, *Ewige Jugend. Konzepte der Verjüngung vom Späten 19. Jahrhundert bis ins Dritte Reich* (Cologne: Böhlau-Verlag, 2004). On the Kammerer scandal as part of the history of anti-semitism at the University of Vienna, see Klaus Taschwer, *Hochburg des Antisemitismus. Der Niedergang der Universität Wien im 20. Jahrhundert* (Vienna: Czernin Verlag, 2015).

4   Maria Rentetzi, *Trafficking Materials and Gendered Experimental Practices: Radium Research in Early Twentieth-Century Vienna* (New York: Columbia University Press, 2008).

5   Leslie Topp, *Architecture and Truth in Fin-de-Siècle Vienna* (Cambridge: Cambridge University Press, 2004). For recent studies of psychopathology and Vienna modernity, see Gemma Blackshaw and Sabine Wieber, eds., *Journeys into Madness: Mapping Mental Illness in the Austro-Hungarian Empire* (New York and Oxford: Berghahn Books, 2012).

6   Sophie Ledebur, *Das Wissen der Anstaltspsychiatrie in der Moderne. Zur Geschichte der Heil- und Pflegeanstalten am Steinhof in Wien* (Vienna: Böhlau Verlag, 2015), 264.

7   Leslie Topp, 'Otto Wagner and the *Steinhof* Psychiatric Hospital: Architecture as Misunderstanding', *Art Bulletin*, 87 (2005), 130–56, here: 151, cited in Ledebur, *Das Wissen der Anstaltspsychiatrie*, 75–76.

8   Ledebur, *Das Wissen der Anstaltspsychiatrie*, 76.

9   See Emily Braun, 'Ornament and Evolution: Gustav Klimt and Berta Zuckerkandl', in *Gustav Klimt: The Ronald S. Lauder and Serge Sabarsky Collections*, ed. Renée Price (New York: Neue Galerie, 2007), 144–69; Scott F. Gilbert and Sabine Brauckmann, 'Fertilization Narratives in the Art of Gustav Klimt, Diego Rivera and Frida Kahlo: Repression, Domination and Eros among Cells', *Leonardo* 44 (2011): 221–27. Both cited in Tatjana Buklijas, 'The Politics of Fin-de-siècle Anatomy', in *The Nationalization of Scientific Knowledge in the Habsburg Empire (1848–1918)*, eds. Mitchell G. Ash and Jan Surman (Basingstoke: Palgrave Macmillan, 2012), 209–44, here: 232.

10 Deborah R. Coen, *Vienna in the Age of Uncertainty: Science, Liberalism and Private Life* (Chicago: University of Chicago Press, 2007). For critical discussion of Coen's thesis, see Veronika Hofer and Michael Stöltzner, 'What is the Legacy of Austrian Academic Liberalism?' *NTM: International Journal of History and Ethics of Natural Science, Technology and Medicine* 20 (2012): 31–42.

11 Michael Stöltzner, 'Vienna Indeterminism: Mach, Boltzmann, Exner', *Synthèse* 119 (1999): 85–111; Stöltzner, 'Franz Serafin Exner's Indeterminist Theory of Culture', *Physics in Perspective*, 7 (2002): 267–319.

12 Ruth Oldenziel, 'Whose Modernism, Whose Speed? Designing Mobility for the Future, 1880s–1945', in this volume.

13 For further discussion and an extension of the argument beyond the turn of the twentieth century, see Mitchell G. Ash, 'Krise der Moderne oder Modernität als Krise? Stimmen aus der Akademie', in *Die Preußische Akademie der Wissenschaften zu Berlin in Krieg und Frieden, in Republik*

und *Diktatur 1914–1945*, ed. Wolfram
Fischer, with the assistance of Rainer
Hohlfeld and Peter Nötzoldt (Berlin:
Akademie-Verlag, 2000), 121–42; Ash,
'Multiple Modernisms? Episodes from
the Sciences as Cultures, 1900–1945', in
Jewish Musical Modernism, Old and New,
ed. Philip V. Bohlman (Chicago:
University of Chicago Press, 2008),
31–54. For a more unitary view of
science, modernity and postmodernity,
respectively, see Paul Forman, 'The
Primacy of Science in Modernity, of
Technology in Postmodernity, and of
Ideology in the History of Technology',
History and Technology, 23 (2007):
1–152.

14   Herbert Mehrtens, *Moderne Sprache
Mathematik* (Frankfurt am Main:
Suhrkamp, 1990).

15   Mehrtens, *Moderne Sprache Mathematik*.
See also Mertens, 'Kontrolltechnik
Normalisierung. Einführende
Überlegungen', in *Normalität und
Abweichung. Studien zu Theorie und
Geschichte der Normalisierungsgesellschaft*,
eds. B. Sohn and Herbert Mehrtens
(Opladen: Westdeutscher Verlag, 1999),
44–64.

16   On eugenics in Vienna, see Gerhard
Baader, Veronika Hofer und Thomas
Mayer, eds., *Eugenik in Österreich.
Biopolitische Strukturen von 1900 bis 1945*
(Vienna: Czernin-Verlag, 2007). For
bodily substances as causal agents, see
Stoff, *Ewige Jugend* and Stoff, *Wirkstoffe.
Eine Wissenschaftsgeschichte der Hormone,
Vitamine und Enzyme 1920–1980*
(Stuttgart: Franz Steiner-Verlag, 2014).
For technological thinking in the social
sciences, see Kerstin Brückweh, Dirk
Schumann, Richard F. Wetzall and
Benjamin Ziemann. eds., *Engineering
Society: The Role of the Human and Social
Sciences in Modern Societies* (Basingstoke:
Palgrave Macmillan, 2012).

17   The following case studies have been
revised from Ash, 'Multiple Modernisms?'.

18   On relativity and the culture(s) of
modernity, see especially Peter L. Galison,
*Einstein's Clocks and Poincaré's Maps:
Empires of Time* (New York: W.W. Norton,
2003).

19   On Mach's appointment and impact in
Vienna, see Josef Mayerhöfer, 'Ernst Mach
in Wien', *Mitteilungen der Österreichischen
Gesellschaft für Wissenschaftsgeschichte* 9
(1989): 19–42. On Mach as predecessor
and in some respects foundational thinker

of the Vienna Circle, see Friedrich Stadler,
*The Vienna Circle: Studies in the Origins,
Development and Influence of Logical
Empiricism*, trans. S. Schmidt, J. Golb,
T. Ernst and C. Nielsen (Vienna: Springer
Verlag, 2001).

20   Ernst Mach, *The Analysis of Sensations
and the Relation of the Physical and the
Psychical*, trans. C. M. Williams, rev. and
suppl. Sydney Waterlow (La Salle, IL:
Open Court 1914 [1886]), 24. Translation
revised by MGA.

21   Philip Blom, *Der Taumelnde Kontinent.
Europa 1900–1914*, seventh edition
(Munich: deutsche taschenbuchverlag,
2015 [2011]), 82. Quotation trans.
MGA.

22   Hermann Bahr, 'Das unrettbare Ich', in
Dialog vom Tragischen, ed. Hermann
Bahr (Berlin: S. Fischer, 1904), 79–101,
here: 101. Originally published in *Neues
Wiener Tageblatt*, 37, no. 90 (10 April
1903), 1–4, here: 4. Blom (cit. n. 20)
erroneously states that the essay appeared
in 1907.

23   Ernst Mach, *The Science of Mechanics:
A Critical and Historical Account of Its
Development*, sixth edition, trans. Thomas
J. McCormack (LaSalle, Ill.: Open
Court Press, 1960), here: 612–13.

24   Ernst Mach, 'The Economical Nature of
Physical Inquiry', in *Popular Scientific
Lectures*, fifth edition, trans. Thomas J.
McCormack (LaSalle, Ill.: Open Court
Press, 1943 [1882]), 186–213.

25   Christoph Hoffmann, ed., *Über Schall:
Ernst Mach's und Peter Salchers
Geschoßfotografien* (Göttingen:
Wallstein Verlag, 2001).

26   Ernst Mach, 'Über Erscheinungen
an fliegenden Projektilen', in
Populärwissenschaftliche Vorträge,
fourth edition, (Leipzig: Barth [1897]),
356–83, here: 357. Quotation trans MGA.

27   Christoph Hoffmann, *Der Dichter am
Apparat*, 45. See also Bernd Stiegler,
'Ernst Machs 'Philosophie des
Impressionismus' und die
Momentphotographie', *Hofmannsthal:
Jahrbuch zur europäischen Moderne*, 6
(1998): 257–80.

28   Lorraine Daston and Peter Galison,
Objectivity (Cambridge: Zone Books,
2007), 157, fig. 3.22.

29   On Boltzmann's career in Vienna and his
later writings, see John Blackmore, ed.,
*Ludwig Boltzmann – His Later Life and
Philosophy 1900 – 1906*, 2. vols.
(Dordrecht: Kluwer, 1995).

30 Ludwig Boltzmann, 'Über die Methoden der theoretischen Physik', in *Populäre Schriften*, (Leipzig: Barth, 1905 [1892]), 1–10, here: 1. Quotation trans. MGA.

31 Boltzmann, 'Über die Methoden der theoretischen Physik', 2.

32 Ludwig Boltzmann, 'Über die Unentbehrlichkeit der Atomistik in der Naturwissenschaft', in *Populäre Schriften*, (Leipzig: Barth, 1905 [1897]), 141–57, here: 144. Quotation trans. MGA.

33 Boltzmann, 'Über die Unentbehrlichkeit der Atomistik in der Naturwissenschaft', 142.

34 Arnold Schoenberg, 'Composition in Twelve Tones', in *Style and Idea: Selected Writings of Arnold Schoenberg*, ed. Leonard Stein (Berkeley: University of California Press, 1984), 217–18.

35 e.g., Arnold Schoenberg, *Theory of Harmony*, trans. Roy E. Carter (Berkeley: University of California Press, 1978 [1911]), 19 ff.

36 Hermann Helmholtz, *Die Lehre von den Tonempfindungen als physiologische Grundlage für die Theorie der Musik* (Braunschweig: Vieweg, 1862); Carl Stumpf, *Tonpsychologie*, 2 vols. (Leipzig: Hirzel, 1883, 1890).

37 Sydney M. Kwiram, 'Tones for Thought: Arnold Schoenberg and the Culture of Scientific Modernism in *Fin-de-siècle* Vienna'. (Honors thesis, Harvard University, 1999). Copy in possession of MGA.

38 See, among many others, Wolfgang Schivelbusch, *Disenchanted Night: The Industrialization of Light in the Nineteenth Century*, trans. Angela Davis (Berkeley: University of California Press, 1995).

39 This remark suggests an expansion of the perspective formulated by Thomas P. Hughes, 'The Evolution of Large Technological Systems', in *The Social Construction of Technical Systems: New Directions in the Sociology and History of Technology*, ed. Wiebe E. Beijker, Thomas P. Hughes and Trevor Pinch (Cambridge, MA: MIT Press, 1987), 51–82. A pioneering older work in this direction is Siegfried Giedeon, *Mechanization Takes Command: A Contribution to Anonymous History* (New York: Oxford University Press, 1948).

2
The cinematic sound of industrial modernity: first notes

Tim Boon

This chapter is about the ways in which filmmakers in the first decade of sound on film used sound to represent industrial modernity. The value in undertaking such a study relates to the importance of sound on both sides of an equation: as a key interwar focus for debate about modernity and as a means of representation in the new media of the time. Emily Thompson's use of Murray Schafer's term 'soundscape' is valuable here: 'Like a landscape, a soundscape is simultaneously a physical environment and a way of perceiving that environment; it is both a world and a culture constructed to make sense of that world'.[1] This central reflexivity of the period – that industrial technology was both author and witness, producer and subject – I offer as an exemplification of Marshall Berman's neat encapsulation that modern*isms* of various kinds may be seen as responses to the modern*ity* their authors experienced.[2] In suggesting that it will be particularly interesting to consider how new sonic technologies used sound itself to represent modernity, I will limit myself here to cinema in the decade from 1927. The virtue of this time frame is that by choosing it we can see a modernistic form in its infancy developing techniques and a language that were later codified or, in some cases, left behind.

First, I consider the sound-on-film practice embodied in two specific films that, in one way or another, are concerned with the nature of industrial modernity, one documentary and one feature. My examples are Paul Rotha's 1935 documentary *The Face of Britain* (British) and René Clair's 1931 feature, *À Nous la Liberté* (French). The directors of these two films both chose industrial modernity as their subject, reflecting back to their societies their own particular responses to it, using modernistic conventions that were necessarily still in the course of being formulated. Both films also have particularly interesting, and self-consciously made, soundtracks, which their authors also discussed in print. Each of these self-conscious cinematic auteurs had highly engaged views on how sound should work within the cinema; their views revolved around the creative possibilities of rejecting for various reasons the synchronised sound of the talkies, where lips are seen to speak words, and instead focussed on the *asynchronous* use of the soundtrack, where sound is used as part of the montage to convey meaning additional to the vision track. I mean by taking these differing examples to suggest a field of cinematic and sonic study that could be populated with many further examples, a possibility which I discuss briefly in the concluding section of this chapter.

There is a further way in which this chapter seeks a synthesis of work pursued in separate fields. Scholars in science and technology studies (STS) will be aware of 'sound studies', the recently opened-up interdisciplinary area between STS, cultural studies and musicology, as represented by Trevor Pinch and Karin Bijsterveld's *Oxford Handbook of Sound Studies* or the 2013 issue of *Osiris* on the topic.[3] The discipline of film studies also made a turn to the sonic a generation ago with foundational publications in 1980 that included Rick Altman's special issue of *Yale French Studies* entitled 'Cinema/Sound' and in 1986 with Elisabeth Weis and John Belton's compendium, *Film Sound*.[4] This essay bridges these two disciplinary areas.

Noise and modernity

As Rick Altman has argued, the world into which cinema was born was one of growing sonic variety, largely the product of new technologies not only in manufacture and transport, but also in sound technologies themselves, including phonographs and telephones. As he says, 'at the turn of the twentieth century, no enterprise involving sound could possibly develop independently of this newly complex soundscape. The sound "vocabulary" of every era depends on the ways in which contemporary media present sound …'. His argument about the changing sonic qualities of so-called 'silent' films goes on to stress the wide range of sources for this vocabulary as it developed in the first 30 years of cinema.[5] It is important to remember that this pre-1927 prehistory of the sound-on-film era was part of the sonic world that the filmmakers under discussion here inhabited, part of the 'vocabulary' put to work via the new grammars of sound on film that they were actively developing in the 1930s.

The significance of noise in interwar modernity is clear on the celebratory side, for example in the noise music of the Futurists.[6] The composer Francesco Balilla Pratella's 1911 *Technical Manifesto of Futurist Music* argued that 'all forces of nature, tamed by man through his continued scientific discoveries, must find their reflection in composition – the musical soul of the crowds, of great industrial plants, of trains, of transatlantic liners, of armoured warships, of automobiles, of aeroplanes. This will unite the great central motives of a musical poem with the power of the machine and the victorious reign of electricity.'[7] Luigi Russolo took up the refrain; his manifesto *The Art of Noises* gloried in 'the muttering of motors, ... the throbbing of valves, the bustle of pistons, the shriek of mechanical saws [and] the starting of the tram on its tracks'.[8] His composition *Awakening of a City*, as scored for an ensemble of *Intonamauri* – or 'noise-intoners' – which he had designed to reproduce machine noises, was heard in London in 1915.[9]

On the distaff side, several authors in sound studies, and especially Karin Bijsterveld, have drawn readers' attention to the negative perception of noise in the emergence of noise abatement movements on both sides of the Atlantic.[10] In New York, Julia Barnett Rice formed a Society for the Suppression of Unnecessary Noise in 1906.[11] Britain had its Anti-Noise League, typical of the interwar polity of voluntary health associations.[12] Active debates were staged in public using contemporary media; the Science Museum in London was, for example, home to a temporary exhibition in 1935 on the theme. The catalogue essay by the League's president, the eminent physician Lord Tommy Horder, shows one of the ways that anti-noise sentiment was presented in public:

> Doctors are definitely convinced that noise wears down the human nervous system, so that both the natural resistance to disease, and the natural power of recovery from disease, are lowered ... Some people say that our nerves are so flexible, and have got such great reserves of energy, that they can adapt themselves without difficulty to noise ... It is true that our nerves have got this power of adjustment. If they had not we could not stand up to the conditions of modern life without becoming hopeless neurasthenics.[13]

James Mansell has shown that Horder's interpretation of the harm of noise via neurasthenia was already rather archaic by the mid-1930s.[14] But this should not cause us to underestimate the significance of anti-noise discourse – understood via several pragmatic and theoretical approaches – in the period. Notably, anti-noise discourse was threaded through with a dialectic of noise and peace, or noise and silence. That dialectic provided one of the ways in which the problem of noise was re-represented to people in the 1930s, via products, advertising and films, as will be clear below.

Paul Rotha's 1935 documentary *The Face of Britain*

Rotha had gained permission to make *The Face of Britain* as a side project, taking advantage of the monthly cross-country trips by car to Barrow-in-Furness where he was filming the construction of the cruise liner *Orion*, the subject of its twin film *Shipyard* for the company Gaumont-British Instructional.[15] Rotha structured *The Face of Britain* as a cinematic and historical dialectic about the impact of the first and second industrial revolutions on the British landscape.[16] In terms of its argument, in its sections on the industrial revolution and its outcomes, Rotha's film is highly critical of the social impact of *laissez-faire* capitalism, as was conventional for anyone on the left at this time. In terms of film technique, as I have described elsewhere, he derived his understanding from his close study at the editing bench of Russian films as well as such written texts as had by that point emerged from Russia, including almost certainly Eisenstein, Pudovkin and Alexandrov's joint 'Statement on sound'.[17] This influential couple of pages, published in English translation in the art film magazine *Close Up* in 1928, clearly expressed an anxiety that sound on film's technical capacity to synchronise speech would ineluctably result in the cinema becoming more theatrical, privileging more 'proscenium arch' styles of filmmaking that would tend to stress speech at the expense of anything more intrinsically cinematic. By contrast, the view of the authors was that the essence of film was that:

> The fundamental (and only) means, by which cinema has been able to attain such a high degree of effectiveness is montage[18] (or cutting). The improvement of montage, as the principal means for producing an effect, was the undisputed axiom on which was based the development of cinematography all over the world.[19]

It followed for them that:

> Only utilisation of sound in counterpoint relation to the visual montage affords new possibilities of developing and perfecting

montage. *The first experiments with sound must be directed towards its pronounced non-coincidence with the visual images.* This method of attack only will produce the requisite sensation, which will lead in course of time to the creation of a new *orchestral counterpoint* of sight-images and sound-images.[20]

In making these arguments, the authors were addressing more than simply aesthetic concerns; silent films – with some translation of intertitles – sold into an international market, whereas talkies were specific to particular language areas.[21]

Rotha proceeded on the basis of contrapuntal sound and vision, and in *The Face of Britain*, each of the dialectical sections – 'heritage of the past', 'the smoke age', 'the new power', 'the new age' – has a different sonic quality as well as differing pace in the editing and different typography.[22] Sonically, the first section, presenting a golden pre-industrial age, combines birdsong and pastoral-style string music with sparse narration. The second 'smoke age' section, on which I will concentrate here, has a much more experimental soundtrack, including a wordless rhythmic, percussive and metallic sequence representing industrial work in steelmaking, ceramic production, mechanical engineering and their industrial locales, that lasts for a whole minute within the overall 20-minute duration of the film. 'The new power', which introduces hydro-electricity and its transmission via the grid, mixes birdsong, water sounds, sparking noises, the whine of a test oscillator and music. 'The new age' starts with a montage of construction noises, which give way to abstract rhythmic sounds and short passages of music accompanying short scenes in the vision and commentary. No composer is credited for the music, but it is likely that Jack Beaver was responsible; he was working for Gaumont-British Instructional at the time, and is credited with the music for *Shipyard* and provided scores for many of their *Secrets of Nature* films.[23]

In 1937 the sound recordist William Frances Elliott[24] brought out a book, *Sound Recording for Films*, which he believed to be the first such publication.[25] His description of the *Face of Britain* soundtrack conveys the character of the 'smoke age' section of the film; his words – almost certainly unknowingly – echo the Futurists:

> One cannot but commend the freshness of invention as led to the use, in Paul Rotha's 'Face of Britain', of orchestrated *noises* … with the rhythm provided by practically every form of machine, it is not a very long step to taking some of the less agreeable noises, each invested with marked significance, and each with its individual

rhythm, and orchestrating them all into a symphony illustrative of the kaleidoscopic panorama of industrial background treated in the film.[26]

He went on to explain that 'actually a "musical" score was written for each of several unconventional "instruments". Hammers were beaten upon empty, resonating, iron tanks in exact tempo to the score, and so on.'[27] One of the two sound scripts for the film preserved in the Rotha Archive gives preliminary instructions:

> Then let sound build to a crescendo – using every type of machine, steam, forges, hooters, wind on mike, rhythmic pistons etc. Punctuate here and there with trams and hooters over the town shots. Climax comes on shot of Angel with Industrial Landscape.[28]

These were very early days in the documentarists' use of sound; the ubiquity of sound films in commercial cinemas made it a necessity for documentarists to have soundtracks for their films to stand any chance of gaining theatrical showings for them.[29] Rotha described the making of the soundtrack of *Face of Britain* in his autobiography, explaining that it would have been far too costly to take a 1930s sound truck with them on their journeys. Here he also revealed that Elliott, recordist on the film, was co-author of the soundtrack:

> The sound tracks … apart from a minimum of music, were fabricated in the back of the studio. With the imaginative help of the recordist W.F. Elliott, every kind of sound effect was conjured up by synthetic means, including a waterfall (the toilet was useful here), a flashover of a million-volt spark, shipyard riveting and so on. We bought a number of old disused cisterns and water tanks; they made a wonderful variety of sounds. A rawlplug driller made an excellent riveting sound. And so on … such synthetic fabrication of sound and the intermixing and overlaying of sound tracks (for some sequences six separate tracks were mixed together into one master track, to be mixed again with voices) were imaginative, stimulating and provided amusement.[30]

In sum, we see in Rotha's use of sound in *The Face of Britain* a commitment specifically to asynchronous sound, partially because of his conviction as a devotee of Russian filmmaking technique, and partially because the novelty of the technique and the unwieldiness of the equipment made any other technique impossible. As he said in his 1936 primer, *Documentary Film*, which he was writing as he edited the film:

> In the same way that we learnt how to create on the cutting-bench, to use the god-like vision of the camera to express in terms of relation and conflict, to dissolve our images one into another, to create tension and suspense by the juxtaposition of shot against shot, so we must employ the cutting-bench and the re-recording panel to give meaning and dramatic power to our sounds.[31]

Given all that can be said about Rotha's approach to the soundtrack of *The Face of Britain*, there remains the question of what kind of account of industrial society the film presents, and what can be said about the contribution of sound to that account. *Documentary Film*'s twenty pages devoted to sound technique vividly conveys his excitement about how sound montage could combine the 'raw materials' of location recordings, studio approximations and music, using the cutting bench and the re-recording panel to layer sounds creatively. He writes warmly about the potential of 'imagistic' use of sounds separated from synchrony with the objects that create them; he gives one explicit example from this film: 'in *The Face of Britain* the plea for slum clearance is ironically commented upon by shots of slums overlaid with the sound of explosions; but the slums remain unchanged.'[32] The sequence in the film runs thus:

Vision	Sound
Caption: 'The New Age'	Commentary: 'A great new world lies ready to be created. There is much to be done.'
Stoke on Trent landscape	[explosion]
Slum demolition	[shovelling sounds, explosion]
Slum demolition, including pneumatic drill	[pneumatic drill, explosion]
Slum demolition	Commentary: 'The heritage of the smoke age, the ghastly squalor, brought about by the uncontrolled spread of industry,
graveyard monument	chaos and filth of an obsolete age
Slums in Coatbridge (Lanarkshire coalfield)[33]	… the scarred and derelict ruins that today are seats of unemployment and misery
Worker on construction site riding on crane	can have no place in the new face of the land.'

This sequence provides a clear example of the 'contrapuntal' technique for sound and vision recommended by the Russian authors of the 'Statement on Sound'. The 'smoke age' section that I have concentrated on earlier uses precisely the kinds of sounds that excited the Futurists put to work in counterpoint to the vision track to represent industrial modernity. Furthermore, although the film is not directly concerned with noise – the commentary nowhere mentions it – the noise abaters' contrast between modern industrial or urban noise and rural peace is built into the film's dialectic, with the first and third sections conveying the comparative peace of the rural past and the rural source of hydroelectricity, and the second and last sections encompassing and representing the noise of modernity; first in industrialisation and latterly in the then current 'age of scientific planning, organisation of cooperation and collective working'. In this final section, the sounds include music, resonant water tanks struck with soft mallets, and commentary. The film ends with resolved string chords and fanfares accompanying views from the brow of a hill, echoing its first section.

Rene Clair's 1931 *À Nous la Liberté*

Bringing my second main example, *À Nous la Liberté*, into comparison indicates the potential breadth of study in this subject of the cinematic representation of sonic modernity, as this film, whilst also containing a critique of industrial modernity, could scarcely differ more from the earnestness of Rotha's documentary. Clair's film is a satire on the dehumanising nature of Taylorised work, as he stated in a 1952 interview: 'At the time I was closest to the extreme left, and I wanted to combat the machine when it becomes an enslaver of man rather than contribute, as it should, to his happiness … Above all [the film is directed] against the idea of the sanctity of work when it is uninteresting and non-individual'.[34]

À Nous la Liberté tells the story of Emile and Louis, two ex-convicts who, at the beginning of the film are experiencing the drudgery of a Fordist production line making toys in a prison. Early in the action, Louis manages to escape the prison, after which he is seen working his way up from selling gramophones to owning and running a gramophone factory that ultimately becomes entirely mechanised. Throughout the film, the prison metaphor for modern work is reinforced, with the same sets, lightly redressed, doing service for prison workshop, factory and cafeteria. The characters of the two main protagonists reinforce the anarchistic motif, typical of the surrealists: Louis is preternaturally lucky and – it seems – cannot help himself from becoming a successful capitalist, despite his ultimate disregard for the trappings of wealth. Emile, meanwhile, is a dreamer reluctant to succumb to the Taylorist yoke, and with a tendency to fall in love – his pursuit of Jeanne is one of the narrative drivers of the film's second half. In the end it is Emile's determination to seize the freedom of the vagrant's life that drives the film to its denouement in which the pals leave behind the world of capitalism and head for the open road.

Considered an exponent of 'pure cinema', Clair was one of those, like Rotha versed in silent cinema, who initially resisted the transition to sound; in common with the authors of the 'statement on sound' he disliked the tendency for 'talking films' to become wordy and theatrical, to become 'canned theatre', as he called it.[35] He promoted instead 'sound films', making use of asynchronous sound – a careful selection and organisation of sounds – 'to recapture some of the poetic energy that animated the silent cinema'.[36] Generically, Clair's sonic solution was musical; as he recalled:

> in order to avoid everything that might make it look like a message picture, I retained the operetta formula. I thought that À Nous la Liberté risked being heavy if treated realistically. I hoped that characters who expressed themselves in song would help put over the satirical nature of the film. And then also I wanted to get to the audience, and I thought that the bitter pill I was preparing would be more easily swallowed if it was coated in amusing music.[37]

We can recruit the author Kurt London to provide a definition of this archaic category of 'operetta film'; in his 1936 *Film Music* (which Elliott had read[38]), he explained the role of the score:

> the music accompanying the scenes which are without dialogue in a sound film is neither illustrative nor mimetic. It is an altogether new mixture of musical elements. It has to connect dialogue scenes without friction; it has to establish association of ideas and carry in developments of thought; and, above all this, it has to intensify the incidence of climax and prepare for further dramatic action.[39]

He explained that Clair had studied sound-film technique in Berlin at a time when Wilhelm von Thiele – director of *Chemin de Paradis* (1930), a French remake of a German operetta film – was experimenting with film operetta technique, matching the film's plot to musical rhythms.[40] According to London, this film 'was one of the first examples of what a modern mime, transferred to the medium of the sound-film operetta, should be'.[41] He also cites the composer Friedrich Holländer (composer for Hanns Schwarz's 1930 film *Einbrecher*) who had 'shown in some experiments how the music has to grow organically out of the rhythm of the pictures and their action. If it then expands into a song … then one can endorse the raison d'être of the theme song, because it is dramatically premised'.[42] This is a technique used at several points in *À Nous la Liberté*.

Georges Auric, the composer Clair chose to write the score for *À Nous la Liberté*, was at that stage beginning to become well established. Along with Louis Durey, Arthur Honegger, Darius Milhaud, Francis Poulenc and Germaine Tailleferre, he was counted one of 'Les Six'.[43] *À Nous la Liberté* was the second of his scores, after Cocteau's *Blood of a Poet* composed the year before, and he went on to write the music for around 130 films.[44] As Colin Roust has explained, Auric and Clair had moved in the same Parisian avant-garde circles since at least 1924, meeting at the Boeuf sur le Toit restaurant. Auric was a convert to Guillaume Apollinaire and Jean Cocteau's *esprit nouveau*, which in music rejected German Romanticism in favour of 'a new and distinctly French musical aesthetic inspired by the popular music found in Paris – that of the circus, the music-hall, the café-concert, and the street fair'.[45] Roust finds the spirit of Apollinaire and Cocteau abroad in Auric's published music criticism from 1922 to 1924, which rejected the confines of the Parisian musical establishment of the Conservatoire and Opéra in favour of kinds of music that would be *net* (plain) and *dépouillé* (stripped-down). In these reviews, Auric also made a play for full collaboration between film directors and composers so as to create a unified aesthetic. It seems that *À Nous la Liberté* became the exemplification of these principles; musically the soundtrack is chanson-like, tonal and melodic, making use of familiar musical forms including march, waltz and foxtrot. Equally, the collaboration with Clair was strong, with Auric gaining second billing ahead of the designer and actors; his presence on set during most of the filming; and the replacement of various originally scripted sound effects with his musical cues.[46]

In his survey of film music composers of different nations, Kurt London speaks warmly of Auric's music for the film: 'in collaboration with the gifted director, he succeeded not only in capturing the desired musical atmosphere, but … in giving a fitting musical finish to the type of musical film developed by Clair – following the self-same principles which we laid down in the section on the various forms of musical sound-film'.[47] In London's account, these principles relate to the opportunity within the true sound film for composers to attain 'an artistic uniformity of style', some of whose components included 'the detached item for a wordless scene (to-day naturalistic noises are dropping-out more and more)' and 'the musical number based on rhythm, with its preparatory stages', as in the example of Friedrich Holländer's technique described above.[48]

Roust identifies two major musical components within the film's semiotics; a march – first heard during the opening titles – represents the heroes' friendship and the freedom of life beyond capitalism. A waltz stands for an idealised vision of love; each has a full set of lyrics that are repeated complete or in excerpt at various points in the film.[49] But here, as my concern is with the film's use of contrapuntal technique in putting across its critique of industrial modernity, I will take two other examples.

Following the sequence that shows Louis's rise to capitalistic success, there is a two-and-a-quarter minute establishing sequence portraying work at his gramophone factory (starts around 14′30″). We see a shot of an industrial works, settling on the gramophone company's logotype displayed on a chimney; a dissolve (accompanied by a drum roll) takes us to a clocking-in machine and workers arriving at the factory, clocking in at a phalanx of machines. The only sonic accompaniment is wordless music closely matched to individual shots, with drums, triangle and percussion joined by sparse woodwind and brass phrases suggesting a march, at first tentatively then definitely, as the workers make their way to the production line. The march motif in a laborious form underpins the mechanised and repetitive work shown throughout the sequence. Workers are seen at conveyor belts attending to gramophones in several stages of production. We see a change of shift, with the workers being frisked to make sure they have removed no tools as in the earlier prison sequence, then taking their lunch at an identical conveyor belt carrying food (with the sparse piano and percussion motif, the same as accompanies the prisoners' meal in the opening sequence). The camera returns to external shots, with workers continuing to march in. The sequence ends on an exterior shot of the factory with the logotype in the foreground and another drum roll. In all, this exemplifies Holländer's technique of growing the music organically out of the rhythm of the pictures and their action. The images and music work in counterpoint to convey a critique of Taylorised work under somewhat militaristic conditions.

The concluding sequence of the film starts with the factory after it has become fully mechanised (around 76′50″). Two workers playing cards sit at a table with a bottle of wine whilst two conveyor belts disgorge complete gramophones behind them, to the accompaniment of a dance tune melody played by woodwind, brass and triangle. A drum roll introduces a march motif somewhat in circus style and the vision shows a panning shot of a geometrically arranged field of the machines. A succeeding pan takes the viewer out of the factory to show the workers, no longer needed in the factory, fishing in the river. Here a chorus singing a refrain about liberty replaces the march. The pan continues to reveal an outside dance floor where characters from the film are waltzing, including Jeanne with her lover.[50] This leads to the denouement where the heroes – one leaving behind a business, the other a sweetheart – make for the road. The music returns to the pals' march theme for the end title.

Arthur Knight has commented – echoing Rotha's contemporary judgement that 'we must employ the cutting-bench and the re-recording panel to give meaning and dramatic power to our sounds' – that 'what Clair had done, what creative directors everywhere were trying to do at the same time, was to discover how to control all the elements that went into the making of a sound film as completely as, in the simpler days of silence, one could control everything that went before the camera'.[51] *À Nous la Liberté* was typical of a group of films whose achievement, in Noel Burch's words, derives from 'a penchant for asynchronous sound based on a paradigm of montage juxtaposition as a means to manipulate, to interpret, and to reconstitute pro-filmic events'.[52]

There is more than coincidence to the fact that both Rotha and Clair selected the asynchronous approaches to the film soundtracks I have described; in a strong sense they were fellow travellers as cineastes. Rotha in particular saw himself as a filmmaker working within an understanding of cinematic technique and its history more broadly, as is evident in his critical writings; before he ever made a documentary he had written *The Film Till Now*, a major work on the history of world cinema.[53] His published criticism continued in parallel with his filmmaking, and Clair was one of the directors he admired. In the third edition of this book, he described *À Nous la Liberté* as 'Clair's masterpiece' in the context of his two previous sound films, *Sous les Toits de Paris* (1929) and *Le Million* (1930). Echoing the point in the 'Statement on Sound' that the talkies constrained international exhibition, Rotha enthused: 'at a time when the language barrier had bereft every European film industry of its foreign market, Clair's marvellous comedies with music were achieving huge success in New York, London and in every capital of Europe … It seemed that Clair had solved the problem of sound film form, and that in doing so he had restored the international appeal of national films.'[54] In a 1931 review of Clair's earlier work, he had stressed what Clair owed in his sound films to the film operetta tradition of Wilhelm Thiele (as we saw in Kurt London's account) and Ernst Lubitsch.[55]

Like *The Face of Britain*, *À Nous la Liberté* does not directly represent nor comment on the noise of industrial modernity. Rather, it uses sound, and explicitly music, to underpin the critique of uninteresting and non-individual work that is carried by the story. Musical themes are used throughout in counterpoint to the action of the film. The use of repeating motifs to signify the prison-like nature of Taylorised factories, compared with choral, chanson-like passages in march time to underpin liberation from the tyranny of work, is one of the main sonic devices of the film. It is worth stressing that, where Rotha and Elliott used industrial sounds as a key component of their soundtrack, Clair permitted Auric to replace various intended naturalistic sound effects with musical cues instead. Recall that Kurt London's view had been that 'to-day naturalistic noises are dropping-out more and more'. In one of his characteristically provocative pieces of journalism, Clair had complained in 1929 of the tedium of synchronised and banal sound effects, proceeding to argue 'we must draw a distinction here between those sound effects which are amusing only by virtue of their novelty (which soon wears off), and those that help one understand the action, and which excite emotions which could not have been roused by the sight of the pictures alone'.[56] It is easy to see that Auric might quite readily have persuaded him to make the substitution.

Conclusion: the empty category

Rotha and Clair in these very different films used contrapuntal sound and vision tracks in distinct, but not unrelated, ways to convey their critiques of aspects of modernity. Honouring the principle of following historical actors' categories, we may follow the already quoted 1930s authors W.F. Elliott and Kurt London to explore how our differing examples can help us to understand the broader picture of the sonic representation of industrial modernity. Elliott's account gives us access to contemporary debates because, writing as a professional sound recordist for films, as early as ten years after *The Jazz Singer* he already had a sense of changing practice; in this chapter's argument, he stresses particularly sound and music for industrial scenes.[57] Elliott's discussion presents *The Face of Britain* as a contrast with an already discarded tradition. Giving the example of Fedor Ozep's 1931 film of *The Brothers Karamazov*, he describes good soundtrack practice as the use of music to suggest all the sonic aspects of a scene (as we have also seen in the case of Clair's film). Bringing to mind Auric's technique in *À Nous la Liberté*, he comments that 'scenes in railway stations, with departures of trains, were expressed entirely by orchestration, yet the spectator remained with the impression of having heard all the sounds incidental to such scenes'.[58] Kurt London was in agreement; for him, Ozep's *Karamazov* was 'exemplary in its form, its rhythm which follows the course of the picture, and its success in seizing the atmosphere of Dostoevsky's masterpiece'.[59] The success in his view derived from the fact that the director cut his visuals to the completed score, 'the composer [Karl Rathaus] could create a form complete in itself on the basis of the pictures already taken, and this form was regarded as the foundation of the final structure to be attained after cutting'.[60] We may see this as a different route to the kind of result achieved by close working between Clair and Auric.

But Elliott went on to decry the over-application of the technique as it developed in the hands of Ozep's and Rathaus's imitators:

> Unfortunately, as sometimes happens, the lesson was learned all too fully by all and sundry, and thereafter the ordinary cinema-goer learned that the appearance of machinery upon the screen was the cue for "machine music", a convention that has persisted almost to the same degree as the theme-song.[61]

Elliott's endorsement of the technique used on the *Face of Britain* soundtrack follows immediately after this statement, with no attempt made to populate this fascinating category with any examples. There is not the opportunity here to seek or to discuss any further examples, but it is all the same a fascinating proposition for further research: a long forgotten overworked cliché of 'machine music' developed between the fourth and the tenth year of the 'talkies' and already superseded by the later date (1931–1937).[62] Elliott's argument proceeds instead to endorse the benefit of the recordist understanding 'the artistic concept of the entire film in advance' so as to 'be able to produce a more harmonious and well-balanced result than his less-penetrating *confrères*', an opinion that aligns well with the views on film music of several authors discussed above.[63] This endorsement on behalf of recordists of the kind of composer-director relationship promoted by Auric provides good evidence that interest in sound technique at that point extended beyond montage-trained directors to sound experts too. But this also may not have had currency for long, if at all; it is striking that John Huntley's influential *British Film Music*, published a decade later, did not cite *Sound Recording for Films*. This suggests that the informed marriage of music and sound recording technique recommended by Elliott in 1937 did not become the normal way of thinking about the soundtrack, any more than the operetta style triumphed in the international market much beyond the relative success of *À Nous la Liberté,* or than post-synchronisation became the normal recourse for all documentary sound.[64]

It is clear that once film on sound became a possibility, any filmmaker telling a story about, representing or critiquing industrial society would have a whole new sense dimension to employ. The evidence presented in this essay shows that industrial society provided particular opportunities for those self-conscious filmmakers who wanted to experiment with the soundtrack, but who wanted to do so in a way that extended the techniques of juxtaposition and rhythm that were the basis of film construction in the silent cinema. These are very different examples: *The Face of Britain* is a serious-minded documentary, using historical dialectic to describe the changing landscape and society of Britain transformed by the first and second industrial revolutions, whilst *À Nous la Liberté*, by contrast, is a romp, an anti-capitalist squib against the tyranny of work. As objects of reflection and study – both in the 1930s and today – they provide vivid examples of how filmmakers could use sounds and noises to represent modern society back to itself. At the same time as they suggest categories of sonic object – including 'machine music' – worthy of extended consideration, they also act as testaments to the sound worlds of our antecedents in High Modernity.

Notes

1 Emily Thompson, *The Soundscape of Modernity: Architectural Acoustics and the Culture of Listening in America, 1900–1933* (Cambridge, MA: MIT Press, 2004), 1.

2 Marshall Berman, *All That is Solid Melts into Air: The Experience of Modernity* (New York London: Viking Penguin, 1988), 36.

3 See for an assessment: T.M. Boon, 'Sounding the Field: New Work in Sound Studies (Essay Review)', *British Journal for the History of Science*, 3 (2015): 493–502.

4 Rick Altman, ed. *Yale French Studies*, 60, Cinema/Sound (Yale French Studies, 1980); Elisabeth Weis and John Belton,

eds. *Film Sound: Theory and Practice* (New York: Columbia University Press, 1985). See discussion in *Sound, Speech, Music in Soviet and Post-Soviet Cinema*, eds. Lilya Kaganovsky and Masha Salazkina (Bloomington; Indianapolis: Indiana University Press, 2014), 2.

5 Rick Altman, *Silent Film Sound* (New York: Columbia University Press, 2004), 28. For a valuable collection devoted to the British context, see Annette Davison and Julie Brown, eds., *The Sounds of the Silents in Britain* (New York: OUP, 2012).

6 A somewhat broader account is given by Karin Bijsterveld in *Mechanical Sound: Technology, Culture, and Public Problems of*

Noise in the Twentieth Century (Cambridge, MA: MIT Press, 2008), especially 137–58.

7   Quoted in Thompson, *Soundscape*, 135–36.

8   Thompson, *Soundscape*, 136.

9   For a more extensive discussion, see Luciano Chessa, *Luigi Russolo, Futurist: Noise, Visual Arts, and the Occult* (Berkeley: University of California Press, 2012).

10 Karin Bijsterveld, *Mechanical Sound: Technology, Culture, and Public Problems of Noise in the Twentieth Century* (Cambridge, MA: MIT Press, 2008), 91–136.

11 Thompson, *Soundscape*, 121.

12 Bijsterveld, *Mechanical Sound*, especially 91–136; James Mansell, *Hearing Modernity: Sound and Selfhood in Early Twentieth-Century Britain* (Urbana, IL: University of Illinois Press, 2016).

13 Noise Abatement Exhibition Catalogue, 1–2. See also James Mansell's dissection of these claims around neurasthenia in Mansell, *Hearing Modernity*.

14 Mansell, *Hearing Modernity*.

15 Paul Rotha, *Documentry Diary: An Informal History of the British Documentary Film, 1928–1939* (London: Secker and Warburg, 1973), 102.

16 For a detailed investigation of *The Face of Britain*, see T.M. Boon, '"The Shell of a Prosperous Age": History, Landscape and the Modern in Paul Rotha's *The Face of Britain* (1935)', in *Regenerating England: Science, Medicine and Culture in the Interwar Years*, eds. C.J. Lawrence and Anna Mayer, *Clio Medica*, vol. 60 (Amsterdam: Rodopi, 2000), 107–48.

17 T.M. Boon, '"The Shell of a Prosperous Age", 107–48; S. Eisenstein et al., 'Statement on sound', which had been published in the film art magazine *Close Up* in 1928. See James Donald, Anne Friedberg and Laura Marcus, *Close Up 1927–1933: Cinema and Modernism* (Princeton University Press, 1999). I have not found a specific citation to the essay in Rotha's published work. In *The Film Till Now*, in a passage alluding to Pudovkin's views, he states a similar conclusion to theirs: 'the sound of rain, leaves, animals, and birds; of trains, cars, machines and ships. These are to be woven into a unity in counterpoint with their visual images, but never in direct conjunction with them' (London: Jonathan Cape: 1930), 310; see 303–11. He was certainly acquainted

with V.I. Pudovkin, *Pudovkin on Film Technique: Five Essays and Two Addresses*, trans. Ivor Montagu, second edition (London: George Newnes, Ltd., 1933 [1929]), which contains two chapters on sound.

18 The original uses the term 'mounting'; for the purposes of clarity, I have reverted to the more normal term.

19 Eisenstein et al., 'Statement on Sound', 83.

20 Eisenstein et al., 'Statement on Sound', 84. Italics in the original.

21 Eisenstein et al., 'Statement on Sound', 84.

22 James Mansell provides a parallel commentary on the sonic quality of the GPO Film Unit's films, and especially under the influence of Alberto Cavalcanti in his essay, 'Rhythm, Modernity and the Politics of Sound', in *The Projection of Britain: A History of the GPO Film Unit*, eds. S. Anthony and J. Mansell (Basingstoke: BFI/Palgrave, 2011), 161–67.

23 The BFI database entry for *Shipyard* notes Beaver as the uncredited composer for that film. There is certainly a strong stylistic similarity in the writing for strings between Rotha's two 1935 films. Beaver was one of the coterie of composers who worked under Gaumont-British's resident music director, Louis Levy; John Huntley, *British Film Music* (S. Robinson, 1947), 33. Huntley notes that Beaver composed all the music for *Secrets of Life* from 1934–46 (104–05).

24 Forenames according to one BL Library entry.

25 William Frances Elliott, *Sound Recording for Films: A Review of Modern Methods* (London: Sir Isaac Pitman and Sons, 1937), vii.

26 Elliott, *Sound Recording for Films*, 77.

27 He continues: 'This is but one example of the relation of music to other sounds, quoted to point the way for individuals who will be able to evolve their own ideas and elaborate upon them'; Elliott, *Sound Recording for Films*, 77. James Kennaway has pointed out to me that a similar sound is audible in Wagner's Rheingold, which can be viewed here: www.youtube.com/watch?v=3ZP-yXsNV2E at 1hr 08'10", accessed June 2018.

28 'Sound Script for "The Face of Britain"', undated [1935], *Face of Britain* documents, Box 12 Rotha Papers, (Collection 2001). Department of Special Collections, Charles E. Young Research Library, University of California, Los Angeles.

29 Conventionally, *Workers and Jobs* (1935) and *Housing Problems* (1935) are presented as the first documentaries to use location sound. See, for instance, Rachael Low, *Documentary and Educational Films of the 1930s*. (London: George Allen and Unwin, 1979), 120–01.

30 Rotha, *Documentry Diary*, 104–05.

31 See Rotha, *Documentary Film*, 1st edition (London: Faber, 1936), 203–04, and, more extensively the section 198–223.

32 Rotha, *Documentary Film*, 222.

33 Identified in plate opposite 276 in Clough Williams-Ellis, ed., *Britain and the Beast* (London: J.M. Dent & Sons, 1937). The caption reads 'Unplanned – a glimpse of Coatbridge'.

34 Quoted in: R.C. Dale, *The Films of René Clair, Vo1.1: Exposition and Analysis* (Metuchen, NJ: Scarecrow Press, 1985), 187, quoted in turn from Georges Charensol and Roger Regent, *Un Maitre Du Cinema, René Clair* (Paris: La Table Ronde, 1952). Also in René Clair, *À nous la liberté and Entr'acte. Films by René Clair*, trans. and description of the action by Richard Jacques and Nicola Hayden (Classic and Modern Film Scripts. no. 22) (London: Lorrimer Publishing, 1970), 9.

35 Arthur Knight, 'The Movies Learn to Talk: Ernst Lubitsch, René Clair, and Rouben Mamoulian', in *Film Sound: Theory and Practice*, eds. Elisabeth Weis and John Belton (New York: Columbia University Press, 1985), 213–20, 216.

36 Elisabeth Weis and John Belton, *Film Sound : Theory and Practice* (New York ; Guildford: Columbia University Press, 1985), 77.

37 Quoted in R.C. Dale, *The Films of René Clair* (Metuchen, NJ: Scarecrow Press, 1985), 187, quoted in Charensol and Regent, *Un Maître Du Cinema, René Clair*.

38 Elliott, *Sound Recording for Films*, 71.

39 Kurt London, *Film Music: A Summary of the Characteristic Features of Its History, Aesthetics, Technique; and Possible Developments*, trans. Eric Sigmund Bensinger (London: Faber & Faber, 1936), 135.

40 Charles O'Brien describes this film as a French alternative to the theatrical style of *The Jazz Singer*; Charles O'Brien, *Cinema's Conversion to Sound: Technology and Film Style in France and the U.S.* (Bloomington: Indiana University Press, 2004), 73–77.

41 London, *Film Music*, 129.

42 London, *Film Music*, 130.

43 Richard Taruskin, *Music in the Early Twentieth Century. The Oxford History of Western Music*, vol. 4. (Oxford: Oxford University Press, 2010), 588.

44 Colin Roust, '"Say It With Georges Auric": Film Music and the Esprit Nouveau', *Twentieth Century Music* 6, no. 2 (September 2009): 133–53, 134.

45 Roust, '"Say It With Georges Auric"', 135. For Auric see also Colin Roust, 'Sounding French: The Film Music and Criticism of Georges Auric' (PhD thesis, University of Michigan, 2007); François Amy De la Bretèque, 'Des Compositeurs de Musique Viennent Au Cinéma : Le "Groupe Des Six"', 1895, 38 (2002): 2, para 12.

46 Roust, '"Say It With Georges Auric"', 138–41, 43.

47 London, *Film Music*, 238.

48 London, *Film Music*, 157.

49 Roust, '"Say It With Georges Auric"', 146–50.

50 A script for the film, reconstructed from viewings has been published: *René Clair. À nous la Liberté and entracte.*

51 Knight, 'The Movies Learn to Talk', 217.

52 Noel Burch, 'On the Structural Use of Sound', in Weis and Belton, eds., *Film Sound*, 200–09, 266.

53 Paul Rotha, *The Film Till Now* (London: Jonathan Cape, 1930).

54 Paul Rotha and Richard Griffiths, *The Film Till Now: A Survey of World Cinema*, rev. and enl. ed. (New York: Vision Press, 1949), 526.

55 Paul Rotha, *Celluloid; The Film To-Day* (London; New York: Longmans, Green and Co., 1931), 181–95. Clair stated that he had learned from Chaplin in his visual comedy (a debt repaid by Chaplin's borrowing of themes from *A Nous la Liberté* for *Modern Times*); for the influence between Clair and Chaplin, see Rotha, *The Film Till Now* (1949 ed.), 526 and René Clair and Georges Charensol, 'A Note on this Edition', *À Nous la Liberté and Entr'acte: Films by René Clair,* Classic and Modern Film Scripts. no. 22. (London: Lorrimer Publishing, 1970), 10.

56 René Clair, 'The Art of Sound', in Weis and Belton, eds. *Film Sound*, 92–95.

57 I have located no other written work by Elliott, despite a search of likely periodicals (*Cinema Quarterly, World Film News, Cine Technician*). Biographical information on him is sparse; he had started in radio, is said to have gained a breadth of outlook on film sound from his work on post-synchronising and adapting

English and American films into other languages, had worked abroad and, by the time of the book's publication, was in charge of the production recording department of British Acoustic Films; Rotha, foreword to Elliott, *Sound Recording for Films*, vi.; Anon. 'Sound Recording for Films (review)'. *World Film News* 2, no. 9 (December 1937): 39.

58   Elliott, *Sound Recording for Films,* 76. The film is not currently available on DVD, but is on YouTube with Spanish subtitles (accessed 19 Aug 2015): see www.youtube.com/watch?v=A98tbzesh E0&feature=youtu.be, accessed June 2018.

59   London, *Film Music,* 225. See also 157 for a similar point.

60   London, *Film Music,* 226.

61   Elliott, *Sound Recording for Films,* 76.

62   The most prominent 'machine' film of the previous year (1936) was undoubtedly William Cameron Menzies's film of H.G. Wells's *Things to Come*. Is this a candidate for the laborious use of 'machine music'? Certainly Kurt London criticised Arthur Bliss's symphonic style in the film's score: 'his orchestra, a big symphony orchestra, has not yet managed to free itself from the symphonic tradition. But the microphone is indeed a problem which even the most prominent musicians have to solve for themselves in practice. In future scores Bliss too will no doubt revise his style'; London, 217–18. See John Huntley, *British Film Music,* 39–40 for an ambivalent account.

63   Elliott, *Sound Recording for Films,* 77.

64   On the reception of the film, see Charles O'Brien, *Cinema's Conversion to Sound: Technology and Film Style in France and the U.S.* (Bloomington: Indiana University Press, 2004).

3

Woolf's atom, Eliot's catalyst and Richardson's waves of light: science and modernism in 1919

Morag Shiach

Introduction

In this chapter, I explore the ways in which three modernist writers engaged with scientific ideas and deployed explicitly scientific metaphors in the year 1919. The aim is to generate new insights into the extent to which, at this particular historical moment, the theorisation and the creation of what was understood as 'modern' writing happened in the interstices between science and literature.

The writers discussed in this chapter are T.S. Eliot, Virginia Woolf and Dorothy Richardson. These writers have been chosen firstly because they are central to diverse accounts of the specific characteristics of modernist writing and the analysis of their work may thus suggest some wider applicability of the arguments to 'modernism' more broadly; and secondly because they each drew on and reworked the languages of science in important ways in this period. The specific metaphors of 'the atom', 'the catalyst' and 'waves of light' have been chosen because they enable a reading that is attentive both to the theoretical and the literary works of these writers. And finally, the year 1919 is chosen because it offers a particularly rich moment within the broad and complex processes of scientific and literary exchange in the early twentieth century. Focussing specifically on this year, it is possible to map with some precision networks of transmission between scientific and literary writings through attention to journals, little magazines and a range of literary and scientific publications from that year, as well as drawing on diaries, letters and other forms of historical evidence.[1]

Woolf's Atom

Virginia Woolf published her essay 'Modern Novels' in the *Times Literary Supplement* on 10 April 1919.[2] The essay (better known in its revised version, 'Modern Fiction', which appeared in *The Common Reader* in 1925[3]) offers a critique of naturalism, or of what Woolf calls 'materialist' modes of fiction, and also develops a theory of a radically different type of 'modern' writing. Woolf's claim in the essay is that early-twentieth-century naturalist fictional writings such as those of Arnold Bennett, John Galsworthy or H.G. Wells are damaged by excessive materialism: 'the great clod of clay that has got itself mixed up with the purity of ... [their] inspiration'.[4] She further argues that:

> Mr. Bennett is perhaps the worst culprit of the three, inasmuch as he is by far the best workman. He can make a book so well constructed and solid in its craftsmanship that it is difficult for the most exacting of critics to see through what chink or crevice decay can creep in. There is not so much as a draught between the frames of the windows, or a crack in the boards.[5]

But why does this absolute solidity ('solid in its craftsmanship') and hermetic sealing (no 'chink' or 'crevice') seem to Woolf so undesirable, and so inappropriate for a 'modern' novel? Why is she concerned about a style of fiction that does not allow for decay, for draughts, or for gaps in the fictional fabric ('a crack in the board')?

One sort of answer to this might be found through a consideration of her fictional writings. *Night and Day* (1919) is commonly thought to be one of Woolf's least experimental or 'modern' novels, and in its narrative frame and structure this is surely true. But Woolf does nonetheless show even within this novel a commitment to exploring the less solid and more mobile aspects of human subjectivity ('the faintly lit vastness of another mind, stirring with shapes, so large, so dim, unveiling themselves only in flashes and moving away again into the darkness') and of the external world ('Moments, fragments, a second of vision, and then the flying waters, the winds dissipating and dissolving').[6] The novel is overall certainly rather solid, and even well made, with a series of chapters setting characters firmly in their social contexts and also with plotting that works through the possibilities of different forms of romantic coupling and ends (rather less than more conventionally) with the prospect of marriage. But the evocation of the transient, of the fragment and of the wind that 'dissipates' suggests a searching after less solid and less materialist foundations. Woolf is thus in *Night and Day* (which was being finalised the month before she was to publish 'Modern Novels')[7] already developing elements of what would come to be seen as characteristically 'modern' aspects of her fiction.

Such aspiration to unravel the apparent solidity and materiality of the external world as well as of human character was certainly to remain important for Woolf's fictional writing. Looking some years ahead, in *To the Lighthouse* (1927), we find in the central section 'Time Passes' writing that seems to echo the language she used in 'Modern Novels', as she evokes the importance of the cumulative creative impact of small intrusions and perturbations thus:

> So loveliness reigned and stillness … among the shrouded jugs and sheeted chairs even the prying of the wind, and the soft nose of the clammy sea airs, rubbing, snuffling, … the falling cries of birds, ships hooting, the drone and hum of the fields, a dog's bark, a man's shout… . Once only a board sprang on the landing.[8]

The 'crack in the boards' and 'draught between the frames of the windows' that Woolf was advocating in 'Modern Novels' here find much fuller literary expression and allow for a condensed and intense exploration of the passing of time.

Woolf's explanation of what was at stake in her overall argument in 'Modern Novels' is related to her claim in the essay that a 'materialist' style of fiction necessarily misses the core of human experience: 'Life escapes; and perhaps without life nothing else is worth while … Let us hazard the opinion that … the form of fiction most in vogue more often misses than secures the thing we seek.' This invocation by Woolf of a collective subject, a 'we' (though a common essayistic device), can be experienced as a coercive form of rhetoric that assumes a shared purpose and set of values amongst the essay's readers. But it is also an attempt at radical thinking, which aims to produce as well as to theorise a new audience for the 'modern' novel. If we are able to accept that there is indeed a shared 'thing we seek', then Woolf's argument is that this thing cannot be captured in a naturalist mode of literary representation. Over-emphasis on plot and on the naturalistic details of everyday life in early-twentieth-century novels have, she argues, created a barrier to the fictional representation of the most important aspects of life.

In seeking to explain in 'Modern Novels' the key features of a modern style of fiction that might enable the richer representation of 'life', Woolf chose to draw on concepts borrowed from the physical sciences. Specifically, in this essay she used the image of the atom as part of her argument that the most compelling aspects of life that the novelist should capture are best understood as the cumulative effect of multiple minute sensations and experiences rather than as a solid and readily quantifiable experience of the material. She further deployed the metaphor of the atom to suggest the importance of abstract pattern 'however disconnected and incoherent in appearance' to the processes underpinning human cognition:

> The mind, exposed to the ordinary course of life, receives upon its surface a myriad impressions – trivial, fantastic, evanescent, or engraved with the sharpness of steel. From all sides they come, an incessant shower of innumerable atoms, composing in their sum what we might venture to call life itself… . Let us record the atoms as they fall upon the mind in the order in which they fall, let us trace the pattern, however disconnected and incoherent in appearance, which each sight or incident scores upon the consciousness.[9]

'Life', the thing that for Woolf it is the business of fiction to capture, is here represented as an 'incessant shower of innumerable atoms' that generate impressions in the mind. The image is perhaps a strange one, suggesting that the atom is being imagined as having something of the substance of water, which does not fit closely with the theories being developed by physicists in 1919. But actually Woolf's emphasis here is on the 'incessant' nature of the impressions being described, and on their 'evanescent' quality, which does resonate with some of the characteristics of the atom as described, for example, by Rutherford as argued below. These 'innumerable atoms' also, Woolf argues, generate a complex pattern that can be grasped, represented and communicated in a way that enables modern fiction to represent modern subjectivity as well as fundamental elements of the modern world. The pattern generated by atoms becomes the substance of the cognitive process Woolf is trying to describe.

Craig A. Gordon, in *Literary Modernism, Bioscience, and Community in Early 20th Century Britain* (2007), has suggested that turning to atomic theory at this stage in Woolf's argument involves 'a surprisingly materialist language' that 'sought to apply the principles of physical science to the human psyche', and that 'having set up the "spiritual" project of modern fiction, Woolf chooses to approach the space of psychological interiority … in terms not simply of the mind, but of the atomic interactions that constitute the bodily processes upon which consciousness depends'.[10] Gordon is arguing here that Woolf's critique of materialist writing takes her finally only to another form of materialism, through the image of the atom to the physical and then the psychological sciences. I argue below, however, that the metaphorical associations of 'the atom' for Woolf or indeed for any other writer in 1919 were not so resolutely materialist, and that use of the atom as a metaphor was intended rather to suggest gaps, 'crevices' and unstable elements in the fabric of everyday life.

A number of recent historians and theorists of modernism have examined the ways in which modernist poets and novelists might have come to be aware of developments in contemporary scientific thought, from psychology to linguistics, physics, chemistry or biology.[11] In *Science For All* (2009), Peter J. Bowler maps in great detail the construction of different audiences for scientific knowledge, noting the development of popular science magazines, the role of public science lectures, and the coverage given to scientific topics within the popular and broadsheet press. He discusses a range of publications that sought to build understanding of developments within atomic physics in the early years of the twentieth century, including Charles R. Gibson's *Autobiography of an Electron* (1911) and Frederick Soddy's *Interpretation of Radium* (1909) and argues that 'almost any educated person was expected to know at least the terminology of the new science and something of its application'.[12] It is also clear from the evidence of these various studies that journals such as *The Athenaeum* and little magazines such as *The Egoist* played a major role in enabling the exchange of information about scientific discoveries, methods and ideas across a broad literate public. I will thus draw on popular science publications as well as on material published in both these journals in 1919 in order to analyse more precisely how modernist writers came to access, and to work creatively with, images and ideas from contemporary science, and in particular to understand what they might have known about the atom.

Woolf began contributing to *The Athenaeum* in 1919, when J. Middleton Murry took over the editorship, noting in her diary on 21 February that 'I am asked to write for the Athenaeum, so that little scratch in my vanity is healed'.[13] Murry, as well as his wife Katherine Mansfield, were personal friends as well as professional colleagues of both Leonard and Virginia Woolf, and the exchange of ideas and texts between them was regular and frequent. Further references to *The Athenaeum* in her diary suggest that Woolf was a close reader of it throughout 1919. It is thus not unreasonable to suggest that she would have seen the notice of Ernest Rutherford's lecture on 'Atomic Projectiles and their Collisions with Light Atoms' that appeared in the June 1919 issue, and that she was likely to have been familiar with key aspects of Rutherford's work on the atom in this period. Rutherford was by 1919 based in the Cavendish Laboratory in Cambridge. The intellectual and scientific world of Cambridge was well known to Woolf through many friends and relations, and was also on her mind in early 1920 (a few months after 'Modern Novels' and Rutherford's lecture) when she began to write *Jacob's Room*, with its key sections set in Cambridge: 'If any light burns above Cambridge, it must be from three such rooms: Greek burns here; science there; philosophy on the ground floor.'[14]

Rutherford's lecture began with the assertion that: 'the discovery of radio-activity … has provided us with the most powerful natural agencies for probing the inner structure of the atoms of all the elements' and he then argued that his experimental design was able to 'obtain important evidence on the strength and distribution of the electric fields near the centre or nucleus of the atom'.[15] This search for 'inner structure' and the pattern of distribution of electric fields resonates interestingly with Woolf's search for 'pattern, however disconnected and incoherent in appearance' in 'Modern Novels'.

Research into the instability of radioactive atoms, and the associated development of new models of the atom, were key elements of modern physics. The scientific understanding of the structure of the atom was undergoing significant revision and the atom was representable only through explication of the outcomes of specific experiments or through the development of abstract 'models' (the 'plum pudding' model of J.J. Thomson [1904]; the 'planetary' model of Rutherford [1911]; the 'Bohr' model [1913]).

Rutherford's 'Gold Foil Experiment' was associated with the development of the 'planetary model' of the atom, and demonstrated that the vast majority of the volume of an atom is empty space (Figure 3.1).

Rutherford published his research on the disintegration of the nitrogen atom by alpha particle bombardment, and the related discovery of the proton, in the *Philosophical Magazine* in June 1919 (though the work had been done some two years earlier).[16] The Rutherford lecture on 'Atomic Projectiles' advertised in *The Athenaeum* in that same month shows that Rutherford's discoveries in this period were being actively disseminated, something that is also clear from Peter J. Bowler's studies of the popularisation of science.

This certainly suggests that the complexity and immateriality of the atom were very likely to have been part of Woolf's thinking as she wrote her essay on 'Modern Novels', and also that her turn to the physical

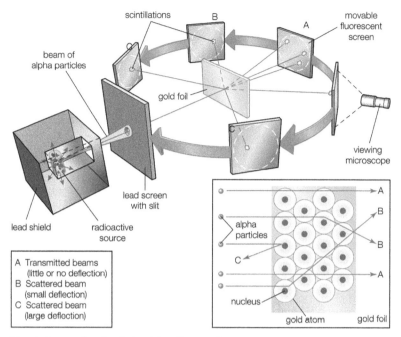

Fig. 3.1 Rutherford's 'Gold Foil Experiment'.

Universal Images Group Ltd/Science & Society Picture Library.

sciences was intended to support her understanding of modernist fiction as a dynamic and abstract capturing of 'life' with its gaps and crevices, rather than being a return to the kind of materialism she saw as so limiting. Further evidence for such a reading can be found in other fictional texts where Woolf draws on the image of the 'atom'. In *The Years* (1937), in a section of the novel set in 1908, we can find the following passage:

> Was someone coming in? She listened. No, it was the wind. The wind was terrific. It pressed on the house; gripped it tight, then let it fall apart... . How little she knew about anything. Take this cup for instance; she held it out in front of her. What was it made of? Atoms? And what were atoms, and how did they stick together? The smooth, hard surface of the china with its red flowers seemed to her for a second a marvellous mystery.[17]

The initial reference to the wind in this passage connects it to other passages of Woolf's writing discussed above, and generates a sense of change, vulnerability and lack of permanence. This powerful wind is in some kind of tension with the 'marvellous mystery' of the smooth hard surface of the china cup, but this solid, hard surface is then in turn undercut by the allusion to the atoms that constitute it, but hold together in such an unconvincing way ('how did they stick together?'). Solidity fights with impermanence here, and imaginatively the latter prevails.

The semantic associations of the atom in Woolf's writing do vary. We have seen that in 'Modern Novels' it was associated with incessant and innumerable sensations, while in *The Years* it signalled a sense of unreality when confronted with the solidity of material objects. In *The Waves* another set of associations with the atom are generated when Bernard's sense of self is articulated as follows: 'I remarked with what magnificent vitality the atoms of my attention dispersed, swarmed round the interruption ... and had created, by the time I put back the receiver, a richer, stronger, a more complicated world'.[18] The atom is here psychologised, and is part of the dynamic construction of Bernard's personality over time and across the novel, but it retains a link to ideas of energy, of dispersal and of complexity. The same kinds of association can indeed by found on the first page of the novel, where we read of 'one incandescence which lifted the weight of the woollen grey sky on top of it and turned it into a million atoms of soft blue'.[19] The atom has layers of semantic possibility for Woolf, but it would seem that it is always associated with elements of the world that are experienced as immaterial and impermanent, as 'a million atoms of soft blue'.

One final example that further suggests the metaphorical richness of the atom for Woolf can be found in her short story 'An Unwritten Novel', which we know from her diary she was working on at the very beginning of 1920. The story's narrator is on a train, and starts to build an imaginary narrative around a woman sitting opposite. Parts of the story are then told from the point of view of this imagined woman (the end of the story will reveal that the narrator's speculations are far from accurate). In this story the atom retains its association with pattern and deep internal structure, but is used to suggest the potential power of such pattern (or one might say of the aesthetic) to resist the 'great clod of clay' that Woolf was so concerned by in 'Modern Novels': 'how the mud goes round in the mind ... till by degrees the atoms reassemble, the deposit shifts itself, and again through the eyes one sees clear and still'.[20]

This interest in the 'atom' in the early twentieth century was not in any sense an idiosyncrasy of Woolf's, though the metaphorical richness generated across her various texts is unusual. Looking at Woolf's own account of the intellectual life of the 1920s in *A Room of One's Own*, we find a representation of the typical areas of debate within a well-endowed university as including: 'archaeology, botany, anthropology, physics, the nature of the atom'.[21] Interest in developments in atomic theory was, as we have seen, generated through a variety of means including lectures, press articles and popular books. One of these, *The ABC of Atoms*, was written by Bertrand Russell, friend of Woolf and associate of the Bloomsbury group, in 1923. Russell begins by suggesting the extent to which atomic theory is a challenge to the perception of the world as solid and material: 'To the eye or to the touch, ordinary matter appears to be continuous … science, however, compels us to accept quite a different conception of what we are pleased to call "solid" matter'.[22] He then goes on to develop a metaphorical account of the structure of the atom that would surely intrigue any novelist, 'the nucleus of any atom except hydrogen is a tight little system, which may be compared to a family of energetic people engaged in a perpetual family quarrel'.[23]

In conclusion, the atom was clearly an object of scientific interest and inquiry in 1919, but it was also an entity of much wider cultural interest. Figure 3.2 maps the frequency of usage of the word 'atom' across all English-language texts since the eighteenth century.[24] One can see a very clear increase in usage of the word 'atom' in the early years of the twentieth century, with the highest 'spike' occurring just around 1919. This was, it would seem, a year of the atom.

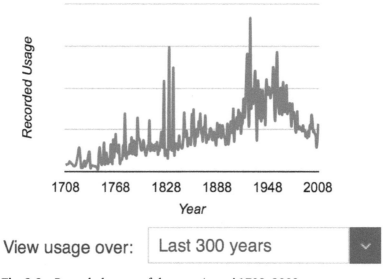

Fig. 3.2 Recorded usage of the term 'atom' 1708–2008.

Eliot's catalyst

As well as publishing various pieces by Woolf and announcing a lecture by Rutherford, in 1919 *The Athenaeum* was also promoting the work of T.S. Eliot, including advertising his lecture on 'Modern Tendencies in Poetry' given to the Arts League of Service in October 1919.[25] In the same year, the journal also published a review of Eliot's *Poems*, which had been recently published by the Hogarth Press (run by Leonard and Virginia Woolf) entitled 'Is this Poetry?'. This anonymous review was in fact written by Virginia and Leonard Woolf. In seeking to capture what they saw as Eliot's novelty and modernity, they asserted that 'Mr Eliot is always quite consciously "trying for" something, and something which has grown out of and developed beyond all the poems of all the dead poets'. What was new about Eliot's poetic method was explicitly associated in the review with the methods of a scientist: 'poetry to him seems to be not so much an art as a science'. The Woolfs also concluded that Eliot has the attitude 'not of the conventional poet, but of the scientist who with the help of working hypotheses hopes to add something, a theory perhaps or a new microbe, to the corpus of human knowledge'.[26]

Given this 'attitude of the scientist', it is perhaps not surprising that Eliot was as alert as Virginia Woolf to the metaphorical richness of the atom in 1919. In 'Gerontion', which was written in that year, the penultimate stanza invokes:

> De Bailhache, Fresca, Mrs Cammel, whirled
> Beyond the circuit of the shuddering Bear
> In fractured atoms. Gull against the wind, in the windy straits
> Of Belle Isle, or running on the Horn.

The gull flying against the wind, the fractured atoms and the shuddering of the Great Bear generate an intense sense of energy, fragmentation, opposing forces and impending collapse that resonates interestingly with the ways in which Woolf used 'the atom' as a figure for the aesthetic practice of modernism in her essay and in her fiction.

However it is not with Eliot's metaphorical use of the atom that this chapter is concerned, but with his use of the analogy of the 'catalyst' as a way of articulating his poetics of impersonality and also grounding his theory of the relations between the individual poet and the tradition(s) of poetic writing he or she inherits. The metaphor of the catalyst is deployed in Eliot's essay 'Tradition and the Individual Talent', which was originally published in *The Egoist* in 1919 (Figure 3.3).[27] The essay was

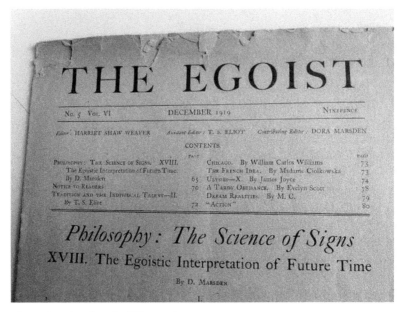

Fig. 3.3 Masthead of *The Egoist*, 1919.

Image Courtesy Michael Sherman

published across two issues, appearing in September and December, in each case accompanied by extracts from Dora Marsden's massive study, 'Philosophy: The Science of Signs' (which had been serialised over the preceding three and a half years of the journal's life) and extracts from James Joyce's *Ulysses*. The juxtaposition of science and high modernism here is striking.

Eliot's argument in this essay is by now a very familiar one, and it is indeed seen as a foundational theory of modernist poetics. Eliot argues both for the creative power of an inherited tradition of poetic writing and for the importance of the individual mind of the poet in enabling the generation of new poetic forms of expression. He begins with the assertion that: 'we shall often find that not only the best, but the most individual parts of his [i.e. the poet's] work may be those in which the dead poets, his ancestors, assert their immortality most vigorously'. This necessary creative relation to the past entails, Eliot further argues, a surrender of selfhood: 'the progress of an artist is a continual self-sacrifice, a continual extinction of personality'.[28]

Michael Whitworth has written very interestingly about the ways in which Eliot's mapping of a relation between the enabling inheritance of previous poetic writing and the simultaneous need for a poet to develop a genuinely new poetic voice resonates with debates about the nature of scientific inquiry that were taking place in 1919, focussed specifically on the role of the individual scientist in generating radically new ideas.[29] Whitworth suggests the relevance to Eliot's thinking here of a series of articles published in *The Athenaeum* by J.W.N. Sullivan in 1919, which examined the consequences for scientific 'tradition' of the radical shift of scientific paradigm implicit, for example, in Einstein's general theory of relativity. Sullivan was both deputy editor and science correspondent of *The Athenaeum*, and Eliot became friends with him through his own friendship and professional links with Middleton Murry.[30] The two essays by Sullivan that Whitworth draws on particularly are 'The Justification of the Scientific Method' (May 1919) and 'Science and Personality' (July 1919), where Sullivan argued that newer scientific theories always incorporated elements of older models (just like Eliot's modern poet is necessarily engaged in and by tradition). Sullivan further argued that the motives that guide the scientist are in fact 'manifestations of the aesthetic impulse', and proposed that despite the general assumption of the impersonality of scientific method a deeper analysis would uncover 'the personal element in the great scientific work'. Overall, Whitworth makes a convincing case that Sullivan was one of Eliot's implied interlocutors as he developed his arguments about tradition, innovation and the mind of the individual poet during these months.[31]

Within the overall argument of 'Tradition and the Individual Talent' it is the assertion that 'the progress of an artist is a continual self-sacrifice, a continual extinction of personality' that leads Eliot to think about the relations between poetic writing and science, and he goes on to say that 'it is in this depersonalization that art may be said to approach the condition of science'.[32] The relations between art and science were indeed preoccupying Eliot throughout 1919, as can be seen both from the content of his lecture on 'Modern Tendencies in Poetry' mentioned above and the argument of his essay 'Humanist, Artist, Scientist', published in *The Athenaeum* in October of that year.[33]

In 'Tradition and the Individual Talent', Eliot's insight that a theorisation of poetic method as impersonal 'approaches the condition of science' leads him immediately to ask his reader to consider a 'suggestive analogy of the action that takes place when a piece of platinum is put into a chamber containing oxygen and sulphur dioxide'.[34] This brief scientific analogy ends the first part of the essay. When Eliot returns to the argument in the December issue and focusses particularly on the relation of a poem to its author, he argues that the mind of the mature poet is a special kind of medium in which feelings are able to enter into new combinations, and the metaphor of the catalyst is now made explicit:

> The analogy was that of the catalyst. When the two gases previously mentioned are mixed in the presence of a filament of platinum, they form sulphurous acid.[35] This combination takes place only if the platinum is present; nevertheless the newly formed acid contains no trace of platinum, and the platinum itself is apparently unaffected; has remained inert, passive and unchanged. The mind of the poet is the shred of platinum … The more perfect that artist, the more completely separate in him will be the man who suffers and the mind which creates.[36]

The metaphor is certainly suggestive, and productive in terms of Eliot's overall argument: it provides an account of how one element within a creative process can both be essential but also 'inert, passive and unchanged'. It resonates with the language Eliot used in his essay on 'Hamlet and His Problems', written in 1919, where he postulated the idea that 'the only way of expressing emotion in the form of art is by finding an "objective correlative"; in other words, a set of objects, a situation, a chain of events which shall be the formula of that *particular* emotion.'[37] The argument here is also based on a theory of poetic impersonality and draws on scientific metaphor, as Eliot goes on to argue that if the 'formula' is enacted, the associated emotion will immediately be evoked, drawing on the idea of an objective formula which can determine the relation between emotion and its expression.

But closer analysis of Eliot's argument about the mind of the poet being understood as a catalyst shows that the metaphor does not in fact enable everything Eliot needs conceptually. His final remark about 'the mind which creates' within the more perfect artist surely goes beyond the 'passive' (or, as he would revise it in 1920, 'neutral')[38] quality of a catalyst, which can only ultimately enable a chemical reaction to happen faster, and with less energy, and thus cannot properly be said to 'create' anything. Eliot has found the limits of his scientific metaphor of the catalyst, but he has also found in contemporary scientific discourse a language that can allow him to pose distinctive and new questions about what it might mean to be a modern poet.

Richardson's waves of light

Dorothy Richardson's major prose work, *Pilgrimage*, consists of thirteen 'chapters' exploring the life of its protagonist, Miriam Henderson, from early childhood to the moment she becomes a writer. The first of these chapters was published in 1915, and the last (posthumously) in 1967. These 'chapters' are each substantial prose works, and are in turn further divided into chapters, so there has been a critical tendency to treat each of them as separate novels, even while recognising the integrity of *Pilgrimage* as a whole. By the end of 1919, Richardson had published five of these novels, and the innovations of her writing had been the subject of significant critical interest, relating particularly to the modernity of her narrative style.

The best-known engagement with Richardson's writing as an example of 'modern' fiction can be found in a review by May Sinclair, which was published in *The Egoist*, in 1918. Here Sinclair was to deploy for the first time the idea of 'stream of consciousness' (borrowed from psychology) to discuss literary writing. Sinclair argued that in *Pilgrimage* there 'is no drama, no situation, no set scene. Nothing happens. It is just life going on and on. It is Miriam Henderson's stream of consciousness going on and on'.[39] The idea that modernist fiction, particularly modernist fiction by women, deployed a 'stream of consciousness method' was to prove a persistent and powerful one. In 1919 Woolf also reviewed *Pilgrimage*, and argued that Richardson displayed 'a genuine conviction of the discrepancy between what she has to say and the form provided by tradition for her to say it', thus recognising the extent to which Richardson was seeking to create a new modern form for the novel.[40] Woolf went on to suggest that the experience of reading *Pilgrimage* requires the reader 'to follow impressions as they flicker through Miriam's mind, waking incongruously other thoughts, and plaiting incessantly the many coloured and innumerable threads of life', and also to argue that this led to a certain disappointment for the reader who was 'kept too near the surface'. This image of many-coloured and innumerable threads is one to which we will return at the end of this chapter.

In chapter five of *The Tunnel*[41] (the fourth volume of *Pilgrimage*, which was published in 1919 but set in 1896), Miriam Henderson attends a lecture at the Royal Institution with her employer, Mr Hancock, who is a dentist.[42] The thought of this event precipitates angry reflection by Miriam on the fact that although her father regularly attended such lectures he had never chosen to take her there: '*never* saying that members could take friends or that there were special lectures for children … it seemed cruel … deprivation … all those years; all that wonderful knowledge just at hand'.[43] Access to knowledge about science seems to Miriam fundamental to being a well-educated person, or an intellectual. She muses on the significance of 'hearing the very best in the intellectual life of London, the very best science there was' and reflects that this opportunity perhaps 'made up for only being able to say one was a secretary to a dentist at a pound a week'.[44] A personal sense of disappointment in her father; the ambition to access the fullest possible range of contemporary scientific ideas; the uncomfortable sense that after twelve years of education she had achieved too little; and a strong sense of having been excluded from the institutions and opportunities that might have helped her to address these issues are all set in play as she contemplates attending a Royal Institution lecture.

Miriam's first experience of the Royal Institution is dominated by a sense of its scale and power and of her own marginalisation. The use of free indirect discourse allows the reader to experience this from Miriam's point of view but with a certain critical distance, which is important to an understanding of the complexity of the responses to science that will follow over the next few pages of the novel:

> It seemed a vast room – rooms leading one out of the other, lit with soft red lights and giving a general effect of redness, dull crimson velvet in a dull red glow and people, standing in groups and walking about – a quite new kind of people … He [Mr Hancock] looked in place; he was in his right place; these were his people … they were all part of science.[45]

'Science' is not at this point primarily a question of ideas or of methods, but is rather experienced by Miriam as a discursive and physical space to which only a few have access. Miriam's thoughts build on this initial position to construct a version of science that might be embraced by such audience: '"Science is always right and the same … the methods of science are one and unvarying"'.[46] This resonates with other moments in *The Tunnel*, and indeed in *Pilgrimage* as a whole, where 'science' is represented as overbearing, rigid and even dangerous: 'The wonders of science for women are nothing but gynaecology – all those frightful operations in the *British Medical Journal*'; 'science is true and will find out more and more, and things will grow more and more horrible'; and 'there's no answering science'.[47] These references point to an imaginative abstraction of 'science' rather than a detailed engagement with scientific inquiry, but they do frame a number of moments in Miriam's life where she makes significant existential choices.

When the Royal Institution lecturer[48] begins to address his audience, Miriam quickly becomes excited by the ideas and arguments being proposed: 'the thrill of truth and revelation running alive and life-giving through every word'. The lecturer talks of Louis Daguerre, of stopping sunlight and breaking it up and Miriam is initially absorbed. But she is quickly distracted by thoughts about the ways in which men perceive and represent women. Then the lecturer talks of 'waves of light' that would rush through the film at an enormous speed, but 'were stopped by some special kind of film and went surging up and down in confinement'.[49] The text of the lecture by Gabriel Lippmann on which Richardson is drawing makes the point rather differently. He begins with the observation that 'when a ray of light falls on a sensitive film, this train of waves simply rushes through the film with a velocity of about 300,000 kilometres per second', notes that 'things change, however, as soon as we pour in mercury behind the plates' and concludes that 'the result is a set of standing waves – that is, of waves surging up and down, each in a fixed plane'.[50] The image of confinement is clearly there in Lippmann's image of 'standing waves' in a 'fixed plane', but for Lippmann the argument then focusses rather drily on the precise technical conditions that generate the experience of colour in a photograph, while for Miriam Henderson the metaphor of confinement leads to a much more personal set of reflections.

Miriam is conscious throughout the lecture both of the ideas being presented and the pressing weight of the subjectivities of other people in the lecture room. Thus, for example, Miriam muses that the chemical properties of 'violet subchloride of silver' mentioned by the lecturer are likely to be of interest to Mr Hancock. But when she turns to glance at him she notices he is asleep. At this point Miriam's thoughts 'recoiled from the platform and bent inwards, circling on their miseries' as she can no longer sustain the tension between the objective and subjective experiences of the lecture she is experiencing.[51]

This mapping of Miriam's inner state of mind through the deployment of scientific images and ideas in the fifth chapter of *The Tunnel* is intriguing. Miriam's fascination with the idea of 'waves of light' whose energy can be diverted so that they are 'surging up and down in confinement' can be read as a metaphor for her own sense of her life as blocked, her ambition as diverted, and her creativity as confined. Miriam's inner life and the constraints of her identity as 'a secretary to a dentist at a pound a week' can be read in and through those confined 'waves of light'. The fact that the presentation of these scientific ideas by the lecturer is both preceded and succeeded by moments of intense and painful reflection by Miriam on her own inadequacies, particularly in relation to the 'proper' performance of her role as a woman, gives greater weight to this reading of the metaphorical importance of 'waves of light' as a representation of Miriam's inner state.

Finally, however, the scientific ideas and methods that have been explicated in the lecture provide the opportunity for a moment of epiphany in the novel, which suggests the possibility both of aesthetic transcendence and a momentary overcoming of the painful divisions that have structured Miriam's sense of self.[52] This, I would suggest, is one of the moments in the novel where its stylistic innovation as 'modern' fiction becomes most compelling. The lecturer distributes colour photographs, 'pictures of stained glass, hard crude clear brilliant opaque flat colour' and the epiphanic moment follows:

> She could not tell him what she felt. There was something in this intense hard rich colour like something one sometimes *saw* when it wasn't there, a sudden brightening and brightening of all colours till you felt something must break if they grew any brighter – or in the dark, or in one's mind, suddenly, at any time, unearthly brilliance.[53]

The many coloured and innumerable threads that Woolf had noted with some concern in her 1919 review of *Pilgrimage* seem here to have achieved a (momentary) integration, an 'unearthly brilliance', which has been possible only through an imaginative exploration of what 'science' could or should mean to an aspiring modern writer.

Notes

1   A rich example of the critical and historical insights that can be generated using the method of the 'year study' can be found in Michael North, *Reading 1922: A Return to the Scene of the Modern* (Oxford: Oxford University Press, 2001).

2   Virginia Woolf, 'Modern Novels', *Times Literary Supplement*, 10 April 1919.

3   Virginia Woolf, 'Modern Fiction' in *The Common Reader* (London: Hogarth Press, 1925), 146–154.

4   Woolf, 'Modern Fiction', 189.

5   Woolf, 'Modern Fiction', 190.

6   Virginia Woolf, *Night and Day*, ed. Suzanne Raitt (Oxford: Oxford University Press, 1992), 531, 534.

7   'My poor old sluggard, Night & Day is to be taken in a parcel to Gerald [Duckworth]. As soon as I can get through with these niggling, bothersome corrections', Virginia Woolf, 7 March 1919, in *The Diary of Virginia Woolf: Volume 1, 1915–19*, eds. Quentin Bell and Anne Olivier Bell (London: Penguin, 1979), 250.

8   Virginia Woolf, 'Time Passes', *To the Lighthouse* (London: Hogarth Press, 1927).

9 Woolf, 'Modern Novels', 189.

10 Craig A. Gordon, *Literary Modernism, Bioscience, and Community in Early 20th Century Britain* (New York: Palgrave Macmillan, 2007), 138.

11 See, for example, Jeffrey S. Drouin, *James Joyce, Science and Modernist Print Culture: 'The Einstein of English Fiction'* (New York: Routledge, 2015); Holly Henry, *Virginia Woolf and the Discourse of Science: The Aesthetics of Astronomy* (Cambridge: Cambridge University Press, 2003); and Michael Whitworth, 'Science in the Age of Modernism' in Peter Brooker, Andrzej Gasiorek, Deborah Longworth and Andrew Thacker, eds., *The Oxford Handbook of Modernisms* (Oxford: Oxford University Press, 2010), 445–460.

12 Peter J. Bowler, *Science for All: The Popularization of Science in Early Twentieth-Century Britain* (London: University of Chicago Press, 2009), 35.

13 Virginia Woolf, *The Diary of Virginia Woolf: Volume 1, 1915–19*, eds. Quentin Bell and Anne Olivier Bell (London: Penguin, 1979), 243. Woolf's frustration was related to the fact that many of her friends had been approached to contribute by Murry before she had.

14 Virginia Woolf, *Jacob's Room*, ed. Kate Flint (Oxford: Oxford University Press, 1992), 49.

15 Sir Ernest Rutherford, 'Atomic Projectiles and their Collisions with Light Atoms' (an address before the Royal Institution of Great Britain, June 1919), *Science*, 21 November 1919, 467–472.

16 Ernest Rutherford, 'Collision of α particles with light atoms, IV: An anomalous effect in nitrogen,' *Philosophical Magazine* 37 (1919): 581–587.

17 Virginia Woolf, *The Years*, ed. Sue Asbee (Oxford: Oxford University Press, 1992), 148.

18 Virginia Woolf, *The Waves* (London: Collins, 1987), 176.

19 Woolf, *The Waves*, 5.

20 Virginia Woolf, 'An Unwritten Novel', in *Monday or Tuesday: Eight Stories* (Digireads.com, 2009), 28. See also Virginia Woolf, 26 January 1920, *The Diary of Virginia Woolf: Volume 1, 1920–24*, eds. Anne Olivier Bell and Andrew McNeillie (London: Penguin, 1981), 13.

21 Virginia Woolf, *A Room of One's Own and Three Guineas*, ed. Morag Shiach (Oxford: Oxford University Press, 1992), 27.

22 Bertrand Russell, *The ABC of Atoms* (London: Kegan Paul, Trench, Trubner & Co, 1923), 7.

23 Russell, *The ABC of Atoms*, 14.

24 'Definition of "atom"', *Collins Dictionary*, see www.collinsdictionary.com/dictionary/English/atom accessed June 2018.

25 Henry, *Virginia Woolf and the Discourse of Science*, 18.

26 Virginia Woolf and Leonard Woolf, 'Is this Poetry?', *The Athenaeum* (29 June 1919): 491.

27 T.S. Eliot, 'Tradition and the Individual Talent', *The Egoist*, vol. 6 (September and December, 1919).

28 Eliot, 'Tradition and the Individual Talent', 55.

29 Michael Whitworth, 'Natural Science', in *T.S. Eliot in Context*, ed. Jason Harding (Cambridge: Cambridge University Press, 2011), 336–345.

30 On this point see Christina Walter, *Optical Impersonality: Science, Images, and Literary Modernism* (Baltimore: Johns Hopkins University Press, 2014), 218.

31 J.W.N. Sullivan, 'The Justification of the Scientific Method', *The Athenaeum* (2 May 1919): 276; and 'Science and Personality' (18 July 1919),): 624–625.

32 Eliot, 'Tradition and the Individual Talent', 55.

33 T.S. Eliot, 'Humanist, Artist, Scientist', *The Athenaeum*, 10 October 1919, 1015.

34 Eliot, 'Tradition and the Individual Talent', 55.

35 I am grateful to Peter Morris who provided me, in a personal email, an explanation of exactly why Eliot's account of catalysis here is inadequate: 'Sulphur dioxide is sulphurous acid (once you add water), the whole point of his catalyst is that it converts sulphur dioxide to sulphur trioxide (a brilliantly white compound which forms delicate and elegant needles). Although on paper adding water to sulphur trioxide produces sulphuric acid, in practice it is very difficult to dissolve sulphur trioxide in water partly because of the heat produced. One adds sulphur trioxide to conc. sulphuric acid and then cautiously dilutes it. This is one reason why the catalytic process took so long to surpass the lead chamber process. The key point here is that Eliot did not realise that sulphurous acid and sulphuric acid were two wholly different compounds.'

36 Eliot, 'Tradition and the Individual Talent', 72.

37 T.S. Eliot, 'Hamlet and his Problems', in *The Sacred Wood* (London: Methuen, 1950), 100. Emphasis in original.

38 See 'Tradition and the Individual Talent', in *The Sacred Wood*, 54.

39 May Sinclair, 'The Novels of Dorothy Richardson', *The Egoist*, vol. 5 (April 1918): 57–59.

40 Virginia Woolf, 'Review of Dorothy Richardson's The Tunnel', *Times Literary Supplement* (13 February 1919): 81

41 Dorothy Richardson, *The Tunnel*, eds. Stephen Ross and Tara Thomson (Peterborough, Ontario: Broadview Press, 2014).

42 Peter J. Bowler makes the important role of the Royal Institution clear in his *Science For All*: see for example 194–198.

43 Richardson, *The Tunnel*, 143. Emphasis in original.

44 Richardson, *The Tunnel*, 144.

45 Richardson, *The Tunnel*, 144–145.

46 Richardson, *The Tunnel*, 145.

47 Richardson, *The Tunnel*, 256, 258, 274.

48 The lecturer is never named, but the snippets from the lecture that appear in the novel can be found in the text of a lecture by Gabriel Lippmann on 'Colour Photography' given at the Royal Institution on 17 April 1896. See *Notices of the Proceedings at the Meetings of the Members of the Royal Institution of Great Britain*, vol. XV, 1896–1898 (London, 1899), 151–156. I am grateful to Frank James for his help in locating this lecture.

49 Richardson, *The Tunnel*, 148, 149.

50 *Proceedings of the Royal Institution 1896–98*, 153–154.

51 Richardson, *The Tunnel*, 149.

52 I differ here from Stephen Ross, who argues in his 'Introduction' to *The Tunnel* that 'Richardson's novels lack a moment of epiphany', Richardson, *The Tunnel*, 38. For Ross, Richardson does not create 'conventional epiphanies' that offer the reader moment of insight. While accepting that this is a reasonable generalisation about Richardson's writing as a whole, this moment in *The Tunnel* is, I would argue, an exception.

53 Richardson, *The Tunnel*, 150. Emphasis in original.

4
T.S. Eliot: modernist literature, disciplines and the systematic pursuit of knowledge

Kevin Brazil

What did it mean for T.S. Eliot to claim in 1919 that 'poetry *is* a science'? This claim was made in a lecture on 'Modern Tendencies in Poetry', given at the Conference Hall in Westminster on 28 October.[1] This lecture has assumed an important place in studies of the relationship between literary modernism and science, for in it Eliot developed many of the ideas around innovation and tradition, and many of the striking metaphors with which to express these ideas, that would also be used in his enormously influential essay, 'Tradition and the Individual Talent'.[2] Most famously, in that essay Eliot compared the process of poetic creation to a catalysing reaction, 'the action which takes place when a bit of finely filiated platinum is introduced into a chamber containing oxygen and sulphur dioxide', the mind of the poet being the shred of platinum that undergoes no change in catalysing poetry out of pre-existing material.[3] Perhaps more infamously, Eliot revealed a less than perfect knowledge of chemistry by failing to mention the necessity of water in forming the resulting sulphuric acid.[4]

Eliot also got his chemistry wrong in the lecture. However, neither the potential of science to supply metaphors for the processes of artistic creation or for the imagination, nor indeed metaphors that were to be included within poems themselves, are the reasons that Eliot proposes that 'poetry *is* a science'. For Eliot, 'to say that poetry is a science is in the first place to say that poetry is a serious study, a life-time's work'. Like the scientist, the poet requires 'training and equipment', and his equipment here is not an instrument like a telescope, Geiger counter, or pipette, but 'his knowledge of what has been done in the past'. Only after a long apprenticeship in and knowledge of his subject is the poet, like the scientist, able to make new discoveries. And extrapolating from this process of assimilation of the work of the past, Eliot claims science offers a model for how the contemporary poet should relate to the poetic tradition. In Eliot's understanding, any individual scientist 'accomplished what he did not through a desire to express his personality, but by a complete surrender of himself to the work in which he was absorbed', a work which is necessarily the task of more than one individual, both across time, and within the present moment of a research community. Yet paradoxically, the more the scientist 'submerges himself in what he has to do', the more there will be 'a cachet of the man all over it'. Submersion into the work of the past reveals to the scientist the almost inevitable discoveries to be made in the present: like oxygen and sulphur dioxide, 'the elements were there to be combined, the work to be done'.[5]

Here 'a science' is first and foremost the systematic pursuit of knowledge, consisting of a programme of education and research; it is a social practice which confers a personal identity; and it is a means of relating a community in possession of specialised expertise to the wider body politic. Only latterly is it a particular cognitive style, whether that is a propensity for 'perceiving new relations' or a gift for 'the analytic, the observing, or the constructive work of science'.[6] Not at all – at least here – is it a particularly privileged epistemology. The term 'science' in this context is being used not to exclusively refer to the natural sciences and their particular methodology, as has become more common in English, but in an older and broader sense of any systematic body of knowledge that is subjected to specialised inquiry. It is a meaning closer to the German term *Wissenschaft*, which can refer to any systematic pursuit of knowledge, carried out through academic scholarship and instruction and organised into a plurality of circumscribed domains; hence why Eliot says 'a science' rather than science as a collective singular.[7]

We can see Eliot using 'science' to mean this broader sense of any systematic body of knowledge in a review, written the following year, of Gilbert Murray's translations of Euripides, in which he reflected upon how the study of Greek had been transformed by the academic research of 'the present day':

> This day began, in a sense, with Tylor and a few German anthro-pologists; since then we have acquired sociology and psychology, we have watched the clinics of Ribot and Janet, we have read books from Vienna and heard a discourse of Bergson; a philosophy arose

at Cambridge; social emancipation crawled abroad; our historical knowledge has of course increased, and we have a curious Freudian-social-mystic-rationalistic-higher-critical interpretation of the Classics and what used to be called the Scriptures. I do not deny the very great value of all work by scientists in their own departments, the great interest also of this work in detail and in its consequences. Few books are more fascinating than those of Miss Harrison, Mr. Cornford, or Mr. Cooke, when they burrow in the origins of Greek myths and rites; M. Durkheim, with his social consciousness, and M. Levy-Bruhl, with his Bororo Indians who convince themselves that they are parroquets, are delightful writers. A number of sciences have sprung up in an almost tropical exuberance which undoubtedly excites our admiration, and the garden, not unnaturally, has come to resemble a jungle.[8]

Again the 'sciences' of Eliot's jungle are not the natural sciences, but the disciplines which have come to be called in the English-speaking world the social sciences. For Eliot, it is their rising prominence that defines 'the present day': their existence characterises what it means to be modern. And yet, as hinted by Eliot's initial move in his lecture from a 'science' as any discipline based on training, apprenticeship and systematic research, to a classically empirical chemistry experiment, the modernity of these new sciences was itself claimed in relation to the sciences of chemistry and physics. The weary and knowing tone of this review essay, as if he had seen it all before, is not in this instance the carefully managed pose of the poet, adopted to manage these proliferating encroachments on his territories of Greek myth, the desires of the 'primitive', or spiritual insight: Eliot really had seen this all before, as we shall shortly see.

These observations by Eliot point to important aspects of the relationship between literature and science in the early twentieth century. First, in addition to offering an ideal form of authoritative knowledge, or a source of metaphors, science in this period offered literary writers a model of social organisation: education, accreditation, social identity and a way to legitimise their knowledge claims. Secondly, long before C.P. Snow's particular institutional context gave him a rather limited view of academic knowledge (sociology was not taught at Cambridge until 1961; the first full professor was not appointed until 1983), Eliot's observations show that for literary modernists there were – to adopt a phrase from Jerome Kagan – not two cultures but three, the social sciences also having a clearly established and publicly visible identity.[9] And thirdly, the emergence of this third culture altered the dynamics of the relationships between literature and science, recalibrating what made them different whilst opening up newer and more complicated modes of interaction.

These interactions were of particular interest to Eliot, and given the role he played as an editor and publisher from the 1920s onwards, his responses to them had particular consequence for the production, circulation and consumption of modernist literature. But Eliot was far from the only writer in the period to reflect upon the relationships between poetry and science. In 1913 Ezra Pound would also declare that '[t]he arts, literature, poesy, are a science, just as chemistry is a science'.[10] Again, a 'science' here is something closer to the meaning of a discipline or a *Wissenschaft*, a term with which Pound was certainly familiar, as shown by his attacks during the First World War on the dominance of German research methods in the American university.[11] Scholars of modernism have studied the relationship between the scientific education and later poetry of writers such as Gertrude Stein and William Carlos Williams, and others critics have traced the knowledge of and reflections on scientific discourses more broadly speaking by writers such as Ezra Pound, Virginia Woolf, D.H. Lawrence and W.B. Yeats.[12]

The claims made by modernist writers that poetry is a science or a specialised subject of study have often been placed within the context of the rise of the ideology of professionalism in the second half of the nineteenth century. According to this line of argument, Eliot's declaration that 'professionalism in art is work on style with singleness of purpose' was made to preserve, as Louis Menand has written, 'the social status of the literary vocation, and thus to some extent the perceived value of literature itself'.[13] Eliot is here read as staking a claim for the prestige of a professional identity. Yet, as a number of historians of science have argued, the sociological theory of professionalisation does not accurately describe the development of science in this period.[14] There is something more specific at stake in these claims, more atuned to developments in science than to professionalisation in general.

Ann Ardis has argued for the importance of considering modernism's relation to mass culture 'in relationship to the pursuit of disciplinary specificity and integrity driving the (re)organisation of the human and natural sciences at the turn of the twentieth-century'.[15] As we have seen, this was a pursuit of which Eliot was well aware. However, for Ardis this process of disciplinary formation, and its ambiguous attraction for literary modernists, is by and large a negative development, one which forecloses the voices of marginalised and middlebrow writers – indeed it is the process that creates the 'marginalised' and the 'middlebrow' themselves. This association between discipline in the sense of an academic subject and discipline in the sense of stifling control has been central both to key arguments within the field of cultural studies, and also to the promotion of interdisciplinary research in the sciences. That the rhetoric related to ideas of 'discipline' of a project born out of post-1968 left-wing activism has now been taken up by governmental and corporate sponsors of scientific innovation is an ironic but typical example of Luc Boltanski and Eve Chiapello's new spirit of capitalism at work just as diligently as Weber's Protestant conscience.[16]

The omnipresent promotion of interdisciplinarity in our present day can, perhaps surprisingly, make it difficult to study the history of disciplinary formation in a way that does not see a disciplinary past ceding to a more liberated interdisciplinary present. This view is the mirror image of the equally teleological vision offered by functionalists or systems theorists such as Niklas Luhmann, in which modernity is an inevitable process of specialisation.[17] Work in the history of science can offer a different perspective on the historical formation of disciplines, avoiding teleologies of both kinds, and avoiding the reductive opposition of disciplines to interdisciplinarity; as Simon Schaffer has written: 'if, as the philosophers of the *fin-de-siècle* notoriously argued, truths are dead metaphors and scientific instruments are boxed experiments about which one has forgotten that this is what they are, then disciplines are interdisciplines about which the same kind of amnesia has occurred'.[18]

Bringing such work into dialogue with studies of literary modernism can expand our understanding of the ways in which science – as a form of social organisation, and as a practice that constitutes an academic discipline – offered a model of being modern in the early twentieth century. The individual case of Eliot can offer an individual point of comparison as well as an especially informed body of commentary. The instinct to associate disciplines with control and to critique both accordingly can also make it difficult to appreciate the appeal clearly evident in claims by poets that poetry should be a science, and to appreciate that appeal as being to do with more than appeals for authority, hierarchy and control, or as masculine reaction to a feminised *belles lettres* – although it was of course all of these things, as the case of Eliot shows more than any. Yet only if we can understand the appeal and prestige of the social organisation of science into disciplines for literary writers can we understand the conflict and ambiguity that accompanied claims regarding literature's status as a modern 'science'.

The most important context for Eliot's encounter with debates around the disciplinary nature of science was his time spent as a graduate student at Harvard University from 1911 to 1914. Gail McDonald has written that, '[i]t is axiomatic that the poetry of high modernism was composed by the educated for the educated, and that Eliot in particular had a central role in shaping literary taste in universities'; this cannot be overstated, so long as one remains attentive to the different national systems that educated the poets of Anglophone modernism.[19] McDonald and more recently Robert Crawford have written about the ways in which Eliot's education was shaped by the characteristics of the elite US university, and other scholars have documented his learning in specific subjects such as philosophy, theology and anthropology.[20] But it was also while as a graduate student that Eliot engaged with a particular moment in a wider series of *fin-de-siècle* debates on the nature of a scientific discipline, of which the debate between Wilhelm Dilthey and Wilhelm Windelband is only the most well-known instance.[21]

These debates were especially intense at Harvard. Its elective system for undergraduate degrees fostered recurring complaints that students lacked a grounding in a specific discipline; the predominantly Unitarian ethos of its philosophy and theology Faculties made the reconciliation of science and Christianity a task as dominant as it was becoming impossible; and many of the professors who taught Eliot were at the forefront of debates about the recognition of new disciplines beyond the academy. Hugo Münsterberg's organisation of a Congress of Arts and Sciences as part of the St Louis World Fair in 1904 aimed at providing a 'synthesis' of the disconnected knowledge produced by the 'specialisation which makes our modern science and scholarship solid and strong' (and at showing the public this was possible), while Josiah Royce and William James advised the Federal Government on the strongly contested recognition of the social sciences by the National Academy of Science, which promoted debates among policy makers about the status and therefore eligibility for national funding of the social sciences and of psychology.[22]

Of course, there is nothing particularly modern nor American about debates about how to classify knowledge into different domains. It has been a central aim of Western philosophy from Plato and Aristotle onwards, and the hierarchical classification of knowledge was central to Francis Bacon's new science as well as to the project of the Enlightenment *Encyclopedié*. But according to Peter Weingart and Rudolph Sitchweh, there are a number of distinctive features of a scientific discipline as a type of classification that emerge in the early nineteenth century.[23] Firstly, disciplines tend to be based not on objects observed in the environment, but on concepts and theoretical objects. Secondly, they are defined by a guiding set of questions and problems put to these concepts and theoretical objects, in which novelty is a necessary condition for contribution to the discipline; the questions and the contributions are determined by researchers themselves, rather than by external actors. As a consequence, disciplines can function without being unified into a hierarchy, but rather interacting horizontally. Thirdly, and most importantly, what now distinguishes a discipline is that communities organised around applying a shared set of approaches to a shared set of objects were institutionalised in the first half of the nineteenth century in a shared set of social practices based in the university: the research programme, the department or faculty, the PhD degree, the specialised journal, a professional association and an educational subject.

Such a compilation of shared features is made possible through detailed historical studies of the formation of individual disciplines. But the striking 'commonality of disciplinary forms at different locations', as Jan Golinski has argued, demands the study of what he terms 'disciplinarity' as a process itself – the embeddedness of disciplinary formation within other cultural formations: changes in educational practice, architecture, visuality, styles of self-fashioning – and in this instance, literature.[24] Yet as Robert Frodeman has recently observed, in contrast to extensive histories and theorisations of interdisciplinarity, 'there exists no substantial body of literature that focuses on the intellectual history of disciplinarity'.[25] Work towards such a history by Timothy Lenoir and Simon Schaffer has engaged particularly with the writings of Michel Foucault, and for the purposes of considering the relationship of modernism as a cultural form to disciplinarity, this can be extended to define a discipline as a *dispositif* – a system of relations between discourses, practices and objects, for the production and regulation of power/knowledge.[26] Such a definition is useful for understanding how science as a body of discourse and as a practice interacted with literary modernism because of the way in which it pulls these two elements – discourse and practice – into a single concept, thus avoiding the dichotomy set up in some work on modernist literature and science between science as a source of metaphors and science as practice.[27] Part of writing the intellectual history of disciplinarity involves recovering the contemporaneous commentary on the nature of disciplines that accompanied their formation; when that commentary is produced by literary modernists, disciplinarity becomes part of the intellectual history of modernism.

Eliot began to engage with the disciplinary nature of knowledge while taking a seminar in logic in 1913–1914 led by Josiah Royce, focussing on the topic of 'A Comparative Study of Different Types of

Scientific Method'. In a paper on 'The Interpretation of Primitive Ritual,' Eliot opened with the questions underlying the seminar: 'On what terms is a science of religion possible? Can it be treated wholly according to the methods of sociology? And are these methods ever wholly "scientific"'.[28] In answering these questions Eliot discussed and rejected Émile Durkheim's claim in *The Rules of Sociological Method* (1895) that this was possible, but in doing so, he also reflected on Durkheim's account of disciplinary formation. *The Rules* was written to prove that 'sociology is not an appendage of any other science; it is itself a distinct and autonomous science'.[29] 'Every science,' Durkheim wrote, 'consists of a specific group of phenomena which are subsumed under the same definition. The sociologist's first step must therefore be to define the things he treats, so that we may know – he as well – exactly what his subject matter is'.

As Durkheim later wrote, '[f]or sociology to be possible, it must above all have an object all of its own ... [a] reality which is not in the domain of the other sciences'.[30] But while the social fact guaranteed the autonomy of sociology as a science, Durkheim defined this autonomy by drawing on – what he understood to be – the methods of the natural sciences, their concepts of mechanical causality and energy, as well as their ethos and practice: '[s]ociologists must adopt the state of mind of physicists, chemists, and physiologists, when they venture into as yet unexplored areas of their scientific field'.[31] *The Rules* is an exemplary case of what Amanda Anderson and Joseph Valente have called the ways in which social science performed a 'contestatory emulation of the scientific disciplines' in the period.[32] As much as *The Rules* was a theoretical manifesto – in this sense an exemplary modernist text – it also reveals the practices through which such definitions were realised: the manifesto itself, the specialised journal, the *Année Sociologique*, the foundation of specialised university departments and the training of graduate students.

Eliot's primary critique of such a method in explaining religious ritual was its failure to consider as relevant, or as having causal force, the intentions and interpretations of actors themselves, and even more myopically, treated its own interpretations as facts: 'What, in short, is the scientific status of a definition which is the description of the meaning of other agents?'[33] Focussing on the procedure by which the social scientist groups phenomena under a definition, he merged this with the problem of the difference between a fact and an interpretation: 'I do not think that any definition of religious behavior can be satisfactory, and yet you must assume, if you are to make a start at all, that all these phenomena have a common meaning; you must postulate your own attitude and interpret your so-called facts into it, and how can this be science? And yet there is the material, and there must be a science of it'.[34] Eliot defined a fact as a 'point of attention which has only one aspect or [can be treated] under a certain definite aspect which places it in a system', and different sciences for explaining religion were an example of such systems.[35] These systems defined their object: if one no longer accepts their explanations, one requires 'a different standpoint; in short, a different science'.[36] In translating the problem of disciplinary definition into that of the hermeneutic circle, Eliot set out his characteristic approach to the question of disciplinarity as a graduate student.

In his final paper written for Royce's seminar, Eliot returned to the question, leaving it unanswered, as to '[h]ow far the so-called social sciences are sciences at all, how far, that is, their objects can be handled as things and the higher object causality imposed upon them, this is a question which would demand careful investigation'. Translated out of the idiosyncratic terminology drawn from the 'Theory of Objects' Eliot was developing, this sentence asks how far the descriptions and definitions of social science refer to things existing in the world, as opposed to relations or ideas, and how far the mechanical notion of causality, proposed by sociologists like Durkheim, can be used to explain them. And this 'Theory of Objects' was pragmatic: whatever the kind of object, 'we do not explain, we only describe: an explanation, that is, always for the purpose of practice', disciplines being one such form of practice.[37] Eliot's turn to the question of how the sciences as practices produce their objects is part of the broader pragmatism Walter Benn Michaels has argued appears in Eliot's doctoral dissertation, and is unsettled by the same conclusions to which this pragmatism led.[38]

This pragmatic approach surfaced again, when Eliot returned to the question of the definition of a science when continuing to develop his 'Theory of Objects' while studying at Oxford in 1914–15. Here Eliot posed the paradox of his own pragmatism. 'There is a point of view', he wrote in sympathy, 'from which it is said that the sciences are mutilated and imperfect parts of reality, the creations of a valuation which takes their objects out of the complete context in which alone they are wholly real'. But he was also in sympathy with this view turned on its head: that '[r]eality is the one thing which doesn't exist', and as such, 'the tenuated positions of reality' presented by the different sciences are 'a partially successful attempt to constitute reality'. The difficulty with this approach was, however, as follows:

> (1) These scientific theories, if each is presented as a final account of its own group of objects, may conflict with each other: the mental

sciences, e.g., anthropology, economics, psychology, may conflict with the vital (physiology), these with the physical. (2) They are obliged to affirm that the existences which they abstract as objects exist just so, as objects, quite apart from the limited universe of discourse in which they are known as objects. That this should be so is inconveniently enough inconceivable.[39]

The dissertation to which all these essays were leading, *Knowledge and Experience in the Philosophy of Bradley*, did not get much further on the question of how these different sciences related to each other than this inconvenient inconceivability, turning its attention instead to a critique of psychology and epistemology. Emphasising a certain relativism, Eliot wrote that '[t]here is a sense, then, in which any science – natural or social – is a priori: in that it satisfies the needs of a particular point of view, a point of view which may be said to be more original than any of the facts that are referred to that science'. 'The attitude of science, then, involves the construction of a larger and larger limbo of appearance – a larger field of reality which is referred to as the subjective side of appearance. Economics is appearance for the biologist, biology for the chemist'. And so 'we have not one science, but a whole universe of sciences' – a foreshadowing of the jungle that would appear to define our present day in 1920.[40]

If this was the thinking that lay behind Eliot's definition of a 'science,' grounded in wider *fin-de-siècle* debates about disciplinarity prompted by the development of the social sciences, positivism and the practices of the American research university, what did it mean for Eliot to propose some five years in later in 1919 that poetry *is* a science? Perhaps a better way of approaching this question would be: how did the definitions of poetry Eliot was offering at this time relate to these previous definitions of a science: one which took in what was common between the natural and social sciences; which was skeptical of reducing the latter to the former; and which saw a science as a practice which defined its own objects, but which paid for this relative autonomy by becoming just another tenuated view of reality.

Central to Eliot's definition of poetry at this time was a version of literary history, one which sought to situate modern poetry in relation to what he called, in a 1920 essay on Dante, 'the greater specialisation of the modern world'. In this history, the metaphysical poets of the seventeenth century – above all John Donne – possessed a unity of thought and feeling, and 'incorporated their erudition into their sensibility: their mode of feeling was directly and freshly altered by their reading and thought'. After this, a 'dissociation of sensibility set in, from which we have never recovered', resulting in a 'reflective' poetry which cannot amalgamate, only merely ruminate upon the fragmented nature of experience.[41] The modernist poet would reverse this dissociation and return to erudition in service of their sensibility.

This historico-poetic myth has been taken as an attack on Romanticism, a manifesto for modernist poetics, and a larger thesis on the nature of modernity and poetry's place within it. But it is also an argument for poetry's relationship to specialised disciplinary knowledge. In a series of 1926 lectures on Donne, Eliot expanded on the nature of Donne's ideal (and idealised) relationship to erudition: 'In the mind of Donne we find all the ideas of his time co-existent in their most abstract form … [s]ome of these ideas are of contemporary science, some of contemporary theology; but they are all entertained on an equal footing; and this is typical of his time'. Not only could Donne possess the entire 'magazine of science and philosophy', he could experience its varied content as feelings, and find 'the emotional equivalent of highly abstract or general ideas' in poetic expression.[42] Such an understanding of poetry, as that which can express the emotional experience of any kind of thought, scientific, religious or otherwise, is clearly in tension with the view of poetry as a single individual 'science' which defines just another tenuated view of reality. In this view of Donne, which is the programme for the modernist poet, poetry is not so much a science as that which provides a synthesis of all the other sciences.

Yet Eliot also argued that poetry should be a science in the sense of a particular form of social practice and organisation of the production and legitimation of its own kind of knowledge. In 1918, Eliot attacked an attitude 'behind all British slackness for a hundred years or more: the dislike of the specialist. It is behind the British worship of inspiration, which in literature is merely an avoidance of comparison with foreign literatures, a dodging of standards'.[43] Although the review is signed with a pseudonym, the position taken is that of a commentator from a society more advanced in its specialised production of knowledge; in the context of the issue's sniping between British critics and American poets, it is hard not to read this as a voice trumpeting a version of the 'all American-propaganda' of which Eliot – this time publicly – accused Amy Lowell in the very same issue.[44] The claim that poetry could only be produced and understood by a specialist was shared by Wyndham Lewis, and in the short-lived little magazine *Tyro*, edited by Lewis and to which Eliot contributed, Lewis extended the analogy: 'I do not for my part believe that *any painter or sculptor has ever been understood, ever, by anyone except a painter or sculptor*: any more than the astronomical mathematics with which Einstein plays are to be understood by anyone but a specialist in that branch of mathematics'.[45] These claims are about judgement, both of a poet and critic, and about the authority of the poet and critic to define what counts as achievement in their field against the amateur and untrained public.

Comparing the poet or a critic to a specialist in this way, however, brings out the imprecision of the analogy; or at least the imprecision of what these writers understood as the specialist. In Eliot's vision of 'The Perfect Critic', included in his 1920 essay collection *The Sacred Wood*, poetry is no more 'highly organised' an intellectual activity than 'astronomy, physics or pure mathematics' – meaning, implicitly, that it is just as highly organised. If the critic, however, in order to understand the product of the activity that is poetry, is tasked with having the 'scientific mind', this is a kind of mind which is not that possessed by the 'ordinary scientific specialist ... [which is] limited in its interest'. It is the kind of 'universal intelligence' possessed by Aristotle, all the more necessary for understanding the 'so many fields of knowledge' that have been 'deposited by the nineteenth century'.[46] In oscillating between defining poetry and defining the critic, Eliot is trying to navigate through the problems raised by imposing his pragmatic definition of a science on poetry: it will have just as much autonomy and authority as physics, but not with the hedge that its view on reality is only partial.

Navigating between the Scylla and Charybdis of synthesis and specialisation also characterised Eliot's thinking on the role of a literary periodical in the years leading up to the foundation of *The Criterion* in 1922. In 'A Brief Treatise on the Criticism of Poetry', published in March 1920 in Harold Monro's *The Chapbook*, Eliot offered a self-admittedly spasmodic argument for how the 'modern literary system' should be organised in order to best ensure the flourishing of poetry. 'There should be no reviewing of poetry in daily newspapers.' Instead, '[l]et practitioners of any art or several arts who have a sufficient community of interests and standards publish their conversation, their theories and their opinions in periodicals of their own. They should not be afraid of forming "cliques," if their cliques are professional and not personal'. This is one explanation for the necessity of the modernist 'little magazine': just as other professions require forums for discussion that outsiders would not be expected to understand, so too does poetry.[47] But what would then distinguish, structurally, such specialised publications from the academic journals to which Eliot had previously contributed, such as *The International Journal of Ethics* or *The Monist*? In her study of modernist print culture, Ann Ardis has shown the perception of this structural homology among one group of readers and editors: Guild Socialists. The editorial practices of a revolutionary socialist journal such as *The New Age* positioned themselves against the effects of the proliferation of specialised journals, whether of art, science or politics, attempting to 'drive into the mind of its readers that life is not composed of water-tight compartments'.[48] From this perspective, the form of a publication as much as its specific content – its closure to those who are not 'practitioners' – indicates its function in entrenching disciplinary specialisation, resulting in an unexpected comparison between the scholarly journal and little magazine.

The Criterion could be compared to neither. At the end of its first year of publication, Eliot outlined a different vision of that periodical and its purpose in sustaining literature as a 'specialised activity'.

> To maintain the autonomy, and the disinterestedness, of every human activity, and to perceive it in relation to every other, require a considerable discipline. It is the function of a literary review to maintain the autonomy and disinterestedness of literature, and at the same time to exhibit the relations of literature – not to 'life,' as something contrasted to literature, but to all other activities, which, together with literature, are the components of life.

If the solution to the support and production of the kind of literature Eliot deemed modern was to be decisively different from the little magazine and *The Monist* alike, and its defence was to draw on the rhetoric of Kantian and Romantic aesthetics rather than that of modern science and its experiments, the unstable and shifting meaning of discipline in such a defence, shading back through 'specialised activity' into associations of the science that poetry is, or could be, serves to show how deeply, if obliquely, science as a social form had seeped into the practices of literary modernism.

The kinds of discipline that preoccupied T.S. Eliot's later writings have rightly come to be associated with the more unsettling aspects of what it meant to be modern in the early twentieth century: the royalist's attraction to authoritarian government and distrust of democracy; the catholic's submission to dogma out of fear of the flawed nature of humanity; the classicist's constriction of the purchase of poetry made all the more insidious through appeals to taste, intelligence and morality. T.S. Eliot reminds us of the ambiguous inheritance of what it means to be modern: the poet who imagined what it was like to be 'a pair of ragged claws / Scuttling across the floor of silent seas' and the excommunicator of *After Strange Gods* (1933). In the same way, his ambiguous yet deeply informed reflections on whether poetry should become as disciplined in its social organisation as a science force critics to consider the attraction of a discipline in our own contemporary moment. Few accolades are higher than the valediction that a scholar has established a field of study; few can resist – historians of science and practitioners of science studies not least – claims for the disciplinary specificity, self-legitimation and thus value of their academic work. The flight of Eliot's poetic successors in the USA into the strictly organised discipline of university-based creative writing is yet another way in which literature and science share more than is commonly recognised. Modernist writers were still coming to terms with the place of science in what it meant to be modern, and every time we allow ourselves to tidy up the world into fields, studies and specialties, so too are we.

Notes

1   T.S. Eliot, *The Complete Prose of T.S. Eliot: The Critical Edition: The Perfect Critic 1919–1926*, ed. Anthony Cuda (Baltimore: The Johns Hopkins University Press, 2014), 212. Hereafter references to quotations from the volumes of the *Complete Prose of T.S. Eliot* are cited as *CP* followed by the volume and page numbers.

2   Michael H. Whitworth, *Einstein's Wake: Relativity, Metaphor, and Modernist Literature* (Oxford: Oxford University Press, 2001), 140–5; Christina Walter, *Optical Impersonality: Science, Images, and Literary Modernism* (The Johns Hopkins University Press, 2014), 214–57.

3   Eliot, *CP2*, 108.

4   As pointed out by an anonymous reviewer in 'Shorter Notices', *The Criterion* XII, no. 46 (October 1932): 163–7.

5   Eliot, *CP2*, 213–4.

6   Eliot, *CP2*, 214.

7   I. Bernard Cohen, *Interactions: Some Contacts Between the Natural Sciences and the Social Sciences* (Cambridge, Mass.: MIT Press, 1994), 7–8.

8   Eliot, *CP2*, 197–8.

9   A.H. Halsey, *A History of Sociology in Britain: Science, Literature, and Society* (Oxford: Oxford University Press, 2004),

101; Jerome Kagan, *The Three Cultures: Natural Sciences, Social Sciences, and the Humanities in the 21st Century* (Cambridge: Cambridge University Press, 2009).

10 Ezra Pound, 'The Serious Artist I–II', *The New Freewoman* 1, no. 9 (October 1913): 161.

11 Ezra Pound, 'Provincialism the Enemy', *The New Age* XXI, no. 11 (July 1917): 244–45.

12 Steven Meyer, *Irresistible Dictation: Gertrude Stein and the Correlations of Writing and Science* (Stanford: Stanford University Press, 2003); T. Hugh Crawford, *Modernism, Medicine, and William Carlos Williams* (Norman; London: University of Oklahoma Press, 1995); Ian F.A. Bell, *Critic as Scientist: The Modernist Poetics of Ezra Pound* (London; New York: Methuen, 1981); Whitworth, *Einstein's Wake: Relativity, Metaphor, and Modernist Literature*; Daniel Albright, *Quantum Poetics: Yeats, Pound, Eliot, and the Science of Modernism* (Cambridge: Cambridge University Press, 1997).

13 Apteryx [T. S. Eliot], 'Professional, Or ...', *The Egoist* V, no. 4 (April 1918): 61; Louis Menand, *Discovering Modernism: T.S. Eliot and His Context* (New York: Oxford University Press, 1987), 100; Thomas

Strychacz, *Modernism, Mass Culture, and Professionalism* (Cambridge: Cambridge University Press, 1993); David Trotter, *Paranoid Modernism: Literary Experiment, Psychosis, and the Professionalization of English Society* (Oxford: Oxford University Press, 2001); Evelyn Tsz Yan Chan, *Virginia Woolf and the Professions* (Cambridge: Cambridge University Press, 2014).

14 Timothy Lenoir, *Instituting Science: The Cultural Production of Scientific Disciplines* (Stanford: Stanford University Press, 1997), 6–8; Jan Golinski, *Making Natural Knowledge: Constructivism and the History of Science, with a New Preface* (Chicago: University of Chicago Press, 2008), 68; David Cahan, 'Institutions and Communities', in *From Natural Philosophy to the Sciences: Writing the History of Nineteenth-Century Science*, ed. David Cahan (Chicago; London: University of Chicago Press, 2003), 297.

15 Ann L. Ardis, *Modernism and Cultural Conflict, 1880–1922* (Cambridge: Cambridge University Press, 2002), 6.

16 Luc Boltanski and Eve Chiapello, *The New Spirit of Capitalism* (London: Verso, 2005).

17 Helga Nowotny, Peter Scott and Michael Gibbons, 'Rethinking Science: Mode 2 in Societal Context', in *Knowledge Creation, Diffusions and Use in Innovation Networks and Knowledge Clusters*, eds. Elias G. Carayannis and David F.J. Campbell (Westport: Praeger, 2006), 39–51; Niklas Luhmann, *Social Systems*, trans. John Bednarz (Stanford: Stanford University Press, 1995).

18 Simon Schaffer, 'How Disciplines Look', in *Interdisciplinarity: Reconfigurations of the Natural and Social Sciences*, eds. Andrew Barry and Georgina Born (London: Routledge, 2013), 58.

19 Gail McDonald, *Learning to Be Modern: Pound, Eliot, and the American University* (Oxford: Clarendon Press, 1993), vii.

20 Manju Jain, *T. S. Eliot and American Philosophy: The Harvard Years* (Cambridge; Cambridge University Press, 1992); Rafey Habib, *The Early T.S. Eliot and Western Philosophy* (Cambridge: Cambridge University Press, 1999); Robert Crawford, *Young Eliot: From St Louis to The Waste Land* (London: Jonathan Cape, 2015).

21 Wilhelm Dilthey, *Wilhelm Dilthey: Selected Works, Volume I: Introduction to the Human Sciences*, eds. Rudolf A.

Makkreel and Frithjof Rodi (Princeton: Princeton University Press, 1991); Wilhelm Windelband, 'Rectorial Address, Strasbourg, 1894', *History and Theory* 19, no. 2 (1 February 1980): 169–85.

22 Hugo Münsterberg, 'The International Congress of Arts and Sciences', in *Congress of Arts and Science: Universal Exhibition, St. Louis* (Boston: Houghton Miffin, 1905), 2; 1; Cohen, *Interactions*, 168.

23 Rudolph Stichweh, 'The Sociology of Scientific Disciplines: On the Genesis and Stability of the Disciplinary Structure of Modern Science', *Science in Context* 5, no. 1 (1992): 3–15; Rudolph Stichweh, 'History of the Scientific Disciplines', in *The International Encyclopedia of the Social and Behavioral Sciences*, vol. 20 (Oxford: Elsevier, 2001), 13, 727–31; Peter Weingart, 'A Short History of Knowledge Formations', in *The Oxford Handbook of Interdisciplinarity*, eds. Robert Frodeman, Julie Thompson Klein and Carl Mitcham (Oxford: Oxford University Press, 2010), 3–14.

24 Golinski, *Making Natural Knowledge*, 69.

25 Robert Frodeman, 'The End of Disciplinarity', in *University Experiments in Interdisciplinarity: Obstacles and Opportunities*, eds. Britta Padberg and Peter Weingart (Bielefeld: transcript Verlag, 2014), 177.

26 Lenoir, *Instituting Science*; Schaffer, 'How Disciplines Look'; Michel Foucault, *Power/Knowledge: Selected Interviews and Other Writings 1972–1977*, ed. Colin Gordon (Hemel Hempstead: Harvester Wheatsheaf, 1980), 194–5.

27 Paul Peppis, *Sciences of Modernism: Ethnography, Sexology, and Psychology* (Cambridge: Cambridge University Press, 2014), 1–2.

28 Eliot, *CP1*, 106.

29 Émile Durkheim, *The Rules of Sociological Method: And Selected Texts on Sociology and Its Method*, ed. Steven Lukes, trans. W.D. Halls (London: Macmillan, 1982), 162.

30 Quoted in Durkheim, *The Rules of Sociological Method*, 3.

31 Durkheim, *The Rules of Sociological Method*, 37.

32 Amanda Anderson and Joseph Valente, 'Introduction: Discipline and Freedom', in *Disciplinarity at the Fin de Siècle*, eds. Amanda Anderson and Joseph Valente (Princeton: Princeton University Press, 2001), 7.

33 Eliot, *CP1*, 112.
34 Eliot, *CP1*, 115.
35 Eliot, *CP1*, 108.
36 Eliot, *CP1*, 109.
37 Eliot, *CP1*, 113; 130.
38 Walter Benn Michaels, 'Philosophy in Kinkanja: Eliot's Pragmatism', *Glyph: Johns Hopkins Textual Studies*, no. 8 (1981): 170–202.
39 Eliot, *CP1*, 166.
40 Eliot, *CP1*, 284; 295; 295.
41 Eliot, *CP2*, 226; 379; 380.
42 Eliot, *CP2*, 725; 726.
43 Apteryx [T.S. Eliot], 'Professional, Or ...', 61.
44 T.S. Eliot, 'Disjecta Membra', *The Egoist* 4, no. 5 (April 1918): 55.
45 Wyndham Lewis, 'Editorial', *The Tyro*, no. 2 (1922): 7.
46 Eliot, *CP2*, 262; 269; 267.
47 Eliot, *CP2*, 207; 208–9.
48 Quoted in Ardis, *Modernism and Cultural Conflict, 1880–1922*, 162.

Section 2
Tensions over science

5

Modernity and the ambivalent significance of applied science: motors, wireless, telephones and poison gas

Robert Bud

Introduction

On Sunday 14 January 1934, the University of Glasgow solemnly dedicated a suite of stained-glass windows illuminating its great chapel overlooking the city. The signs of the zodiac, representing creation, provided the theme of several designs, but in addition two were dedicated to human endeavour. The subject of one of these chapel windows was theology; the other, which was to shine over the congregants, more surprisingly displayed the prominent title 'Applied Science', and portrayed an heroic muscular workman, surrounded by derricks, machines and bottles of chemicals (Figure 5.1).

The subjects had been freely chosen by the designer, distinguished Scottish artist and nationalist Douglas Strachan. Perhaps his decision had been influenced by the chapel's situation just a few metres from a house, which was once the home of telegraph-pioneer and physicist Lord Kelvin. His thinking, however, was also more profound. At the dedication, the Vice Chancellor quoted the artist's explanation that he had attempted

> to figure man's life, all life, as engaged on a spiritual enterprise: to visualise our little planet moving on through infinite space – or perhaps one ought now to say Finite Space, whatever that may mean: man's unceasing search and endeavour to comprehend the universe and his own spiritual aspirations, and to find one image for both.[1]

Fig. 5.1 Douglas Strachan, 'Applied Science' stained glass window for Glasgow University Chapel, 1934.

Image courtesy Nick Haynes.

Here the term 'applied science' served to link the industrial reality of contemporary work to an awareness of cosmic grandeur.

In contrast to the spiritual connotations in the chapel so eloquently expressed, the same term had strictly bureaucratic meanings. In 1930 a proposal put to the University of London that a degree in Textile Science be made available to external students was rejected strenuously by the Faculty of Pure and Applied Science. The committee dismissed the proposed syllabus as not worthy of recognition, 'a heterogeneous set of applications of a number of sciences' and quashed it by suggesting that the proposers were really raising the question of a separate 'Technology' Faculty; this question was allocated to another ad hoc committee, which simply faded.[2]

Since the 1860s, 'applied science' had been deployed to denote the background knowledge that it would be invaluable for an intending engineer or chemist to possess, without itself being the profession's core knowledge, which could be gained only through apprenticeship to a master and experience on the job. The subject was at the same time oriented to the useful arts and rooted in the science that was part of liberal culture offering character development. For the student, it would combine understanding of the present with flexibility in the face of rapid change. Its study was pursued in universities and was an alternative to the 'technology' which was useful knowledge taught in technical colleges, promoted by their own national examination body, London's 'City and Guilds', and much closer to vocational education.

The bureaucratic connotations of the term in its educational context are indicated by the rigour with which the boundary with technology was meticulously policed through the whole of our period. A committee of Britain's leading politicians in a subcommittee of the Privy Council considered the award of Charters to the North of England's universities in Manchester, Leeds and Liverpool in 1902. They were anxious to counter any tendency to 'degrade University teaching to technical teaching' and to emphasise that university education was concerned with principles rather than their application.[3]

The example of the 'civic universities' had encouraged colleagues in Glasgow. The English Charters having been negotiated, in 1909 the Glasgow and West of Scotland Technical College (now the University of Strathclyde and previously known as the Andersonian Institution) bid to be allowed to offer Glasgow University degrees in applied science. Over the previous few years, the college averred, such rights had been won by lesser colleges in such cities as Manchester, Leeds, Sheffield, Newcastle and Bristol. The College could cite the Glasgow University

Calendar, typically the driest of dry documents, which announced that under the Universities (Scotland) Act of 1889, 'Degrees in Applied Science could be instituted and conferred.' This, claimed the ambitious and distinguished College, justified its proposal to join the University in the domain of 'Applied Science'.[4]

A third much less positive connotation could be instanced by others, particularly the Church, in the wake of the First World War. A Canon of Westminster used a Lenten address in 1927 to warn the congregation in terms that would become familiar: 'in the very hour of his triumph man found himself the slave of the machine he himself had created, as though it was some Frankenstein monster which, because it had no soul, he was unable to control … It [science] might also be used for infinite destruction, and applied science might be the very devilry of mankind as in the great war.'[5] This trope of applied science as the uncontrollable monster made by man, and the model role of Frankenstein's monster, would become very familiar.

In these three examples of the term's use we have seen not just a diversity of connotations but also its deployment in a diversity of discourses, ranging from the existentially profound to the bureaucratic. When the twentieth century had begun, the dominant meaning had related to its pedagogical context.[6] Academic usage was certainly maintained and indeed became more important as it came to encompass research particularly in engineering as well as teaching.[7] Across the first half of the twentieth century, however, use of the term transcended this narrow context to circulate through bureaucracy, literature, art and politics. These usages were no mere popularisations of the academic term. This chapter addresses the questions of how and why this concept became a widely proclaimed reality with broad cultural underpinning. In debates over British modernity, one can see the competing claims of groups of narratives, many of which shared the subject of 'applied science'. Their telling, although intended to have competing effects, collectively assured the concept's vitality and reality. I shall suggest that such stories became 'representative anecdotes' and applied science became an emblematic term, for differing reasons it is true, for opposing social movements dedicated to interpreting change in the early twentieth century.[8]

This was not the only key to modern times on offer at the time, even within English-speaking lands. Lewis Mumford and his mentor Patrick Geddes preferred the more holistic concept of 'Technics', rooted in the means by which energy was produced. 'Machine civilisation' was also a frequent trope of the time, similar to, though different in emphasis from, an interpretation directly rooted in science, never formally distinguished from it. These categorisations were internationally shared but the means by which concepts were discussed were shaped by the reach of mass media, which were still local and national. The narratives and the mechanisms of transmission we will see here were therefore both distinctively British and closely related to parallel and complementary discussions overseas.[9]

Concepts, social movements, narrative and circulation

Neither a solid thing nor merely a linguistic convention, in these narratives 'applied science' came to be widely treated as real and undoubtedly 'there' as any table. It is useful to think of it as a 'concept', a class on which there has been much reflection. Gallie talks of 'essentially contested concepts', McGee of 'Ideographs', Raymond Williams of 'keywords', Blumenberg of 'metaphor'.[10] At Bielefeld, Reinhard Koselleck and his colleagues promoted the study of concepts in political and cultural history and created the school of *Begriffsgeschichte*, normally translated as 'conceptual history'. They argued that a 'concept' differs from a simple 'term' through the richness of meanings associated with it.[11] Though generally the references explored by such historians have tended to be to strictly political issues, that is an artefact of past concerns of the historical discipline. In the case of applied science we see too the activity of movements intent on national, social and cultural change, of press coverage of the institutions they created, and of individual promoters working through rapidly developing mass media including newspapers, exhibitions and the wireless. The term was deployed as part of the process of forming movements by key national and local influential individuals and media, and was itself thereby reshaped.

Analysts have become interested in the ways by which social movements seek to exploit existing culture to transform culture, and the use of language and stories as frameworks of meaning. More generally, distinctive cultures have increasingly been seen as important characteristics of movements. The sociologist Gary Fine indeed has gone on to argue that a social movement can be seen as a bundle of canonical narratives.[12] An American historian of political conservatism has reflected, 'Being conservative is about telling tales. It is a way of speaking. It is about knowing and repeating the stories of the movement'.[13] By analogy, this characterisation would fit the proponents of technical education in the late nineteenth century. Stories of such allegorically significant figures as Humphry Davy, James Watt, Robert Stephenson, Justus Liebig and William Perkin were told and retold.[14] In twentieth-century Britain, tales of German and then American industrial leadership linked to investment in industrial research would become more familiar accounts of the recent past.

In the later-nineteenth century, 'applied science' and its role in replacing 'rule of thumb' had played an important role in the promotion of civic colleges which prepared students for an industrial life. This paper deals with the subsequent period of the early twentieth century in which 'applied science' came to transcend deployment by a single social movement.[15] The utility of the same term to complementary but also competing movements entrenched the reality of the concept in the eyes of many. The term's use did not just 'happen' to circulate across divides, there was neither necessary unity between the movements to which it became important, nor was it necessary that science would be important in any of them. Mass media circulated the term's use across diverse discourses by individuals on the national and local stage, acting as self-conscious agents or as critics of 'modernising' social movements and through the presence of the institutions they created. The prominence of these people ensured their speeches would be reported, and later broadcast. So in the work of users one can see how the term was entrenched in discourses of education, of gadgets, of war, Britain's place in the world and this moment's place in time.

Usage of the term in the press

Before exploring how the narratives were adopted by social movements and how they were transmitted between movements, it is worth seeing how the term was used in the press. This chapter draws on a study of the term's use on 2161 occasions found in the now digitised daily and weekly newspapers from the period 1900 to 1939.[16] Both national and major regional newspapers are included and also the influential *Spectator* and *Economist* magazines. Every use found, whether in advertisement or reported speeches, has been taken into account. Although strict quantitative conclusions would be inappropriate, this examination of the contexts and interpretations in which millions of people would have encountered the term provides a useful counterpart to the study of the thoughts of a few individuals. Coverage in newspapers served to amplify the efforts of agents seeking to use the term to their own ends. For, particularly after the world-widening experience of the Great War, there was a new wider audience for the printed word, and particularly education. John Buchan's estimate of a 40 per cent increase in the reading public due to the war has been widely cited.[17] The widening of the audience also had longer-term causes. In 1922, the journalist Sidney Dark coined the description 'The new reading public' to describe the beneficiaries of three generations of compulsory primary education who were now thirsting for more.[18]

Use of the term served a variety of purposes. It explained the nature of and relationships between organisations: initially principally universities and faculties within them but later industrial laboratories too. It could be used to reflect on change over time and particularly the changing nature of war. Applied science could also be used to explain the more beneficial wonders that were now being made possible and even appearing in daily lives. Rapid developments in such areas as aircraft, agriculture (particularly in the empire) and increasingly medicine were described in these terms. This continued after the First World War. As Peter Bowler has shown, the self-styled 'Professor Low' became a popular journalist describing the latest miracle of applied science.[19] He was not the only one. The feminist journalist Storm Jameson urged her readers in the *Daily Mail* newspaper to look around them: 'since you are compelled to live in a day of noise, speed, central heating, motor-cars, convenient flats, aeroplanes and wireless – instead of in a day of quiet, few and slow trains, inconvenient houses and splendid isolation – make up your mind to use these things for your better advantage … Make modern applied science lighten your work and enrich your leisure.'[20] This genre emphasised stories of past deprivation and increasing convenience. The familiarity of this usage was exploited in the marketing of what otherwise might seem more traditional devices and materials: in 1926, in the era of the gramophone, a large number of advertisements for a new player piano claimed to show what applied science can do for music.[21]

Modernity in general could be identified with the concept as in James Joyce's *Ulysses* of 1922. Two pages are devoted to explaining Stephen Bloom's interest in 'applied science'. Alongside this structuring of time, applied science also structured conceptions of the nature of spaces. It was used in the description of regions of Britain, such as Manchester and Glasgow, and also the country's relations with other countries: particularly Germany but also the USA. These countries were often so closely identified with applied science that their very evocation would be used to colour the term. By extension, other lands could be included to make a surprising statement. In 1912 the Kingdom of Italy took out full-page advertisements in both *The Times* and *The Manchester Guardian* to inform readers that Italy was now a 'new nursery of applied science'.[22]

These usages did not exist in a political vacuum. Rather they signified visible exchanges in debates between active and vocal social movements. In an era of World War and the threat of war, great social change, the Russian Revolution, political upheaval and mass unemployment at home, the concept of applied science was transmitted across a variety of such movements.

National efficiency and the promotion of applied science

The beginning of the period reviewed here coincided with a new era of national debate at the onset of the twentieth century, sparked by the British Empire's difficulties in quelling the resistance of Boer farmers. Much of the criticism was characterised by the catchword 'efficiency'.[23] This represented, as was then pointed out, 'the political gospel opposed to the fine old English doctrine of "muddling through", the phrase in which Lord Rosebery summed up the Boer war'.[24] The invocation of efficiency would continue through and after the First World War, expressed in the 1920s by such events as the 1921 Efficiency Exhibition briefly replacing the Ideal Home Exhibition. However, a single gospel did not a single movement make. For the years before the First World War, the historian G.R. Searle has identified three competing groupings to which 'efficiency' was particularly attractive, the Liberal Imperialists, their Parliamentary opponents reforming Conservatives and the Fabians. These three categories were not themselves coherent, as Searle himself points out in explaining why neither the Liberal nor the Conservative party split more formally.

Proponents of the Army's decisive role in the defence of the nation formed the National Service League in 1901 to pressurise for compulsory national service for all young men between 18 and 21. Meanwhile advocates of the Royal Navy's key role in keeping the enemy from the shores and protecting trade argued for Dreadnoughts, the latest in modern technology, to meet the German threat. 'We want eight and we won't wait' went the song. Others sought a new welfare system such as the Germans had already instituted, with health care and pensions made available to the working classes. The suffragette movement expressed the wish of many women to take a part in public life. Baden

Powell saw the salvation of the nation in the fitness of its youth, while others were motivated to improve education.

Appeals to 'applied science' were frequently associated with reference to efficiency, but this did not mean either that use of the term indicated a single frame of reference. At the highest of levels there was of course a distinction between those who welcomed what they proclaimed to be the progressive benefits of applied science, while others found in the attraction of new gadgets challenges that could distract citizens from combatting wrong or, even, evil ends. These categories were however neither 'naturally' pre-existing nor homogenous. Instead, talk about applied science was part of the creation and maintenance of social movements. Communicating stories about the future of applied science, the formation of social movements and the meaning of the concept were intimately interconnected.[25]

At the very beginning of the century, leaders of the Fabian Society would be particularly influential. In London, from the base of the municipal Technical Education Board (TEB), Sidney and Beatrice Webb created an atmosphere which linked efficiency and science through such publications as the 1902 'Report on the Application of Science to Industry'. They conspired with their friend the lawyer R.B. Haldane to create the federal Teaching University of London (1898), many of whose accredited teachers worked in the city's polytechnics, the London School of Economics and Political Science (1895), which became part of the University of London in 1902, the Day Training College for teachers (now the Institute of Education) in 1902 and finally Imperial College for Science and Technology (1907).[26] Although the College had moved from its originally proposed title, 'The Imperial College of Applied Science', the dean of its constituent Royal College of Science described the College's mission as 'adequate, systematic and superior instruction in applied science – science applied to various branches of industry.'[27] It was in this climate that the suffragist Evelyn Sharp greeted the award of degrees in Home Economics at the neighbouring King's College London with the reflection, 'Five hundred years ago it would have been called witchcraft. Today it is called applied science and all sorts of people who would once have been burnt at the stake are encouraged to pass examinations in it'.[28]

Institutions with authoritative names served to give an air of timeless reality to the outcomes of pressure-group politics. The numerous departmental names that expressed affiliation to applied science drew on nineteenth-century precedent but also made claims about the twentieth century. The highest profile of these models was Sheffield University, the largest part of which was the 'Department of Applied Science'. The Department (actually a faculty) was an artefact of the creation between 1903 and 1905 of the University and seems to have been inspired by Michael Sadler, formerly of the Government's Board of Education.[29] In exile from making national policy, Sadler had been asked by the city council to advise on secondary and higher education in Sheffield. Like all effective leaders of social movements, he drew radical conclusions from widely accepted principles. For him the deployment of a new bureaucratic phenomenon followed from his analysis of the world, both characterised by applied science. He explained his proposals in his 1903 report to the City's Council:

> The development of applied science, and the world-wide growth of commercial and industrial activities, have changed the situation. We can no longer prosper without giving as much thought to education as to markets and machinery. Mother-wit and business-power are as valuable as ever they were, but they cannot dispense with scientific training and a sound education. What we need is to provide ourselves with forms of education which are both thorough and practical, and which fail neither in respect of direct utility or of liberal culture.[30]

From this common-place, Sadler argued that the University College, created a few years earlier by integrating local colleges and including a large technical department, should seek to be upgraded to the status of a University in its own right.[31] Responding to Sadler's recommendations, the head of the Technical Department, writing on behalf of the College, welcomed the proposal and agreed that students would need a stronger theoretical base: 'no really advanced work can be done by the Technical student in applied science until he has first received a sound theoretical training.'[32] When that upgrade happened two years later, the 'Technical Department' was rebadged as the 'Department of Applied Science'. Across the north of Britain, readers of not just the *Yorkshire Post* but also of such papers as *The Manchester Guardian* and *The Scotsman* more remote from Sheffield, encountered the term 'applied science' through references to the celebrated department. This title would be highly visible particularly, but not only, in northern newspapers right up to the Second World War. Several dozen articles from the set mentioned the department in the period even before the First World War.

The celebrity border crosser

If the development of, and news about, educational institutions and technical education provided a continuing public profile to 'applied science', the use of the term was taken far beyond. Before the First World War, the key actor in the public eye was the Liberal politician and Minister of War R.B. Haldane: a man with many faces, philosopher, conspiring politician, modernising Minister of War and the man who coined the term the 'civic university'.[33] Haldane was a friend of the Webbs, and a member of the small club of turn-of-the-century thinkers known as the 'Coefficients'. While such people are typically treated as sources of creativity, here I want to suggest how their celebrity enabled them to retransmit ideas across the public sphere.[34] Haldane's engagement with civic universities as well as the military and the wider public shows how concepts could circulate through the civic prominence of universities and the energy of politicians.

Haldane capitalised on his education in Germany at Göttingen University and his undoubted love of the country. Never himself a minister with responsibility for education, he was nonetheless a promoter of what was a triple helix of science, university education and the armed forces. His impact on the Army as Secretary of State for War from 1905 to 1912 must have been alarming. Asked at his introduction to the Department by the generals what sort of army he wanted, he later recalled replying 'an Hegelian army'.[35] The founder of the Imperial General Staff, he is reported to have said about his work in military reform, 'I have never had a more congenial occupation than this attempt at reorganisation and the introduction of science into the business'.[36] Haldane's influence spread the use of the term across discourses which would otherwise have been quite separate. The preface to his 1902 *Education and Empire* cited the role of higher education as inculcating a love of culture for culture's sake on the one hand and, on the other, expertise in the application of science in the training of 'captains of industry' – a phrase he popularised.[37] His friend, the educational reformer and civil servant Robert Morant wrote to him, 'You have said what needs saying over and over again and you have said it pointedly and with illustrations which ought to make the Englishman listen if anything will.'[38]

The effectiveness of Haldane's effort to emphasise educational competition with Germany was indicated in 1904. On 6 October the Kaiser gave a short speech on the importance of applied science in opening the new Technische Hochschule in Danzig. Quite remarkably, this event in a distant town, whose location would have been unknown to most British people, was covered at length the very next day in both *The Scotsman* and *The Manchester Guardian* newspapers.[39] In 1909, Haldane could explain his decision to depend on a committee of scientists for advice on aeronautical development, by saying that he had learned in his recent trip to the University of Göttingen that the Germans were relying on science for their work on aeroplanes. Although he was blamed by some for being slow to take up the Wright Flyer, he has been shown, instead, to have been waiting for the success of British aviator and designer Lieutenant Dunne.[40] He could, and did, appeal to 'common sense' and say he was merely continuing a campaign begun in the mid-nineteenth century. The context and outcomes in the years leading up to the Great War, the war itself and the post-war years would prove, however, quite different from their Victorian predecessors.

In October 1913, it was Haldane, now Lord Chancellor, who opened new buildings of Sheffield's Department of Applied Science. These were built on a lavish scale to accommodate the two thousand night school students and two hundred day students it was attracting. The excellent facilities were intended to help the local steel and coal industries that had liberally supported their building, along with the government.[41] The local *Sheffield Telegraph* reported that it was impossible to find anyone more distinctively or even 'aggressively modern' to open the department.[42] Haldane's three speeches on this occasion linked the discourse of war and universities. He said that the graduates would be the 'general staff of the empire', and described the men of the Officers' Training Corps, an organisation he had invented and whose ranks at Sheffield he inspected, as his 'spiritual sons'. The speeches were widely quoted *in extenso* and reflected upon in *The Times*, *The Scotsman* and *The Manchester Guardian* as well as in local papers, and the politician's circulation of an educational term into the military sphere was widely reinforced.

In the war that followed shortly, science and its potential were widely discussed.[43] In the words of the engineer and codebreaker Alfred Ewing, 'With the community generally, applied science took on a new significance; till then it had meant little to them; they now saw it as a man struggling in the water sees a plank within his reach.'[44] The Department of Scientific and Industrial Research (DSIR) founded in 1915 reflected in its first annual report that business had firmly adopted the concept of applied science.[45] By this the organisation meant the research that would have benefits in the short term and would be conducted by new industry-wide research associations. In public discourse during the war itself, applied science was discussed remarkably widely as a German strength, even if lack of moral guidance would vitiate its potency in the end. Thus a *Times* editorial in 1915 about DSIR reflected that industry had passed through two phases, 'mechanical invention' and 'applied science'. While Britain had led in the first it was having to catch up in the second.[46] Of course views about the German development of chlorine gas as a weapon also became a symbol of the abuse of applied science as we shall see below.

R.B. Haldane's friend, the one-time Prime Minister and head of the Conservative Party, Arthur Balfour, shared many of his views and was a much-cited commentator on science. This Scottish laird and philosophical idealist might have seemed the archetypical traditionalist, but he was personally fascinated by science and by such novel gadgets as fast cars.[47] He invested his own money heavily in novel technology though he lost a fortune. In the early years of the century, in particular, Balfour's was an important presence in the press. Time and again his speeches, reported in full, emphasised that applied science was based on pure science and hence enthusiasm for the former should not detract from study of the latter. A fond obituary of several pages in the *British Medical Journal* by the Secretary of the Medical Research Council in 1930 testified to the recognition of his role.[48] Balfour's writing and speeches were widely reported to the end of his days, so he was an influential circulator of ideas. When, in 1924, the government's chief chemist told the British Association meeting in Toronto that Britain had an unparalleled record in pure science but weakness in applied science, his comments were noticed in *The Times* but not elsewhere. By contrast, however, when four years later Balfour contributed a single-page preface to a review of the research associations, reflecting that while Britain had made its full share of discoveries in pure science, as for applied science, 'I am not so sure', his comment was widely reported, appearing even in the *Daily Mirror* and the *Daily Mail*.[49]

The widespread reporting of the words of the wise in the press flattened the difference between national politicians, journalists and celebrity writers. Of all such authors of the early twentieth century, H.G. Wells was the most widely cited promoter of applied science. Initially also a member of the Webbs' circle, his devoted fans would be called Wellsians from the First World War onward.[50] An editorial in *The Yorkshire Post* in 1935 posed the issue thus: 'There are still those who talk as if yet more applied science, not a humane way of living, were the ideal to be followed. They it is, with publicists like Mr. H.G. Wells to lead them, who inspire so much of the rabid apocalyptic which disorders the nerves of the simple-minded. If we are not careful our civilisation will collapse; if we are careful, and throw overboard a few immemorial human traditions, we shall immediately achieve the Earthly Paradise of no work to do, lots to eat, and endless toys to play with."[51] As this quotation illustrates, Wells's name would be almost synonymous with a belief in the power of applied science during the interwar years. Although not himself a frequent user of the term, the link between scientific research and technical change lay at the heart of many of Wells's novels, his non-fiction and his prophetic utterances. Wells has unique standing but he was not the only writer to be discussed in terms of applied science. In general, the genre of science fiction would follow the Second World War (though see Charlotte Sleigh, Chapter 7 in this volume). Science was already, however, an important theme in literature, for instance in a number of stories about the consequences of atomic weapons. Not only were these read directly, they were also discussed in the press in terms of applied science.[52]

Politicians on the national stage were complemented by local influentials. In Hornsey, north London, gadgets such as cookers, electric lights and television illustrated national progress (Figure 5.2). The development of the town hall complex, with its pioneering modernist design, coincided with the opening, in Alexander Palace on the overlooking hill, of the world's first public television mast. At the same time the borough decided to switch the municipally run electricity service from DC to AC, which prompted sales effort for new appliances and competition from the privately operated gas company. Showrooms for both the electricity and gas services were built as an integrated part of the whole, through the resolution of the modernising Alderman and committee chair Rudolph Moritz, a chemistry graduate who had studied in Zurich, a published astronomer and successful patent lawyer.[53] Moritz ensured they were embodied in the creation of the complex and the integrating designs imposed across its buildings. The gas showroom was emblazoned with reliefs romanticising energy and the chemists who brought it safely to the customer. Its neighbour featured the splendid Promethean 'Spirit of Electricity' shown on the cover of this book. These futuristic images, now protected as architectural heritage, still hover over the citizens of North London.[54]

The exemplary development

The rapidly changing profusion of gadgets, materials and new processes was symbolised by the most public of large research projects of the

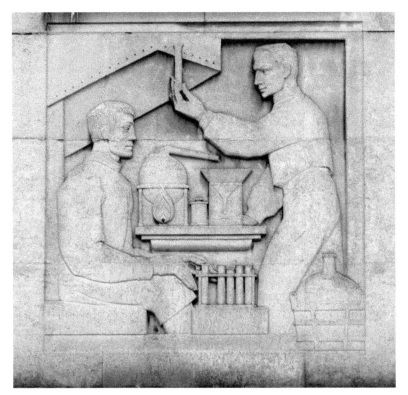

Fig. 5.2 Arthur Ayres, panel from Hornsey Gas Company mural, Hornsey Town Hall Complex, 1937.

Image courtesy Bob Speel.

interwar years: the attempt to turn coal into oil. British coal was in a slump after the First World War as export markets disappeared and its traditional use in transport was threatened by oil, through the greater use of petrol and diesel-driven cars and lorries on the land, and fuel-oil-powered ships on the sea. Even on the railways, diesel and electric traction were making headway. Meanwhile smoke pollution from domestic uses was becoming increasingly unacceptable. Conversion of coal to a low-tar smokeless fuel and creosote which could be hydrogenated seemed a dream quite capable of realisation. The German Alfred Bergius had developed the hydrogenation stage and low-temperature carbonisation of coal implemented, for example, by the Coalite Company to produce smokeless fuel and creosote from coal.[55] The possibility and indeed the prospect of such plants were widely promoted by enthusiasts seeking to rescue the coal industry through applied science. As *The*

Manchester Guardian pointed out in August 1933, '[The average man] holds with an invincible faith that applied science can always work miracles if it is given time to experiment, and he believes that coal spirit may ultimately be producible at competitive prices, a faith which is shared by some eminent chemists'.[56] The politician and first Chairman of the conglomerate Imperial Chemical Industries, Alfred Mond, imported the term 'rationalisation' from Germany and campaigned for the public adoption of this 'scientific' approach to balancing supply and demand. He drove his chemists at ICI to pursue the conversion of British coal into oil with the brief 'cost is immaterial within reasonable limits'.[57] Despite this lack of concern with short-term economics, the ambition was long-term prosperity and a special tax break on synthetic petroleum was sought and won. Technical success was marked by the opening of the world's first plant for converting bituminous coal to oil in Billingham in the north-east of England. Industrialists and trade unionists in both South Wales and Scotland bid for further such works.[58] Industrial espionage around a fictional coal-oil plant in Wales was the subject of a thriller by Hammond Innes in 1939.[59] With greater gravity, at the 1938 Empire Exhibition in Glasgow, the coal exhibit organised in the government building by the DSIR featured at its centre a huge model of the Billingham oil works. Its intended significance went even beyond oil to illustrate 'that to some considerable extent the future of coal lies in its use as a raw material for chemical industry apart from its use as a fuel'.[60] Others were more concerned with cost than Mond, and no other such plant was built in Britain before the Second World War. Despite its lack of economic viability and only marginal effect on fuel production during the 1930s, the conversion of coal to oil had become an important and much cited symbol of applied science as an agent of economic modernisation and the search for new sources of wealth.

The exhibition

The experience of the First World War and the growing American challenge encouraged successive governments to promote the narrative of efficiency and applied science through exhibitions and the support for museums. In 1924, the nation celebrated relief from the war by hosting the British Empire Exhibition in Wembley, attended by 25 million visitors.[61] It fascinated Virginia Woolf.[62] In the Palace of Engineering, the National Physical Laboratory displayed research supporting industry, such as the work on ship design to 'remind the visitor that shipbuilding is an applied science'.[63] Elsewhere in the vicinity enthusiastic demonstrators on the stand of the Vickers company showed how X-rays could be used to test for cracks in metals.

The discourse of applied science in the promotion of popular acceptance of modern civilisation was a means of making sense of such novelties as plastics and a wide range of new electrical goods. The opening of the new building of the Science Museum in 1928 was a catalyst for such a usage. The museum was run by the Department of Education and was the government's foremost investment in interpreting the benefits of science to the people. The replacement for a collection of rundown old buildings on Exhibition Road was acclaimed as 'a cathedral of applied science' and quickly became the first museum in the Empire to attract over a million visitors a year.[64] Deeply influenced by the even larger Deutsches Museum in Munich, the museum consciously played the role of Britain's government-funded and most powerful means of scientific promotion. The Director invited government research associations to promote their work to the public within the museum.[65] Citing the German example, the 1929 Royal Commission on Museums and Galleries emphasised yet further efforts: 'We are impressed by the need for a more adequate representation of current practice in the manifold fields of Applied Science which the Museum is designed to illustrate'.[66] In April 1933 therefore the museum opened its first exhibition promoting an industry, on 'plastic materials and their uses' representing 'eighty industries, most of them of post-war birth'.[67] Following its intended six-month run, and three-quarters of a million visitors, it was extended a further three.[68] Plastics would be the largest such exhibit to be mounted in the entire twentieth century.

Less well funded but with wide appeal, social movements campaigning against excessive noise and excessive smoke respectively adopted applied science. In October 1936 the Smoke Abatement Society organised a special exhibition at the Science Museum. The press release emphasised that not only did the show explain the origins and nature of smoke, it also 'shows how much is being done with the aid of applied science to reduce the smoke nuisance'.[69]

In historical memory, the Science Museum plastics exhibition and its successors have been overshadowed by a shorter nearby exhibition also intended to emphasise the close links between science and practice.[70] The 1931 exhibition in London's vast Albert Hall, a few hundred metres north of the Science Museum, celebrated the centenary of Michael Faraday's discovery of induction. A small central core of the exhibit dealing with the hero himself was surrounded by the multiplicity of modern electrical gadgets that his discovery had made possible and manufacturers were hoping to promote. The message was clear: those surrounding machines embodied the application of the discovery at the centre.[71]

This section has argued that the concept of applied science was circulated across broadly supportive social movements by the direct and reported words of national and local celebrities and the stories of writers, by the prominence of such developments as oil from coal and through the narratives of exhibitions. If, however, the term had been the exclusive property of its proponents and enthusiasts, it would have served as jargon, but not treated as a widely recognised 'real thing'. Instead it did reach wider audiences because equivalent mechanisms worked also to spread the concept among groups much more sceptical of modern life.

Social movements against science

Even as talk about science entered, to an unprecedented degree, the life-worlds of artists, politicians and citizens, opposing narratives also emerged. The experience of the First World War, the diffusion of radically new consumer goods in the 1920s, unemployment associated with new machines, industrial change and the Depression of the early 1930s, combined with challenges to religion, inspired quite different talk about applied science.[72] One approach was to emphasise a distinction between the practitioners of 'pure' or 'fundamental' science and those of 'applied' science. Thus, in preparing for the British Empire Exhibition in 1924, the Royal Society was asked by the DSIR to organise an exhibition on pure science in the government building, separate from the 'applied' exhibits in the Palaces of Engineering and Industry.[73] Alan Morton has shown how the adapted cathode ray tube of J.J. Thomson, originally reported as a means for measuring the velocity of cathode rays and of the m/e ratio of corpuscles, became in 1924 the locus of discovery of the electron. Here, in Morton's words, was 'knowledge without destructive consequences'.[74]

The sensitivity to destruction and pain from it was illustrated by the physicist Oliver Lodge, the best known scientist in Britain in the 1920s.[75] Recalled now probably for his dogged promotion of psychic phenomena, he was one of the founding discoverers of radio waves and inventor of the coherer, an early tuner, and he became a popular broadcaster. A perhaps unlikely follower of the aesthete and socialist John Ruskin, for him science was a calling.[76] Repeatedly, he turned for his theme to such great men of the past as Isaac Newton, the preeminent example of the 'heaven-sent', 'epoch-making' man.[77] Quite different as far as Lodge was concerned were the scientists who pursued their calling like any other tradesmen, under the control of governments. He lost a son in the First World War and increasingly identified the future of applied science with spiritualism and the recovery of contact with lost souls. He condemned the military uses of science. 'Patriotism may run riot, as well as other virtues; and the nations may vie with each other in developing their powers of destruction.'[78] In opening an extension to the Department of Applied Science in Sheffield in 1923 he reflected on the risks of discovering atomic energy, for 'we were getting hopefully or dangerously near it, according as we regarded it.'[79] Such anxieties were multiplied as the traumas of the Great War were amplified by rearmament in the face of the threat from Germany.

The other worry was associated with machinery. Unemployment in the north of England and Scotland was high throughout the interwar years. Many philosophers and writers were concerned about the way machines threatened jobs entirely or about the pleasure they brought. This could be linked to the threat of the abuse of science to make arms. The American writer Raymond Fosdick put the problem as the 'Old savage in the new civilization'.[80] This was not just a matter of philosophy. At the depths of the Depression, the distinguished engineer Alfred Ewing blamed abuse of applied science for weapons, surplus of production and unemployment. In a sad, much-reported speech, he concluded that the optimism view of his youth, that science would be immune to abuse, had been misplaced: 'the command of nature has been put into his [man's] hands before he knows how to command himself'.[81] From the pulpit of the British Association, Ewing was widely heard not just by the participants at the annual meeting and readers of *Nature* and of the *Annual Reports* of the BAAS but also by readers of the daily newspapers to whom both Lodge and Ewing were familiar figures. Although he did not himself specify it, audiences who received his message via *The Observer*, *The Times* or *The Church Times* heard a condemnation specifically of 'applied science'.[82]

Beyond such individuals, Oxford proved a home for concerns about the modern conceptions of science, and this doubt came to be embodied in the Museum of History of Science, formed in the aftermath of the First World War and housed in the old Ashmolean Museum built in the seventeenth century, the oldest dedicated museum building in the world.[83] The three key players were the collector Lewis Evans, the passionate museumophile Robert Gunther, originally a botanist, and his patron the distinguished Regius Professor of Physic Sir William Osler, familiar to Americans as previously a father of the Johns Hopkins school. Osler was horrified by the war and its barbarity.

To Osler the solution to the association of science with barbarism was to associate it rather with civilisation and the humanities. He refused to support colleagues seeking to remove the requirement on all Oxford students to pass an examination in Greek and in 1919 addressed the Classical Association meeting in Oxford. 'The salvation of science lies in a new philosophy – the scientia scientiarum of which Plato speaks'.[84] For Osler the science of science was its history and he was particularly interested in the older history of science in the time of the Greeks up to the Middle Ages. He commended the call for the history of science as the source of a new humanism, which came from George Sarton, the Belgian refugee now at Harvard. Associated with Osler's talk and with the conference was a wonderful exhibition of mediaeval and older instruments. Gunther would argue that scientific instruments of the Renaissance were the northern European counterparts of Italian painting and a worthy model for today.

To many who saw that portion of the collection which was exhibited in the Bodleian, these instruments came as a double revelation; they showed a precision of workmanship that was unexpected in a period as remote as the Elizabethan, and still more so in the days of Roger Bacon; 'they prove, too, that the early masters of the craft of the instrument-maker have been able in their chef-d'oeuvre to combine with scientific accuracy high artistic qualities, in a manner that is entirely foreign to the style adopted by the scientific workers of the present day'.[85]

Gunther's museum would have been entirely different in style to the Science Museum in London, with its celebration of contemporary scientific culture. It was completely appropriate for Gunther to memorialise himself in a stain-glass window in a style contrasting radically to that installed at almost the same time in Glasgow (Figure 5.3). Thus the Museum of History of Science could applaud science in its artistic past, but preserve a telling silence on its modern counterparts.

In Oxford, applied science was damned by implication and the construction of a less damaging interpretation of science.[86] The church proved an influential context for the transmission of questioning views and its leaders were widely reported.[87] In 1924, the independent Anglican newspaper *The Church Times* appointed Sidney Dark as its first lay editor. Dark used his editorial position to introduce readers to debates about the place of science and the balance of the challenges from new devices and opportunities, and the ability of society and the church to respond.

Fig. 5.3 In honour of Robert Gunther, stained glass window, Museum of the History of Science, Oxford, 1938.

Courtesy of the Museum of the History of Science.

Some, such as the Moderator of the Church of Scotland warned of a world civilisation based on applied science but with no place for God.[88] In a 1937 lecture reported by *The Scotsman*, the Professor of Divinity at Edinburgh complained that secularism driven by liberal education and applied science united the modern intelligentsia.[89] He seemed here to be referring to communism and secularism in general, though he also saw other risks in the demonic abuses of Nazi Germany.

Many church thinkers who spoke up cited applied science not as a bad thing in itself but as a challenge that needed to be met. Among

Anglicans, Dean Inge, Professor at Cambridge and then Dean of St Paul's, was an influential figure. An enthusiast for eugenics and an academic as well as perhaps the best theologian of the time, he was an important agency for the promotion of thinking about science within the church. As early as 1910 he was reflecting to the Church Congress on the failure to 'reconcile idealism and naturalism through applied science and co-operative industry'.[90] His 1930 volume *Christian Ethics and Modern Problems* dealt with 'the age of science' in a chapter entitled 'Problems of Social Ethics'. He saw applied science as the inevitable result of the modern confluence of science and industrialism, neither of which were in themselves bad. His influence could be seen in the similar diagnosis of the Christian student movement published and widely reported in 1932.[91] Yet churchmen were often interpreted as challenging applied science, even when that was not their principal concern.

Most often cited was the 1927 intervention, during the British Association meeting in Leeds, by Edward Burroughs, the Bishop of Ripon. Burroughs made science the topic of a sermon in which he argued that so violent were its effects, that man was not yet ready for its impact and indeed our priority now should be to change man rather than gadgets. His call for a ten-year moratorium on science was widely reported even though he was emphatic that he was not attacking science itself and he did not mean to criticise such theories as evolution.[92] The challenge of man needing to conform to the discipline made necessary by modern science had become so widely expressed within talk of modern times that it could be described as a commonplace.

At the end of the 1930s this commonplace would be assaulted by the academic and writer C.S. Lewis, whose religious allegories would be popular into the twenty-first century. He had fought in the First World War, and during the 1930s had been both impressed and frightened by the success of H.G. Wells.[93] His first anti-Wellsian science fiction novel *Perelandra* was published in 1938. The third chapter of his tract *The Abolition of Man* published in 1941 began, '"Man's conquest of Nature" is an expression often used to describe the progress of applied science'. He wrote caustically of three exemplary products of applied science, the aeroplane, the wireless and the contraceptive. Perhaps none was innately 'bad' but each granted power to some men over other men. 'From this point of view', he wrote, 'what we call Man's power over Nature turns out to be a power exercised by some men over other men with Nature as its instrument … Man's conquest of Nature, if the dreams of some scientific planners are realized, means the rule of a few hundreds of men over billions upon billions of men'.[94] Lewis was not alone in Oxford. From the late 1930s he met with a few distinguished likeminded writers including his friend J.R.R. Tolkien at the Eagle and Child pub in the group known as the Inklings.[95] Unsurprisingly, Gunther (curator at the Museum of History of Science), Frank Sherwood Taylor (an expert on alchemy, whom Gunther admired) and Lewis formed an alliance devoted to the tension between naturalism and religion.[96]

Frank Sherwood Taylor, chemist, writer and convert to Roman Catholicism, was deeply aware of the cultural tensions surrounding the science whose history he would write and present. He himself had been a casualty at the battle of Passchendaele. He introduced a 1940 book about successful modern medicines with the words of his ailing brother. Gassed in the First World War and now unable to speak, his brother had scrawled, 'This is what modern science has done for me'.[97] This poignant quotation in a 'popular' science book highlighted the iconic role of poison gas for an entire generation. Beyond its use as a weapon, poison gas had served as the development that was the single most evocative symbol of the dangers of applied science. In 1915, in the wake of its first use on the Western Front, the Canadian Prime Minister was widely quoted bemoaning the application of the instruments of applied science to the work of destruction.[98] After the war, writers and journalists would refer to it time and time again. The charge was, for instance, promoted repeatedly in the pages of the popular *Daily Mirror* newspaper.[99] In 1922 it published a denunciation of 'destructive science' by which the article explained it meant 'applied science'.[100] Its columnist Jennings often wrote under the by-line 'WM', evocative of the Victorian writer William Morris. By the 1930s the newspaper was using the template of destruction wrought with gas to warn of the new threat of atomic bombs created by scientists who had always wanted to destroy the world.[101]

In the wake of the Depression and the even worse unemployment of the early 1930s, 'applied science' and its threat to jobs were often seen as a danger to individual welfare. The term circulated also amongst those who condemned modern changes. The new ethnographic reports of the British, published under the title 'Mass Observation', found by the late 1930s that the public were indeed worried. A 1941 survey summarised its findings as 'People feel that science has got out of control'.[102]

Amongst promoters of the greater knowledge and use of science, we see therefore the term 'applied science' being circulated between discourses about modernity and our place in time, about education, about military preparedness, and relations with Germany and the benefits of research. A few figures addressed the tension between the two movements directly. Most distinguished was the Director of the Royal Institution, Sir William Bragg. His 1928 address as President of the BAAS quite explicitly addressed the concerns of those who had found 'in the last few years, scientific inquiry has advanced at a rate which to all is amazing, and to some is even alarming'.[103] His solution was to interpret the application of science not as something quite new but rather an extension of the reassuringly old, 'craftsmanship'. Industrial research became the process of better harnessing science to craft. In his hands 'The craftsman becomes an association of men, a great manufacturing firm, even, we might say, a nation, if all the members of the nation contribute through government intervention and control to the maintenance of some industry'. Bragg's talk earned the applause of both science and religion. It was published both in *Nature* and in the American journal, *Science;* and it was extracted at length in the American *Journal of Chemical Education.* He also satisfied religious sensitivities and was congratulated by the Bishop of Willesden.[104] *The Church Times*, however, argued that Bragg had failed to resolve the fundamental inhumanity of science: 'Science not only de-personalizes industry, it also de-personalizes the workers in industry. That, indeed, is but one illustration of a fundamental characteristic of science, which explains why the relations between religion and science have never, since the time of Plato and Aristotle, been any better than an armed truce. Science is impersonal, in its methods, its aims, and in its results.'[105] Nonetheless even if Bragg could not answer the fundamental humanist challenge, he became a much-appreciated mediator and sought-after broadcaster.[106]

The radio as a new medium

Radio is the one important form of mass media that has so far not been mentioned. Broadcasting was new in this period. The BBC as a monopolistic public corporation dates from 1923 and its Royal Charter was granted in 1927. There was one national wireless channel, which therefore had immense power. This was circumscribed by political constraints on the treatment of controversial subjects. Politics and strikes were not to be treated. Science however seemed safe and in 1930 a specialist science producer was hired to take responsibility for science talks. Mary Adams was a remarkable producer for she recruited presenters who talked not so much about breakthroughs in science as about science and its broader meanings.[107] The broadcasts made during her tenure brought together the discourses of enthusiasts eager to encourage national efficiency, and sceptics worried about the inhumanity of applied science. For audiences of hundreds of thousands to the radio, and publication in the BBC's *Listener* magazine of tens of thousands of readers, these programmes entrenched the contested concept of applied science.

From Adams' appointment in 1930 until 1935, nearly every week there was a programme at peak time, 7.30pm, about science. Many constituted substantial edited series on such topics as science and religion, science and civilisation and science as a social activity. Contributors were drawn from across a wide spectrum of opinion. Thus the twelve-part series on science and religion broadcast in the Autumn of 1930 included Dean Inge, and the pacifist vicar Dick Shepard as well as the organiser Julian Huxley.[108] The six-part 'Science and Civilisation' broadcast early in 1932 included both the radical J.B.S. Haldane, nephew of 'R.B.' and the Catholic controversialist known for sparring with H.G. Wells, Hilaire Belloc. Julian Huxley's brother Aldous, who would author the novel *Brave New World*, contributed to the series with a broadcast published as 'Science – the double-edged tool'.[109] The next year the BBC decided on national surveys of industry, religion and science. Julian Huxley was recruited to visit laboratories and reflect on his observations with such friends as P.M.S. Blackett, and particularly with the well-known Communist and Professor of Mathematics at Imperial College, Hyman Levi. Concerned that Levi would give the series too radical a tone, Adams invited William Bragg to introduce the series and set the tone.[110] In his preface to the book that came out of the series, Bragg summarised the issues it raised: 'Is scientific research drawing us together or forcing us apart? Is it to be commended for supplying our needs or blamed for causing unemployment? Does it help to bring peace between the nations, or war? Does it add to mankind's vision or restrict it? If it is solving some problems, is it perhaps raising others still more difficult and troublesome?' Huxley, who had been struck by the amount of research conducted in industry, also had a more down-to-earth objective, to tease out the relationships between pure and applied science. Though challenged on his assumption of a 'linear model', and willing to accept the occasional two-way relationship between the two, he came down to a sequence going from basic research to development. In summarising his argument, a *Times* article concluded: 'However even if the distinction between pure and applied breaks down in theory, and should perhaps be replaced by some other terminology – such as basic and long-term research, on the one hand, as against ad hoc and development research on the other – it still remains convenient enough in most cases'.[111]

By 1935 the activities of such producers as Mary Adams were giving rise to concern in the BBC, and she and others left their posts. The radio with its breadth of perspectives but whose broadcasters used the term 'applied science' from a variety of points of view confirmed the reality of the concept.[112]

Conclusion

By the end of the 1930s, applied science was widely cited, and treated as a cultural reality. Whatever their considerable differences on its desirable place in culture, the left-wing materialist physiologist J.B.S. Haldane and the conservative Platonist Dean Inge each gave lectures in 1927/1928, proclaiming the centrality of applied science to the present age.[113] Each asserted the reality of the phenomenon, though one delighted in it and the other proclaimed its inadequacy. In this paper I have suggested that this term was circulated through the narratives of a variety of social movements grappling with modernity, often to achieve commercial, political, institutional or religious ends. Each found uses in the trope of 'applied science'. Before and indeed during the First World War it was a useful way of giving reality to 'efficiency'. It was a term that was circulated from educational circles to the military and to talk about industry and the nation. In the post-war years the term circulated through movements as diverse as the Smoke abatement society and the Anglican Church. Each used 'applied science' as the key to making sense of the welter of strange experiences and gadgets encountered personally by citizens as well as the increasing amount of daily news of unprecedented phenomena. Each deployed the concept of applied science in their visions of past, present and future, of Britain's place in the world, of education and of contemporary experience.

The ironic parallels between those celebrating modernity and those deeply worried by it was exploited by Aldous, the writer brother of Julian Huxley and himself grandson of Thomas Huxley. His novel *Brave New World* was described by H.G. Wells as 'a blasphemy against the religion of science'.[114] Yet the treatment of the character of the benevolent ruler Mustafa Mond, who secretly read the works of Isaac Newton, was sympathetic. Its author was much more torn than might appear, and was a founder member of Political and Economic Planning (PEP).[115] His 1932 broadcast on science and civilisation recounted the victories that had followed from applying science to human affairs: famines had been vanquished and mountains crossed. Now psychology should be applied to government. He argued that 'Ideally, science should be applied by humanists. In this case it would be good'.[116]

As we have seen, scientists themselves were not always happy to be so clearly identified with the characteristics of modernity, negative as well as positive. In the 1920s the terms 'pure science' and 'fundamental science' distanced the academic chemist and physicist from the publicly controversial outcome of their work. Frank A.J.L. James (Chapter 6 in this volume) shows how historians of science reconstructed a history without recourse to the modern traumas of science.

Yet the narrative of a society rooted in science was encountered by the public through the names of departments and degrees in modern subjects, in engagement with exhibitions, in popular books and lectures, in films and on the radio. The consequence of looking at these manifestations of public discourse is to focus the historian's attention on the uses of the term 'applied science' in making sense of changes in society. That has not eliminated the role of the great communicators. Individual tellers of stories about the past and future, R.B. Haldane, Oliver Lodge, H.G. Wells and C.S. Lewis, have had a prominent place in this account. So have places where stories were transmitted, exhibitions and museums; even stained glass in a museum and a chapel, and above all the new space of the BBC's National Channel. The BBC played a distinctive role by bringing together otherwise contrasting and competing discourses and thereby highlighted the interest they shared in using 'applied science' as a means of making sense of a world, not just science. This was the key legacy of the four decades reviewed here. Talk about applied science was too important to be reduced to an interpretation of science alone.

Acknowledgements

This paper was made possible by the award of AHRC Leadership Fellowship award no AH/L014815/1. The author is grateful to the Museum for the photograph for Figure 5.1 specially taken for this volume.

Notes

1   The original correspondence is contained in 'Notes on a stained glass scheme for the chapel Glasgow Univ.' 8 December 1930, OC8/669, Glasgow University Archives. In a previous letter, Douglas Strachan to Principal Rait, dated 8 November 1930, Strachan wrote, 'Subject matter in the decoration of a University Chapel should satisfy – or at any rate not offend – the sections of thought represented within the University & should in a small and unimportant way therefore face the

"Science & Religion" question'. The Vice Chancellor's speech is reproduced in the coverage of the occasion in *The Scotsman* newspaper: 'Glasgow University. Memorial Windows in Chapel. Dedication Ceremony', *The Scotsman* 15 January 1934. See Nick Haynes, '"A Spiritual Enterprise" – Douglas Strachan's Stained Glass in the Memorial Chapel, University of Glasgow', *Historic Churches* (April 2014). On Strachan see Juliette MacDonald, 'Aspects of Identity in the Work of Douglas Strachan (1875–1950)' (PhD thesis, St Andrews, 2003).

2 'Position of Technological Studies in the University', University of London Senate Minutes (20 December 1933), para. 1067, UoL ST 2/2/50, Special Collections, University of London Library.

3 Question from Lord Balfour of Burleigh to R.B. Haldane, 'Transcript of Shorthand Notes of Proceedings of Committee of Council on 18 December 1902 (2nd day)', PC8/605 part 2, 32, The National Archives. 'Applied science' and 'technology' were related but also competing concepts and to avoid confusion I shall avoid the term 'technology' except where it is explicitly intended.

4 'The Glasgow and West of Scotland Technical College. Relations with the University', May 1909, in 'Special Commissions on Relations with the University. Minutes etc.' C1/27/1 University of Strathclyde Archives.

5 'Industry without Morality: Canon Donaldson and More Social Sins', *The Manchester Guardian* (31 March 1927).

6 It is indicative that of the 16 references to applied science found for 1900 in the digitised newspaper selection of over 2,000 papers in all over 40 years (1900–1939) all but two related to academic contexts or associations. In 1939, of the 41 references less than half were academic. Moreover, while in 1900 a reference to the invention of a folding bicycle by the brother of the founder of the scouts Baden Powell in terms of applied science showed how clever he was, in 1939 a report from a meeting of the Institute of Inventors that a splash screen for grapefruit eaters was applied science was clearly light-hearted. The *Yorkshire Post* had been excluded from the set before 1930 because when it was constructed the paper was not available for the entire period. Had it been included, three further articles would have been available for the 1900 set (all academic). In 1939, 18 articles were published in the *Yorkshire Post* mentioning applied science, of which five were not academic, and 13 related to academe.

7 The identification of academic research in engineering with research in applied science at the time was very common. See for instance Michael Sadler, 'Science, Pure and Applied', *The Manchester Guardian* (22 August 1917). [Sidney Dark], 'The Editor's Table. Mechanization', *The Church Times*, 2 April 1931.

8 For popular discussion of science in this period see Peter J. Bowler, *Science for All: The Popularization of Science in Early Twentieth-Century Britain* (Chicago: University of Chicago Press, 2009). The concept of the representative anecdote was explored in K. Burke, *A Grammar of Motives* (Berkeley: University of California Press, 1969, first published 1945), 59–61 and 325–28. Robert Bud, 'Framed in the Public Sphere: Tools for the Conceptual History of "Applied Science" – A Review Paper', *History of Science* 51, no. 4 (December 2013): 413–33. Frank Trentmann explored the problem of the relationships between diverse discourses in popular politics in his working paper 'Civil Society, Commerce, and the "Citizen-Consumer": Popular Meanings of Free Trade in Late Nineteenth-and Early Twentieth-Century Britain', Working Paper No. 66 (Center of European Studies, Harvard University, 2003). For parallels in the way a concept is produced as much by its opponents as proponents see Darrin M. McMahon, *Enemies of the Enlightenment: The French Counter-Enlightenment and the Making of Modernity* (New York; Oxford: Oxford University Press, 2001) and Matthew H. Wisnioski, *Engineers for Change: Competing Visions of Technology in 1960s America* (Cambridge, Mass: MIT Press, 2012).

9 A large US historical literature has now explored the self-conscious manner in which a boosterist literature around the benefits of chemistry in particular was intended to answer doubts raised by the science's use in the development of poison gas. See, for instance, Andrew Ede, 'The Natural Defense of a Scientific People: The Public Debate Over Chemical Warfare in Post-WWI America', *Bulletin for the History of Chemistry* 27, no. 2 (2002): 128–35. However, in American

discourse it was the 'machine' rather than science which was to blame for unemployment. See, for example, the well-known, and widely banned, tract William F. Ogburn, *You and Machines* (Chicago, Ill., 1934). See Arlene L. Barry, 'Censorship during the Depression: The Banning of "You and Machines"', *OAH Magazine of History* 16, no. 1 (October 1, 2001): 56–61. See also Amy Sue Bix, *Inventing Ourselves Out of Jobs? America's Debate over Technological Unemployment, 1929–1981* (Baltimore, MD: Johns Hopkins University Press, 2000). On technological culture see Lowell Tozer, 'American Attitudes toward Machine Technology, 1893–1933' (PhD thesis, University of Minnesota, 1953). On the complex parallels between discourses in British and the more German-influenced Sweden see Emma Eldelin, 'The Cultural Transfer of a Concept: C.P. Snow's "Two Cultures" and the Swedish Debate', in *Inter: A European Cultural Studies Conference* (11–13 June 2007): 185–203. Proceedings published by Linköping University Electronic Press, www.ep.liu. se/ecp/025/, accessed June 2018. See also Martin Kohlrausch and Helmuth Trischler, *Building Europe on Expertise Innovators, Organizers, Networkers* (London: Palgrave Macmillan, 2013). On thinking about machines in Britain see Daniel Wilson, 'Machine Past, Machine Future: Technology in British Thought, c. 1870–1914' (PhD thesis, Birkbeck University of London, 2010).

10   W.B. Gallie, 'Essentially Contested Concepts', *Proceedings of the Aristotelian Society*, New Series, 56 (1 January 1955): 167–98; Michael Calvin McGee, 'The "Ideograph": A Link between Rhetoric and Ideology', *Quarterly Journal of Speech* 66, no. 1 (1980): 1–16; Raymond Williams, *Keywords: A Vocabulary of Culture and Society* (Oxford: Oxford University Press, 1985); Hans Blumenberg and Robert Ian Savage, *Paradigms for a Metaphorology* (New York: Cornell University Press, 2010).

11   Reinhart Koselleck, 'Introduction and Prefaces to the *Geschichtliche Grundbegriffe*', *Contributions to the History of Concepts* 6, no. 1 (2011): 1–37.

12   For Fine see Hank Johnston and Bert Klandermans, eds., 'Public Narration and Group Culture: Discerning Discourse in Social Movements,' in *Social Movements and Culture edited by Hank Johnston and*

Bert Klandermans (London: UCL Press, 1995), 127–43; and see Bert Klandermans and Conny Roggeband, *Handbook of Social Movements Across Disciplines* (Springer Science & Business Media, 2007). See also Teun van Dijk, 'Ideology and Discourse' in *The Oxford Handbook of Political Ideologies*, eds. Michael Freeden, Lyman Tower Sargent, Marc Stears. (Oxford: Oxford University Press, 2013), 191–212.

13   Michael James Lee, 'Creating Conservatism: Postwar Words that Made a Movement' (PhD thesis, University of Minnesota, 2008), 14.

14   See Bud, 'Framed in the Public Sphere', 413–433.

15   There may be a tendency to see such a term as 'applied science' as the catchphrase of a single social movement which today we might identify as 'technocracy'. This 1932 term however had a specific meaning in the USA and its application to an array of British discourses originating earlier is unhelpful. For the history of the technocrat movement see William E. Akin, *Technocracy and the American Dream: The Technocrat Movement, 1900–1941* (Berkley: University of California Press, 1977).

16   Newspapers explored are *The Times* (Gale), *The Manchester Guardian* (Proquest), *The Scotsman* (Scotsman Archive), *The Daily Mail* (Daily Mail Archive), *Daily Mirror*, *The Daily Express* and *The Yorkshire Post* (1930s only) (UK Press Online). Weekly magazines are *The Church Times* (UK Press Online) and *The Spectator* (Spectator Archive).

17   See for example Jonathan Wild, '"Insects in Letters": John O'London's Weekly and the New Reading Public', *Literature and History* 15, no. 2 (Autumn 2006): 50–62; Joseph McAleer, *Popular Reading and Publishing in Britain 1914–1950* (Oxford, Clarendon Press, 1992), 73.

18   Sidney Dark, 'The New Reading Public: A Lecture Delivered Under the Auspices of "The Society of Bookmen"' (London, 1922), 15.

19   See Bowler, *Science for All*.

20   Storm Jameson, 'Get the Best out of Today: Don't Yearn for Yesterday and Get the Best out of Life', *Daily Mail* (4 April 1931).

21   See the advertisements for the Broadwood Expression Player Piano, for instance *The Times* (3 February 1926).

22 'Hustled History. Italy's Triumphant
Progress. New Nursery for Applied
Science', *The Manchester Guardian*
(17 December 1912); and *The Times*
(17 December 1912). A subsequent
advertisement for Italy in similar terms
appeared a few months later, 'Italy's
Industrial Growth', *The Manchester
Guardian* (13 February 1913).

23 Most notably see G.R. Searle, *The Quest
for National Efficiency: A Study in British
Politics and Political Thought, 1899–1914*
(Oxford: Blackwell, 1971).

24 A.G. Gardiner, *Prophets, Priests and Kings*
(London: Alston Rivers, 1908), 211.

25 On the linkage between culture and
communication see James W. Carey,
*Communication as Culture, Revised
Edition, Essays on Media and Society*
(London: Routledge, 2009).

26 These developments have been well
traced and linked in *Education for
National Efficiency: The Contribution of
Sidney and Beatrice Webb*, ed. E.J.T.
Brennan (London: Athlone Press, 1975).
See also R. MacLeod, 'Science for Imperial
Efficiency and Social Change: Reflections
on the British Science Guild, 1905–1936',
Public Understanding of Science 3, no. 2
(1994): 155–193; Linda Simpson,
'Imperialism, National Efficiency and
Education, 1900–1905', *Journal of
Educational Administration and History*
16, no. 1 (1984): 28–36.

27 'Royal College of Science', *The Times*
(4 October 1907).

28 Evelyn Sharp, 'Science in Home Life', *The
Manchester Guardian* (8 July 1909).

29 David Phillips, 'Michael Sadler and
Comparative Education', *Oxford Review of
Education* 32, no. 1 (2006): 39–54; Lynda
Grier, *Achievement in Education: The Work
of Michael Ernest Sadler, 1885–1935*
(London: Constable, 1952).

30 City Council (Sheffield). Education
Committee, *Report on Secondary and
Higher Education* (London: Eyre &
Spottiswoode, 1903), 4. Sadler was
himself a most interesting circulator of
discourse, proceeding to the leadership of
Leeds University in 1911 and then in 1923
becoming Master of University College
Oxford. His responsibility for success
in Leeds' competitor Sheffield was
recognised by that University in 1913
with an Honorary Doctorate of Letters.

31 City Council (Sheffield). Education
Committee, *Report on Secondary and
Higher Education*, 11–12.

32 W. Ripper [Head of the Technical
Department] to John F. Moss [Secretary
to the Education Committee, Sheffield]
(14 September 1903), 170–171, Sheffield
Technical School Minute Book No. 2,
SUA/1/2/, Sheffield University Archives.
Quoted with permission.

33 See Eric Ashby and Mary Anderson,
Portrait of Haldane at Work on Education
(London: Macmillan, 1974).

34 I am not suggesting at all here that
Haldane was original in conflating
of academic and military categories.
In 1903, Lockyer (founder of the
British Science Guild of which
Haldane was chair) told the British
Association for the Advancement of
Science in his presidential address that
'the land unit, like the battleship, will
become a school of applied science,
self-contained, in which the "officers"
will be the efficient teachers'. 'The British
Association. Meeting at Southport', *The
Scotsman* (10 September 1903). In his
talk, 'The Influence of Brainpower on
History', Lockyer acknowledged the
inspiration of Captain Mahan, author of
The Influence of Seapower on History.
Lockyer however was much more
successful at speaking to the faithful
than in reaching out in the way Haldane
was so successful. There is a large
literature on the role today of 'celebrities'
in spreading knowledge of, and interest
in, climate change. See for instance,
Maxwell T. Boykoff and Michael K.
Goodman, 'Conspicuous Redemption?
Reflections on the Promises and Perils of
the "Celebritization" of Climate Change',
Geoforum 40, no. 3 (2009): 395–406.
They argue, 'This increase in media
coverage of celebrities and climate
change/global warming illustrates a
need for an understanding of the new
and changing connections of the
celebritization of climate change where
media, politics and science intersect',
395. See too Alison Anderson, 'Sources,
Media, and Modes of Climate Change
Communication: The Role of Celebrities',
*Wiley Interdisciplinary Reviews: Climate
Change* 2, no. 4 (2011): 535–546.

35 R.B. Haldane, *An Autobiography* (London:
Hodder and Stoughton, 1929), 185. See
also Andrew Vincent, 'German Philosophy
and British Public Policy: Richard Burdon
Haldane in Theory and Practice', *Journal
of the History of Ideas* 68, no. 1 (2007):
157–79.

36 Gardiner, *Prophets, Priests and Kings*, 211.

37 R.B. Haldane, *Education and Empire. Addresses on Certain Topics of the Day.* (London: John Murray, 1902), x. The term 'captain of industry' had been coined by Carlyle in *Past and Present*, published in 1843 but its usage tripled between the late 1890s and 1903, according to Google 'ngrams'. For other literary interpretations of the term see Alan Hunt, 'The Captain of Industry in British Literature, 1904–1920' (PhD thesis, University of Toronto, 1998).

38 Morant to Haldane, 25 May 1902, f. 175 MS 5905, National Library of Scotland.

39 The opening was recorded in Germany by a celebratory booklet: Hans von Mangoldt, 'Denkschrift Über Die Eröffnungsfeier Der Königlichen Technischen Hochschule Zu Danzig: Am 6. Oktober 1904'. 'The German Emperor on Education', *The Scotsman* (7 October 1904); 'The Kaiser on German Education', *The Manchester Guardian* (7 October 1904).

40 Alfred Gollin, 'The Mystery of Lord Haldane and Early British Military Aviation', *Albion* 11, no.1 (April 1979): 46–65.

41 Michael Sanderson, 'The Professor as Industrial Consultant: Oliver Arnold and the British Steel Industry, 1900–14', *The Economic History Review* 31, no. 4 (1978): 585–600. For Jonas see Gerald Newton, 'Joseph Jonas, University Benefactor, Lord Mayor of Sheffield 1904–5 – German Spy?' in *Mutual Exchanges I. Papers presented in Commemoration of the 50th Anniversary of the Sheffield-Münster Academic Links*, ed. R.J. Kavanagh (Frankfurt/Main: Peter Lang, 1999).

42 'Degree Day. Lord Haldane in Sheffield. Applied Science. Opening of the New School. The Nation's Duty', *Sheffield Daily Telegraph* (27 October 1913).

43 See for instance Roy M. MacLeod and E. Kay Andrews, 'Scientific Advice in the War at Sea, 1915–1917: The Board of Invention and Research', *Journal of Contemporary History* 6, no. 2 (1 April 1971): 3–40; Roy MacLeod and E. Andrews, 'The Origins of the DSIR: Reflections on Ideas and Men, 1915–1916', *Public Administration* 48, no. 1 (1970): 23–48; R. MacLeod, 'The Scientists Go to War: Revisiting Precept and Practice, 1914–1919', *Journal of War and Culture Studies* 2, no. 1 (2009):

37–51. Roy MacLeod, 'The Chemists Go to War: The Mobilization of Civilian Chemists and the British War Effort, 1914–1918', *Annals of Science* 50, no. 5 (1993): 455–481. T.R.V. David, 'British Scientists and Soldiers in the First World War (with Special Reference to Ballistics and Chemical Warfare)' (PhD thesis, Imperial College, 2009); A. Hull, 'War of Words: The Public Science of the British Scientific Community and the Origins of the Department of Scientific and Industrial Research, 1914–16', *The British Journal for the History of Science* 32, no. 4 (1999): 461–481. Ian Varcoe, 'The DSIR: A Study in the Growth of Organized Science' (DPhil dissertation, University of Oxford, 1972).

44 James Alfred Ewing, *An Engineer's Outlook* (London: Methuen, 1933), 40. This was originally presented as 'The James Forrest Lecture' at the Institution of Civil Engineers on 4 June 1928.

45 'Science and Industry. First Report of the Advisory Council. Corps of Research Sappers Wanted', *The Manchester Guardian* (1 September 1916). For opposition to such views see Graeme Gooday, '"Vague and Artificial": The Historically Elusive Distinction between Pure and Applied Science', *Isis* 103, no. 3 (1 September 2012): 546–54.

46 'Science and Industry', *The Times* (28 July 1915).

47 John David Root, 'The Philosophical and Religious Thought of Arthur James Balfour (1848–1930)', *Journal of British Studies* 19, no. 2 (January 1980): 120–141; David Knight, 'Arthur James Balfour (1848–1930), Scientism and Scepticism', *Durham University Journal* 87 (1995): 23–30. L.S. Jacyna, 'Science and Social Order in the Thought of A.J. Balfour', *Isis* 71, no. 1 (1980): 11–34. Ruddock Mackay, H.C.G. Matthew, 'Balfour, Arthur James, first earl of Balfour (1848–1930)', *Oxford Dictionary of National Biography* (Oxford: Oxford University Press, 2004) online edition, January 2011, www.oxforddnb.com/view/article/30553, accessed June 2018.

48 Walter Fletcher, 'Lord Balfour and the Progress of Medicine', *British Medical Journal,* no. 3613.1 (5 April 1930): 660–61.

49 'Laundries Big Claim. Customers' Huge Saving as Result of Applied Science' *Daily Mirror* (27 April 1927); 'Research in Industry. A Real Advance', *The Times*

(27 April 1927); 'Science Applied to Industry. Results of Research. Cheapening Costs of Manufacture', *The Manchester Guardian* (27 April 1927); 'Research Policy. Lord Balfour's View', *The Scotsman* (27 April 1927); 'Research Pays. Earl of Balfour on Example of US and Germany', *Daily Mail* (27 April 1927).

50 Robert Crossley, 'The First Wellsians: A Modern Utopia and Its Early Disciples', *English Literature in Transition, 1880–1920* 54, no. 4 (2011), 444–69.

51 'Happiness and Invention', *Yorkshire Post* (25 June 1935).

52 Spencer Weart, *Nuclear Fear* (Cambridge, Mass: Harvard University Press, 1988).

53 See the short obituary of Moritz in *Monthly Notices of the Royal Astronomical Society* 102 (1942), 69.

54 See Bridget Cherry, *Civic Pride in Hornsey: The Town Hall and Its Surrounding Buildings*, new ed. (London: Hornsey Historical Society, 2006). On the dominating role of Moritz see Borough of Hornsey, *Minutes of the Proceedings at the Meetings of the town Council 1935–36*; Borough of Hornsey, *Minutes of the Proceedings at the Meetings of the town Council* 1937–38, Haringey archive service.

55 On the 1920s work of the fuel research station of the DSIR to develop the Bergius process see Anthony N. Stranges, 'From Birmingham to Billingham: High-Pressure Coal Hydrogenation in Great Britain', *Technology and Culture* 26, no. 4 (1 October 1985): 726–57. For a contemporary assessment see W.R. Gordon, 'The Utilisation of Coal', *Journal of the Royal Society of Arts* 82, no. 4255 (1934): 756–80.

56 'Petrol from Coal. The New Scheme and the Ordinary Man', *The Manchester Guardian* (2 August 1933). The author would probably have been the newspaper's science correspondent J.G. Crowther, himself an enthusiast for applied science.

57 'The Present Situation Regarding the Production of Synthetic Petrol', attachment to Coal & Oil Hydrogenation Group Committee and Oil Division Hydrogen group committee, 'Minutes of the Extraordinary Joint Meeting Held on January 05 1929', ICI Papers 17.059 RMC, Teeside Archives.

58 For Scotland see Scottish National Development Council, 'First Report of the Oil from Coal Committee', Economic Series, Scottish Industry (Scottish National Development Council, 1935); for South Wales see 'Report and Findings of Special Coal Research Committee set up under the auspices of the National Industrial Development Council of Wales and Monmouthshire by the Commission for the Special Areas (England and Wales) October 1935/December 1939, DIDC/7 Glamorgan Archives; for the National Development Council of Wales and Monmouthshire see Robert E. Counsel, 'The National Development Council of Wales and Monmouthshire 1932–55' (MA Thesis, University of Wales, Cardiff, 1987).

59 Hammond Innes, *All Roads Lead to Friday* (London: Herbert Jenkins, 1939).

60 'Glasgow Empire Exhibition. Coal Hall in Government Pavilion', Appendix to minutes of the Second Meeting of the Glasgow Empire Exhibition Committee of the Department of Scientific and Industrial Research, January 26 1938, para. 10, 3, BT 60/47/9, National Archives.

61 Donald R. Knight and Alan D. Sabey, *The Lion Roars at Wembley: British Empire Exhibition, 60th Anniversary 1924–1925* (New Barnet: D.R. Knight, 1984).

62 Scott Cohen, 'The Empire from the Street: Virginia Woolf, Wembley, and Imperial Monuments', *Modern Fiction Studies* 50, no. 1 (2004): 85–109.

63 'The Value of Research: Tests and Trials', *The Times* (24 May 1924).

64 On the Science Museum as centre of applied science see J. Bailey to the editor, 'The Science Museum', *The Times* (27 December 1923); 'A New Palace of Machinery. The King opens Science Museum', *The Manchester Guardian* (21 March 1928).

65 Henry Lyons to the Secretary of the Cast Iron Research Association (29 June 1927). File 2609, Z Archive, The Science Museum Library.

66 Royal Commission on National Museums & Galleries, *Final Report, part II. Conclusions and Recommendations relating to Individual Institutions dated 1st January, 1930*, 1929–30 [Cmd. 3463], 47. Note the treatment of the Science Museum was followed immediately and quite anomalously by a page-long treatment of its German counterpart, The Deutsches Museum 'not only because it is in itself a remarkable example of how a modern Museum can be made a great instrument of technical as well as of

popular instruction, but because it is a symbol of national efficiency. It reveals the intense concentration in the Germany of to-day on the scientific means of industrial progress, a concentration which we believe has its sharp significance for this country', 48–49.

67 'Plastics Industrial Exhibition 1933', *The Engineer* (31 March 1933): 319.

68 'Miscellanea', *The Engineer* (20 October 1933).

69 'The Smoke Abatement Exhibition to be held at the Science Museum South Kensington', File 5397AZ Archive, Science Museum Library. On temporary exhibitions at the Science Museum see, Peter Morris, '"An Effective Organ of Public Enlightenment": The Role of Temporary Exhibitions at the Science Museum', in *Science for the Nation* ed. Peter Morris (London: Palgrave, 2010), 228–65.

70 F.A.J.L. James, 'Presidential Address: The Janus Face of Modernity: Michael Faraday in the Twentieth Century', *The British Journal for the History of Science* 41 (2008): 477–516; G. Cantor, 'The Scientist as Hero: Public Images of Michael Faraday', in *Telling Lives in Science: Essays on Scientific Biography*, eds. Michael Shortland and Richard Yeo (Cambridge: Cambridge University Press, 1996), 171–93.

71 *The Times* announced its 'Faraday Number' of 21 September 1931 under the heading 'Applied Science'. See 'Applied Science', *The Times* (18 September 1931).

72 For relations between science and religion early in the twentieth century, see Peter J. Bowler, *Reconciling Science and Religion: The Debate in Early-Twentieth-Century Britain* (Chicago: University of Chicago Press, 2010).

73 Department of Scientific and Industrial Research, Advisory Council, 'Memorandum on the British Empire Exhibition, 1924' (20 June 1922), British Empire Exhibition Committee Papers, CD.470–497, Royal Society Archives.

74 Alan Morton, 'The Electron Made Public: The Exhibition of Pure Science in the British Empire Exhibition, 1924–25,' in *Exposing Electronics*, ed. Bernard Finn, Artefacts 2 (Reading: Harwood Academic, 2000), 25–43, 39. For fundamental science see Sabine Clarke, 'Pure Science with a Practical Aim: The Meanings of

Fundamental Research in Britain, circa 1916–1950', *Isis* 101, no. 2 (1 June 2010): 285–311.

75 Peter Rowlands, *Oliver Lodge and the Liverpool Physical Society* (Liverpool University Press, 1990), 271.

76 See Lodge's introduction to John Ruskin, *Unto This Last: And Other Essays on Art and Political Economy* (London: Dent, 1907), vii–xv.

77 See for example Oliver Lodge, *Pioneers of Science* (London: Macmillan, 1926).

78 Oliver Lodge, 'Science and Human Progress', The Halley Stewart Lectures 1926 (London: George, Allen & Unwin, 1926), 35.

79 'Sir Oliver Lodge and Atomic Energy', *The Manchester Guardian* (4 May 1923).

80 See Peter H. Denton, *The ABC of Armageddon: Bertrand Russell on Science, Religion, and the Next War, 1919–1938* (Albany, NY: State University of New York Press, 2001), Chapter 1, 'Science and the New Civilization'.

81 Alfred Ewing, 'An Engineer's Outlook', *Supplement to Nature* no. 3279 (3 September 1932): 341–350.

82 'Double-Edged Gifts of Science: The Three Riddles. War Economics Leisure', *The Observer* (4 September 1932); 'Mechanical and Moral Progress', *The Times* (1 September 1932); 'British Association at York. Mechanical Progress not altogether a Blessing', *The Church Times* (9 September 1932).

83 A.V. Simcock, *The Ashmolean Museum and Oxford Science, 1683–1983* (Oxford: Museum of the History of Science, 1984); A.V. Simcock, *Robert T. Gunther and the Old Ashmolean* (Oxford: Museum of History of Science, 1985).

84 W. Osler, 'The Old Humanities and the New Science: The Presidential Address Delivered before the Classical Association at Oxford, May, 1919', *BMJ* 2, no. 3053 (5 July 1919): 1–7.

85 'Collection of Historic Scientific Instruments. Public Opening by the Earl of Crawford & Balcarres … May 25 1925', f. 52 Old Ashmolean Letter Book, Volume 1, Archive of the Museum of the History of Science, Oxford.

86 For the many scientists at Oxford who contributed to war work see Jack Morrell, *Science at Oxford, 1914–1939: Transforming an Arts University* (Oxford: Clarendon, 1997), 6–14.

87 See Bowler, *Reconciling Science and Religion*.

88 'World Conference. Stewardship and Church Finance', *The Scotsman* (23 June 1931).

89 'Lee Memorial Lecture. Opposition to Christianity. Prof. Baillie's Analysis', *The Scotsman* (24 May 1937). The title of the lecture was 'The secularistic and demonic forces in the world of today'. Baillie was chiefly concerned with 'the political paganisms of our time, rather than with applied science per se.

90 'The Church Congress', *The Times* (30 September 1910). In 1909, Inge's enthusiasm for the applied science of eugenics had been reported in the *Daily Mirror*. See 'Finest Specimens of Humanity. Cambridge Professor's Eulogy of the Upper Classes of this Country', *Daily Mirror* (10 February 1909).

91 William Ralph Inge, *Christian Ethics and Modern Problems* (London: Hodder & Stoughton, 1930), 204; V.A. Demant, *This Unemployment: Disaster or Opportunity?* (London: Student Christian Movement Press, 1932).

92 'The British Association. Bishop of Ripon's Address, Science and Religion', *The Times* (5 September, 1927); See also 'Bishop of Ripon on the Science Holiday. Science ahead of Man', *Evening Standard* (6 September 1927). The discussion has been analysed by Anna-K. Mayer, '"A Combative Sense of Duty": Englishness and the Scientists', in *Regenerating England: Science, Medicine and Culture in Inter-War Britain*, ed. Christopher Lawrence and Anna-K. Mayer, vol. 60 (Amsterdam: Rodopi, 2000), 67–106; C. Pursell, '"A Savage Struck by Lightning": The Idea of a Research Moratorium, 1927–37', *Lex et Scientia* 10 (1974): 146–58.

93 On C.S. Lewis and H.G. Wells, see David Downing, 'Rehabilitating H.G. Wells: C.S. Lewis's Out of the Silent Planet', in *C.S. Lewis: Lifework and Legacy*, ed. Bruce L. Edwards, 2 (Westport: Praeger, 2007): 13–34.

94 C.S. Lewis, *The Abolition of Man, Or, Reflections on Education, with Special Reference to the Teaching of English in the Upper Forms of Schools*, third edition (Glasgow: Collins, Fount Paperbacks, 1978), 22. First published 1941.

95 Humphrey Carpenter, *The Inklings: C.S. Lewis, J.R.R. Tolkien, Charles Williams and Their Friends* (London: HarperCollins, 2006); Philip Zaleski and Carol Zaleski, *The Fellowship: The Literary Lives of the Inklings: J.R.R. Tolkien, C.S. Lewis, Owen Barfield, Charles Williams* (New York: Farrar, Straus and Giroux, 2015).

96 C. Mitchell, 'Following the Argument Wherever It Leads: C.S. Lewis and the Oxford University Socratic Club, 1942 to 1954', *Inklings* 17 (1999): 172–96. At a meeting of the Socratic Club in 1946 Sherwood Taylor spoke on 'The Scientific World Outlook'. He was bitterly criticised by the biologist Peter Medawar who 'attacked the picture put forward of the scientific materialist. He was confident that there was now no such person. It was the efforts of Dr Sherwood-Taylor and Mr C.S. Lewis to prop the corpse up into a sitting position which gave it a semblance of life', *Socratic Digest* 3 (1946): 155–156.

97 Frank Sherwood Taylor, *The Conquest of Bacteria from 606 to 693* (London: Secker and Warburg, 1940), 13. On Taylor see A.V. Simcock, 'Alchemy and the World of Science: An Intellectual Biography of Frank Sherwood Taylor', *Ambix* 34, no. 3 (1987): 121–39. On the British response to gas see Marion Girard, *A Strange and Formidable Weapon: British Responses to World War I Poison Gas* (University of Nebraska Press, 2008). For Osler's shocked response see William Osler, 'Science and War', *The Lancet*, (9 October 1915), 798.

98 The speech was reported in 'A War of Applied Science', *The Times* (7 August 1915); and 'Sir R. Borden and Warsaw: Courage of the British Empire Undaunted', *The Scotsman* (7 August 1915).

99 The *Daily Mirror* columnist successfully spanned the range between elite, for he had 'discovered' the literary talent of the writer Edith Sitwell, and his mass public. A pen-portrait of Jennings is painted by Jon Pilger in *Hidden Agendas* (London: Vintage, 1988), 381.

100 'Destructive Science', *Daily Mirror* (2 September 1922).

101 WM, 'One Good Bang', *Daily Mirror* (25 October 1935). Also see 'World May Blow Up', *Daily Mirror* (25 October 1935). Sherwood Taylor too would warn of the dangers of cheap nuclear energy in his book on science in 1939: 'If this discovery is made, the world had best be on its guard, for there will be put in the hands of the greedy, ambitious and cruel such a power of accomplishing their ends as men have barely dreamed of'. F. Sherwood Taylor, *Science Front, 1939* (London: Cassell, 1939), 248.

102 Mass Observation, 'Report on Everyday
Feelings about Science' (October 1941),
no. 951. Reproduced with permission of
Curtis Brown Group Ltd, London on
behalf of The Trustees of the Mass
Observation Archive © The Trustees of
the Mass Observation Archive.

103 William Bragg, 'Craftsmanship and
Science', *Science* 68, no. 1758 (7
September 1928): 213–23. See Jeffrey
Hughes, 'Craftsmanship and Social
Service: W.H. Bragg and the Modern
Royal Institution', in *The Common
Purposes of Life: Science and Society at the
Royal Institution of Great Britain*, ed.
Frank A.J.L. James (Hambledon: Ashgate,
2002), 224–47.

104 Bishop of Willesden to W.H. Bragg
(5 September 1928), W.H. Bragg/8A/4,
Royal Institution Archives.

105 'Science and Truth', *The Church Times*
(14 September 1928).

106 James Friday, 'The Braggs and
Broadcasting', *Proceedings of the Royal
Institution of Great Britain*, 47 (1974):
59–85.

107 Allan Jones, 'Mary Adams and the
Producer's Role in Early BBC Science
Broadcasts', *Public Understanding of
Science* 21, no. 8 (2012): 968–83; Ralph
John Desmarais, '"Promoting Science":
The BBC, Scientists, and the British
Public, 1930–1945' (MA thesis, IHR
University of London, 2004).

108 Julian Huxley, ed., *Science and Religion:
A Symposium* (London: G. Howe, 1931).
The Science and Civilisation series was
published as Mary Grace Agnes Adams,
Science in the Changing World. (London:
George Allen & Unwin, 1933).

109 Aldous Huxley, 'Science – The Double-
Edged Tool', *The Listener*, issue 158
(20 January 1932): 77–79, 112.

110 Adams to Bragg (14 September 1933),
BBC Written Archives Centre.

111 'Science', *The Times* (1 January 1936).

112 The disproportionate significance of
Mary Adams's tenure is indicated by
the concentration of references to
'applied science' in the *Listener* during
the years of her posting. In the 11 years
between January 1929 (when the
magazine was launched) and December
1939, there were 76 uses of the term in
the main editorial sections. Of these,
65 are to be found in the five and a
half years from January 1930 to
June 1935.

113 Dean Inge told the Church Congress in
August 1927, 'we were living at the end
of an extraordinary episode in history,
marked especially by the advance of
applied science and a vast increase
of wealth and population', 'Modern
Churchmen in Conference. Dean Inge's
Presidential Address', *The Manchester
Guardian* (23 August 1927). For
Haldane's expression of this belief see
J.B.S. Haldane, 'The Place of Science in
Western Civilisation', *The Realist* 2.2
(November 1929), 49–165, presented to
the Fabian Society (25 October 1928).
The essay begins, 'Western Civilisation
rests on Applied Science', 49.

114 J.R. Hammond, ed., *H G. Wells: Interviews
and Recollections* (London: Macmillan
1980), 95.

115 Robert S. Baker, 'Aldous Huxley:
History and Science between the
Wars', *Clio* 25, no. 3 (1996): 293–300;
John David Bradshaw, 'Aldous Huxley's
Ideological Development 1919–1936'.
(DPhil dissertation, University of Oxford,
1987).

116 Huxley, 'Science – The Double-Edged
Tool', 112.

6
'The springtime of science': modernity and the future and past of science

Frank A.J.L. James

As the different chapters in this volume attest, science was a central feature of modernity in Britain during the period roughly from 1890 to 1950. Many areas of cultural activity, including literature, art, music and some parts of religion, drew heavily on contemporary science and that of the recent past. Its pervasiveness stemmed from images of science developed and cultivated by members of the scientific community and their supporters. One image asserted that scientific knowledge possessed a special epistemological status. That is, its cognitive content depended entirely on understanding what was taken to be the existence of an invariable external natural world, which could be known with increasing certainty through observation, experiment and theory. (This image did start to change in the 1920s and 1930s as the implications of relativity and quantum theories began to be appreciated). A second image was that scientific knowledge could be deployed, through practical application, for the betterment of humanity. Involving devising some sort of under-standing about how science connected to various areas of society and culture, issues relating to the epistemological independence of scientific knowledge might consequently be raised. Thus, it became possible that the two images could, to some degree, be put in opposition.

During the interwar period, two approaches were articulated to avoid addressing such potential problems, and the subject of this paper, were how science might be used in the future and the history of science. The consequences of what emerged from these discourses are still very much with us in guiding generally what we think about the nature, purposes and function of science and how we think about the future. More specifically much of the core content of the contemporary academic discipline of the history of science stems directly from the work undertaken in the interwar period.

Through a few case studies, I will provide some suggestions as to how and why science and its history related so closely to modernity. The examples will include how the future of science was imagined (especially in the writings of William Inge and Aldous Huxley where current scientific trends projected into the far future, came, to some extent, to serve as a proxy for anxieties concerning the potentialities of science) and how the history of science was deployed in the interwar years (as seen through its place at the Royal Institution and the 1931 International Congress). Finally, I will discuss, primarily through the work of Rupert and Marie Hall, how approaches to the history of science were debated thereafter.

During the interwar years, the history of science generally dealt with the development of scientific knowledge and came to concentrate in Britain, for reasons that will be outlined, on the seventeenth century. Although this chapter will centre on Britain, it should be noted that, remarkably, practitioners in countries with very different national and ideological backgrounds such as the USA, France and the Soviet Union also come to focus on the same period. There was thus potential for some scholarly international discussion but its effects were limited, at least so far as Britain was concerned. Indeed, though the history of science has become much more international since 1945, vestiges of these different national pre-war styles are still recognisable in current historiography.

During the interwar years, the history of science in Britain existed as a very marginal academic discipline. The first department, initially of the History and Method of Science at University College London, began work in 1921,[1] while the University of Cambridge established a committee in 1936.[2] Both those groups, and indeed the larger subject generally, were then dominated mostly by scholars originally trained in science or medicine (including Charles Singer, Douglas McKie, Angus Armitage, Joseph Needham and Walter Pagel), although the historian Herbert Butterfield played a significant and, as the scientists went off to war in the 1940s, increasingly dominant role at Cambridge as Professor of Modern History from 1944.[3]

Beyond the narrow confines of the academic discipline, a significant amount of history of science was deployed publicly to support various agendas, including science education.[4] Another public use of history of science was celebrating scientific anniversaries, for example the centenary of the discovery of benzene in 1925[5] and, perhaps most significant, the extensive 1931 celebrations, put on at considerable cost, to mark the centenary of Michael Faraday's discovery of electro-magnetic induction.[6] Events included, amongst much else, an exhibition held at the Albert Hall, which displayed most of Faraday's original apparatus, and a grand commemorative meeting at the Queen's Hall. Though this latter was held at the time of the pound being forced off the gold standard, nevertheless the Prime Minister, Ramsay MacDonald, who had studied science when he first arrived in London, honoured his promise to address the meeting. All these events were heavily funded by the newly created electrical engineering industry, which explicitly portrayed itself, for example through its design language, as modern. But by celebrating a long-dead scientist, the industry invoked some kind of respectable scientific ancestry for their commercial aims presented in the visual language of modernity, thus displaying a curious Janus face of contrasting the modern and the past with a strong eye to the future, a topic which was of profound interest to Inge.

Dean Inge and the future

As other chapters in this volume illustrate, it was not only in the presentation and representation of science and engineering that modernity made an impact. In England the term 'modernist' became closely associated with 'The Churchmen's Union for the Advancement of Liberal Religious Thought', an Anglican organisation founded in 1898. Partially because of the success of contemporary scientific and historical knowledge, The Churchmen's Union sought to oppose high church Tractarian or Anglo-Catholic practices and beliefs (such as in miracles) within the Anglican Church. (There was a similar movement in the Roman Catholic Church, which Pius X attempted to proscribe in his 1907 encyclical *Pascendi dominici gregis*.[7]) Peter Bowler has discussed how some of the leading modernist figures in The Churchmen's Union, for example, William Inge, Ernest Barnes, Edward Burroughs, etc., sought to reconcile Anglican Christianity with scientific knowledge in the 1920s and 1930s.[8] So successful was this that many believed that the conflicts associated with the nineteenth century (Charles Darwin, Thomas Huxley, John Tyndall, etc.) had all but been settled to such an extent that the biologist Julian Huxley published in the journal *The Modern Churchman* an essay on his grandfather and religion to mark the centenary of his birth in 1925. There he claimed that 'to-day the liberal wing of religious thought rejects all naïve creationism; rejects miracles; sees in prayer a special form of mediation, of value for the spiritual results which it engenders in the praying mind'.[9]

The term 'reconciliation', a Christian notion after all, does not seem to do full justice to how one churchman, William Inge, actively deployed science to support his ideological and theological views.[10] Born in 1860 into a strong Tractarian family, who were appalled at his apostasy to the modernist position, Inge attended Eton and King's College, Cambridge, and was later a Fellow of Hertford College, Oxford, and Lady Margaret Professor of Divinity at Cambridge University. In 1911 the Prime Minister, Herbert Asquith, appointed him Dean of St Paul's, a position he held until retirement in 1934. In that role and indeed until his death in 1954 he enjoyed an enormous public reputation as the spokesman for the modernist Anglican position.

In the spring of 1914, Inge delivered his first lecture course at the Royal Institution on the third-century CE Greek philosopher Plotinus, whom he described as a religious teacher and mystic, a view probably similar to how Inge saw himself, especially in his adherence to Platonism. In these lectures, which attracted an average audience of 53, typical for the Royal Institution that spring, Inge tackled themes that became increasingly familiar in his writings, such as racial exhaustion, referring specifically to the third century, but doubtless with 1914 in mind. In 1922 he delivered a second lecture series on 'Theocracy' which had a marginally larger audience.[11] But when he delivered a Friday Evening Discourse at the Royal Institution on 29 May 1931, the fame that he had built up since his appointment as Dean ensured that it was attended by 450 people which he described as 'a record audience'.[12]

Inge entitled his Discourse, 'The Future of the Human Race', which summarised much of his thinking during the previous decade, devoting a large part of it to imagining what England would be like in the year 3000 as he 'should like to see it'.[13] His vision included no armed forces or national debt and low rates of crime, though persistent offenders would be 'painlessly extinguished in a lethal chamber, without any publicity or humiliation to his family'.[14] Indicating the impact of contemporary science, Inge asserted that the twentieth century would still be viewed as the 'century of discovery',[15] which would be exemplified by new books being read on that iconic modern invention, the wireless.

But there would be problems in achieving Inge's vision. A further great war would be the end of civilisation. The use of a new poison gases (the centenary of their first use during the Great War being the occasion for the meeting where these essays were originally presented) or atomic weapons might allow for the extermination of a nation in a day. Inge thought that the Soviet Union would be capable of performing and willing to perform such an act.[16] Not that he was any kinder to America: 'When during the [Great] War they applied to their young conscripts the mental tests of scientific psychology, they discovered that their mental age was only thirteen. Thus, the average American citizen has to be classed as "a high-grade moron," which is jargon for semi-imbecile. The Americans still believe in democracy'.[17]

This last sentence went to the heart of what Inge considered to be the fundamental problem with society. Being a Platonist with a strong anti-democratic authoritarian streak, Inge railed against what he took to be the decline of the race, due to its being 'appallingly dysgenic'.[18] He had coined the latter word in 1915 because the war, in his view, eliminated the strongest and healthiest of the population, 'leaving the weaklings at home to be the fathers of the next generation'.[19] It was a word that Julian Huxley used when endorsing Inge explicitly, though he saw the issue more in class terms, with the upper classes having better control in restricting family size, while the working classes would multiply uncontrollably.[20] By 1931, with the absence of war, Inge had come to see it in much the same way. Taking full advantage of the Anglican Church's approval, less than a year before, of the use of artificial contraception in some circumstances,[21] Inge imagined a future where children could be conceived only by certified permission. This would hold Britain's population to 25,000,000, many of whom would be 'A1' men and women[22] and thus avert the 'biologically disastrous'[23] effects of better healthcare and education of the working classes, which allowed them to enter into the professional classes. In the case of health he cited a piece of research on the transmission of increasing incidence of blindness due to a genetic fault receiving expression over several generations. He asked 'Can any nation which permits such an exercise of the alleged right of procreation be called highly civilised?'[24] Unsurprisingly Inge was an enthusiast for eugenics which he asserted 'Christianity has much in common with. It aims at saving the soul – the man himself'.[25] As evidence for this contention Inge summarised from the Sermon on the Mount: 'Do men gather grapes of thorns, or figs of thistles? A corrupt tree cannot bring forth good fruit, neither can a good tree bring forth evil fruit.'[26]

Inge's view of the future meant that one did not have to look beyond these islands to see a strong strand of authoritarian thought linking science and ideas of modernity with corporate, centralised and co-ordinated states.[27] But, of course, one could. The research on inherited blindness that Inge discussed had been undertaken by Karl Pearson who had been appointed Galton Professor of Eugenics at University College

London, the same year that Inge became Dean. It was Pearson in his speech at a dinner given in his honour in April 1934 who suggested that the culmination of eugenics:

> lies rather in the future, perhaps with Reichskanzler Hitler and his proposals to regenerate the German people. In Germany a vast experiment is in hand, and some of you may live to see its results. If it fails it will not be for want of enthusiasm, but rather because the Germans are just starting the study of mathematical statistics in the modern sense![28]

Pearson did not live long enough to see the consequences of such enthusiasm.

From the examples of the deployment of science by Inge, Huxley, Pearson and others and from the Faraday celebrations, it is abundantly clear that modern science in its applications, both ideological and practical, was viewed as intimately connected with the contemporary and future state of society and culture. Such views were brilliantly satirised by Huxley's brother Aldous Huxley early in 1932 when he published *Brave New World*. Shot through with electrical imagery, such as the game of electro-magnetic golf at St Andrews, and populated with people categorised from Alpha-double-plus to Epsilon-minus, Huxley's novel extrapolated into the far future what life might become if the trajectory proposed by Inge, Pearson and Julian Huxley was followed. But where they believed this would be beneficial to humanity, Aldous Huxley provided, through his dystopia, a salutary warning.

The history of science

The success of *Brave New World* suggests that there existed some significant contemporary repugnance at the possible connections linking science, culture and society in the name of modernity. Science, under such circumstances, would not be able to retain its special epistemological position, an outcome that most practising scientists found highly undesirable. One way of defending the status of science was by turning to its history and this can be examined by the treatment of the subject at the Royal Institution. Of course, the Friday Evening Discourses at the Royal Institution between 1890 and 1950 included many devoted to aspects of the Institution's history and those who worked in it – indeed until the 1960s the vast majority of writing about the Royal Institution was produced either by members of staff or members of the Institution.[29] But there were also Discourses devoted to topics within the history of science that had no connection with the Royal Institution. Thus before the Great War Discourses were delivered by British scientists on the recently deceased German-speaking scientists, Heinrich Hertz, Herman Helmholtz and Robert Bunsen.[30] There were none after, possibly because the Director of the laboratory at the Royal Institution from 1923 to 1942, William Bragg, hated the Germans due to the death in action of his younger son Robert at Gallipoli; he did not even go to Stockholm in 1920 to belatedly collect his 1915 Nobel Prize, since German laureates would also be there.[31] In the 1920s and 1930s Discourse topics moved backwards in time to eighteenth-century figures such as Joseph Priestley or Josiah Wedgwood[32] and in the early 1940s to the seventeenth century. In 1942, the astronomer Henry Plummer lectured on 'Galileo and the Springtime of Science' which he commenced by asserting that the death of Galileo Galilei three centuries before had brought to an end 'the first chapter of modern science'.[33] As yet another global conflict approached and commenced, it would seem that in Bragg's Royal Institution at least there was a retreat to happier scientific times, or so they thought.

The image of that happier time, where scientific knowledge had its own existence separate from society and culture, needed protection from the associations that science now had, thanks to the Great War and the views of Inge and others, in the modern world. That historical image was seriously imperilled at the second International Congress for the History of Science, held in London, between 30 June and 4 July 1931, of which Singer was President, Bragg, Treasurer, and Henry Dickinson (Secretary of the Newcomen Society, the British host society for the Congress), Secretary. The Congress was attended by a delegation of eight Soviet savants, including Nikolai Bukharin and Boris Hessen, who arrived with minimal notice due to Stalin shortly before changing policy by deciding to authorise Soviet participation in foreign international conferences; the Congress provided the first occasion where his wishes could be implemented.[34]

It is perhaps therefore not too surprising, partly because of time constraints and partly because of ideological objections by Singer, that the Soviet participants did not have sufficient time to fully make their presentations.[35] Instead their papers were quickly translated into English and published as *Science at the Cross Roads*. This provided a Marxist interpretation of science in terms of dialectical materialism. Most of the papers were not historical but dealt with contemporary issues, emphasising the scientific and technological progress that the Soviet

Union had made or would be making. There were only two historical papers in this volume. One, a short account of the modern application of Faraday's work, doubtlessly responding to the preparations for the electro-magnetic induction celebrations centenary less than three months hence.[36] The second, by Boris Hessen, entitled 'The Social and Economic Roots of Newton's "Principia"', basically argued that 'the contents of the "Principia" exhibits the complete coincidence of the physical thematics of the period, which arose out of the needs of economics and technique'.[37]

The disjoint between most of the Soviet papers and the subject of the Congress was doubtless partly due to the shortness of the time in which they could prepare them. But there was also a more substantive reason. The papers argued that scientific ideas and knowledge, both in the past and contemporaneously, were a product of the prevailing social structures and the mode of production at any particular point in time and place. Since the Soviet delegation wanted to understand the nature of science and its place in the modern world, this could be achieved as well by contemporary examples as by historical. For them there would have been no incongruity about presenting their work at the Congress, especially as there was no consensus as to where the boundaries of history of science lay; hence the subtitle of *Science at the Cross Roads* referred, incorrectly, to the Congress as being for the history of science and technology.[38] Their argument became a part of the externalist view of the history of science in opposition to the assertion that science developed through its own internal logic, building, largely cumulatively, on previous knowledge.[39]

Most members of the British scientific and historical communities either ignored or reacted negatively to the views of the Soviet delegation, though tinged with a degree of envy about the size of the discipline in the Soviet Union.[40] One historian who responded was George Clark who, newly elected to the Chichele chair of Economic History at the University of Oxford, provided the opening paper at the Congress. Six years later, in a lecture that was quickly published as a paper and reprinted in his *Science and Social Welfare in the Age of Newton*, Clark, Labour left-leaning in his politics, attacked Hessen's thesis which, he opined, from 'the point of view of the historian ... has serious defects'.[41] This view was initially endorsed by the American sociologist Robert Merton who drew heavily on Hessen's work for his long paper, first published in 1938, 'Science, Technology and Society in Seventeenth Century England'. Here he argued, through statistical analysis, that the rise of science in seventeenth-century England could be attributed to technological needs and the prevalence of Puritanism at the time. A copy of Clark's lecture seems to have reached Merton as he was completing his paper since he discussed it briefly in an addendum. Whilst asserting his heavy indebtedness to Hessen, Merton agreed with Clark's criticism that other 'classes of influence outside of science proper were operative' and by implication had not been addressed by Hessen. These included 'medicine, arts, religion and most important of all, the disinterested search for truth'.[42] Very rapidly Merton reversed his position and the following year published a critique of Clark and reasserted his own position.[43] Merton's thesis later reverberated in the historiographical literature as one of the earliest extended articulations of the externalist position. But none of those writings really addressed the issue as to why the sixteenth and seventeenth centuries became identified as the beginning of something called modern science other than it was when the 'Scientific Revolution' happened.[44]

There were two issues at stake which became connected. First, what was the impact, if any, of scientific knowledge on the modern world? And, beyond that, what were the entailed moral and judgemental issues? Both Inge and Aldous Huxley had answered the impact question in remarkably similar ways, but their views on the consequences were entirely at variance. The second issue centred on the cause or causes of scientific developments particularly in the seventeenth century. Did new knowledge cumulate and stem, entirely independently of social, cultural, economic factors etc., from earlier scientific knowledge because of science's internal logic? This position was argued for strongly in 1939 by the Russian-born Platonist historian and philosopher of science Alexandre Koyré who went so far as to suggest that Galileo had not performed many of the experiments he described.[45] Or was science the product of social, economic and ideological structures as proposed by Hessen and Merton? The former position supported science's special epistemological status; the latter might throw it into doubt.

One of those who attended the 1931 Congress was an Anglo-Catholic Cambridge University embryologist, Joseph Needham. His first book, an edited collection of essays, entitled *Science, Religion and Reality* (1925), included contributions from Singer (on science and religion up to the time of Isaac Newton) and Inge, who chaired the editorial committee for the volume, of which Needham was one of the secretaries.[46] This book was part of the reconciliation process between science and Christianity and Bowler was surely right to count Needham as a sort of Modernist.[47] Although Needham drifted away from Anglo-Catholicism, he remained committed to High Church practices, including sexual liberation, and embraced a form of Christian socialism interpreted through Marxism which he pursued following the 1931 Congress. As well as being a first-rate embryologist (he was elected a Fellow of the Royal Society of London in 1941), Needham was highly interested in the history of his subject, hence his involvement with the Congress in the first place. As Gary Werskey pointed out, within three years Needham had moved from the publication of a nearly 200-page internalist account of the development of embryology, which introduced his three-volume treatise on chemical embryology[48] (as well as many historical references scattered throughout the text), to advocating an Hessenist view of how the history of embryology should be written.[49]

Needham and other scientists further to the left of Clark, such as Desmond Bernal and Lancelot Hogben, and slightly later Stephen Mason and Samuel Lilley, argued, following Hessen, that external social factors contributed significantly, if not decisively, towards constructing scientific knowledge, though generally they sought to maintain the special epistemological status of science.[50] It is no surprise then that Needham played a significant role in founding the history of science committee at Cambridge University in the latter half of the 1930s. There he came into conflict with the liberal historian Butterfield who firmly believed that the history of science should be undertaken by trained historians and not by practising or former scientists, such as Needham, Singer or Herbert Dingle, or by the whole host of chemists turned historians of their subject then active, such as Eric Holmyard, James Partington and Douglas McKie.[51] Needham lost this argument largely because of his going to China on behalf of the Royal Society of London in 1942, followed by a spell in Paris, at the invitation of Julian Huxley, the first Director General of UNESCO, to help establish that organisation.[52] When he returned to Cambridge in 1948, he found that Butterfield had ensured that the committees were dominated by historians and Needham's influence within the discipline became marginal, especially as he worked on what was generally regarded as his esoteric multi-volume *Science and Civilisation in China*.[53]

From the position of power that he had acquired, Butterfield ensured that his protégé Rupert Hall (whose doctoral thesis was examined by Singer and Clark) was appointed in 1948, against Needham's wishes on explicitly political grounds, to provide a course of lectures in history of science at Cambridge University and in the early 1950s was appointed assistant lecturer and then lecturer.[54] In 1959 Hall divorced his first wife and married the American historian of chemistry Marie Boas, and shortly afterwards they moved to California. Both had served with their respective armed services during the first half of the 1940s (coincidentally both in signals). Furthermore, immediately after the war Boas had worked under the direction of Henry Guerlac on the very contemporary history (including publishing a paper on the history of naval radar) of the Radiation Laboratory at the Massachusetts Institute of Technology. After the war and both heavily influenced by Koyré's work, Hall wrote his thesis on the theory of seventeenth-century ballistics, while Boas moved with Guerlac to Cornell University to study the seventeenth-century natural philosopher, Robert Boyle.

One thing that is apparent was that for the Halls, and indeed for many of their contemporaries, the history of science they practised avoided addressing the issues raised by science having been so effectively and devastatingly deployed by all sides during the conflict. In the case of electronic communications (and in Boas's case of radar as well), they both had first-hand experience of the power of science, and, one has to say, of its limitations as well. What is significant, indeed striking, is that after those experiences, they and many others of their generation deliberately chose to research what they took to be science in the sixteenth and seventeenth centuries, rather than the more recent past which might have thrown some light on contemporary problems. The approach that the Halls and their colleagues adopted in their work was that science, at least during that period, should be shorn of any links to social, economic, political, even religious considerations.

This view was, of course, challenged by those who wanted to link scientific knowledge to broader social and cultural issues. The bitter debates which ensued centred on what was regarded as the mutually exclusive internal and external approaches to the history of science, but may also have been a consequence of the terms being loosely defined.[55] The depth of feeling engendered is well illustrated by Bob Young's contribution to Needham's Festschrift.[56] In this rather ill-tempered piece, Young accounted for the debate in terms of academic politics at Cambridge University in the 1950s and 1960s. Despite trying to be aware of his own historical construction, Young, possibly because he was American, did not realise that the predominant view in Britain of what was regarded as central to the content of the history of science had emerged, at least in part, from how science and its history was seen to relate to modernity in the interwar years.

But herein lies a paradox in the subject. The reason for studying the history of science in the first place was because of science's modern importance and influence, so clearly on display during the 1920s and 1930s. Perhaps this is one of the reasons why the term 'early modern' began to be increasingly used at this time; scholars of the seventeenth century, such as Plummer, could claim to be studying the modern world which thus began at the same time as modern science. Butterfield, in his *Origins of Modern Science* (1949) believed that it began in the seventeenth century with 'the scientific revolution' which, as his ringing and oft-quoted declaration made clear, 'outshines everything since the rise of Christianity and reduces the Renaissance and Reformation to the rank of mere episodes, mere internal displacements, within the system of medieval Christendom'.[57] Rupert Hall, in his influential book, *The Scientific Revolution 1500–1800: The Formation of the Modern Scientific Attitude* (1954) deliberately echoed Butterfield: '[science] is the one product of the West that has had decisive, probably permanent, impact upon other contemporary civilizations. Compared with modern science, capitalism, the nation-state, art and literature, Christianity and democracy, seem regional idiosyncrasies, whose past is full of vicissitudes and whose future is full of dark uncertainty'.[58] There was no explanation of how or why modern science had achieved this standing or what its implications might be. Furthermore, there was no discussion of either engineering or technology (words noticeably absent from the index), nor of politics, economics or religion.[59]

A sort of strategic consensus had formed in the interwar years which asserted that the crucial development of modern science as a powerful independent epistemic system occurred during the seventeenth century. This was given force by Koyré's work, and by the titles of both Butterfield's and Hall's books containing the word 'modern'. Not discussed was how and why the effects of 'the scientific revolution' and subsequent scientific developments had impacted on the modern world, to make science such an integral part of modernity. Thus while the Halls, Koyré and their colleagues made the claim that figures such as Copernicus, Kepler, Galileo, Boyle and Newton were the founders of modern science and hence supported the Baconian narrative that 'knowledge itself is power', very little of this addressed the place and power of science in modern culture, society and the general polity.

Those who tried to illustrate the contemporary power of science showed every sign of being influenced by the events of the Cold War in one way or another. For example, the Princeton historian Charles Gillispie, in his *Edge of Objectivity* (1960) asserted, in a vein similar to, but presumably unbeknownst to him, Inge's view of the Soviet Union, that 'Men of other traditions can and do appropriate our science and technology, but not our history or values. What will the day hold when China wields the bomb? And Egypt? Will Aurora light a rosy-fingered dawn out of the East? Or will Nemesis?'[60] Hall's mentor at Cambridge in the late 1930s, C.P. Snow, identified in his 1959 Rede lecture, 'The Two Cultures and the Scientific Revolution', a cultural divide between the sciences and the humanities.[61] Although Snow took a very technocratic view of science, strongly antipathetical to modernist writers,[62] this was yet another attempt to address the place of science in modern culture and society, with, despite the title, little reference to history, since history had not tackled the issue. An ironic consequence of the controversy that ensued following Snow's lecture was that the history of science became identified as one of the ways of bridging the cultural divide he had identified. Indeed, the advertisement for the chair of history of science and technology at Imperial College of Science and Technology was framed explicitly in terms of bridging the cultures.[63] Hall, who was appointed to this post after Needham had declined it, said in his inaugural lecture that he preferred leaving bridge building to engineers.[64] Thus Hall and Boas Hall stayed well clear of such uses of the history of science and sought to concentrate, as decent scholars would, on their research. They at least seem to have recognised the impossibility of the history of science, or at least their form of it, delivering what had been asked of it, and quite soon paid a heavy institutional price for that self-awareness.

Conclusion

So, does any of this help in understanding the issues with which I set out? So far as the content of the history of science is concerned, why the seventeenth century became the key area for research and study is, I think, clearer. It was not just to do with Butterfield emphasising something called the scientific revolution and Hall popularising the term. It was closely related to the place of science in the modern world of the 1920s and 1930s, as envisioned by Inge and other modernists. They wished to see science as a force for good in the modern and future worlds and to ensure this happened, it had to be shorn of any unfortunate connotations. Writing the history of modern science, especially chemistry, at that time would not have achieved this. Inge and later Gillispie were quite clear that it was the misuse of scientific knowledge by political systems that did not in their view share Western values or lacked internal moral controls (i.e. countries such as the Soviet Union, China, etc.) that was the problem, rather the science itself. But this was a potentially weak argument for the value of science. Much better to claim that science in the seventeenth century emerged as an independent epistemic entity and so it cannot be anything else today – a view still widely held in today's scientific community, I might add. Hence the regression back to the springtime of science, when it appeared that such claims could be historically substantiated. Those who contradicted this view, who wrote about science from a Marxist or externalist stance, even, perhaps especially, in the seventeenth century, were thus heavily criticised and marginalised. This offers a rather different explanation for such marginalisation which Merton, writing in 1970, suggested only ended 'when science itself came to be widely regarded as something of a social problem', as if science, at least to some, had not been a problem before the 1960s.[65]

While the modern practice of the history of science has become less adversarial[66] and now researches into contemporary science as well as earlier periods (the Halls only had one student who worked on the twentieth century), most scientists (there are some honourable exceptions), view the history of science with some suspicion as potentially undermining what they take to be the special epistemic status of their work. They are of, course, quite right to be suspicious, but the approach I have adopted here implies that some of what was achieved in the name of modernity needs to be undone.

Acknowledgements

I thank Imperial College (IC) and the Royal Institution (RI) for access to their archives as well as an anonymous referee and Robert Bud for helpful comments on earlier versions of this text.

Notes

1   W.A. Smeaton, 'History of Science at University College London: 1919–47', *British Journal for the History of Science*, 30 (1997): 25–28.

2   Anna-K. Mayer, 'Setting up a Discipline: Conflicting Agendas of the Cambridge History of Science Committee, 1936–1950', *Studies in the History and Philosophy of Science*, 31 (2000): 665–89.

3   Michael Bentley, *The Life and Thought of Herbert Butterfield: History, Science and God* (Cambridge: Cambridge University Press, 2011), 188–89.

4   Anna-K. Mayer, 'Moralizing Science: The Uses of Science's Past in the 1920s', *British Journal for the History of Science*, 30 (1997): 51–70 and 'Fatal Mutilations:

Educationism and the British Background to the 1931 International Congress for the History of Science and Technology', *History of Science*, 40 (2002): 445–472.

5   Jennifer Wilson, 'Celebrating Michael Faraday's Discovery of Benzene', *Ambix*, 59 (2012): 241–65.

6   Frank A.J.L. James, 'Presidential Address: The Janus Face of Modernity: Michael Faraday in the Twentieth Century', *British Journal for the History of Science*, 41 (2008): 477–516. Also discussed by Robert Bud in Chapter 5 in this volume, in the context of exhibiting science.

7   Darrell Jodock, ed., *Catholicism Contending with Modernity: Roman Catholic Modernism and Anti-Modernism*

in Historical Context (Cambridge: Cambridge University Press, 2000).

8 Peter Bowler, *Reconciling Science and Religion: The Debate in Early-Twentieth-Century Britain* (Chicago: University of Chicago Press, 2001), 244–86.

9 Julian Huxley, 'Thomas Henry Huxley and Religion', *The Modern Churchman*, 15 (1925): 109–18, 163–71, 167.

10 Also discussed by Robert Bud in Chapter 5 in this volume.

11 Syllabus of lectures in RI MS AD/12/E/1/1914/10. Attendance figures from RI MS AD/6.

12 William Inge, *Diary of a Dean: St Paul's 1911–1934*, (London: Hutchinson, 1949), 159. That year only the archaeologist C. Leonard Woolley attracted a larger audience with his 'Latest Excavations at Ur', *Proceedings of the Royal Institution*, 26 (1931): 553–62. Attendance figures from RI MS AD/6.

13 William Inge, 'The Future of the Human Race', *Proceedings of the Royal Institution*, 26 (1931): 494–515, 500.

14 Inge, 'Human Race', 503.

15 Inge, 'Human Race', 504.

16 Inge, 'Human Race', 505–06.

17 Inge, 'Human Race', 510.

18 Inge, 'Human Race', 507.

19 William Inge, 'Patriotism', *Quarterly Review*, 224 (1915): 71–93, 77.

20 Julian Huxley, 'Birth Control and Hypocrisy', *The Spectator* (19 April 1924): 630–31.

21 'Resolutions of the [Lambeth] Conference', *The Times* (15 August 1930): 6f–g.

22 Inge, 'Human Race', 500–01.

23 Inge, 'Human Race', 508.

24 Inge, 'Human Race', 509.

25 Inge, 'Human Race', 514.

26 Inge, 'Human Race', 514–15. The full quotation from Matthew 7: 1620 (KJV) is 'Do men gather grapes of thorns, or figs of thistles? Even so every good tree bringeth forth good fruit; but a corrupt tree bringeth forth evil fruit. A good tree cannot bring forth evil fruit, neither *can* a corrupt tree bring forth good fruit. Every tree that bringeth not forth good fruit is hewn down, and cast into the fire. Wherefore by their fruits ye shall know them' (italics in original).

27 R. Griffin, *Modernism and Fascism: The Sense of a Beginning under Mussolini and Hitler* (Basingstoke: Palgrave Macmillan, 2007).

28 *Speeches Delivered at a Dinner Held in University College, London in Honour of Professor Karl Pearson 23 April 1934* (Cambridge: Cambridge University Press, 1934), 20.

29 Frank A.J.L. James, ed., 'The Common Purposes of Life': Science and Society at the Royal Institution of Great Britain* (Aldershot: Ashgate, 2002).

30 Oliver Lodge, 'The Work of Hertz', *Proceedings of the Royal Institution*, 14 (1894): 321–49; Arthur Rücker, 'The Physical Work of von Helmholtz', *Proceedings of the Royal Institution* 14 (1895): 481–96; Henry Roscoe, 'Bunsen', *Proceedings of the Royal Institution*, 16 (1900): 437–47.

31 John Jenkin, *William and Lawrence Bragg, Father and Son: The Most Extraordinary Collaboration in Science* (Oxford: Oxford University Press, 2008), 400–01.

32 Philip Hartog, 'Joseph Priestley and His Place in the History of Science', *Proceedings of the Royal Institution*, 26 (1931): 395–430; John Thomas, 'Josiah Wedgewood and His Portraits of 18th Century Men of Science', *Proceedings of the Royal Institution*, 30 (1938): 497–517.

33 Henry Plummer, 'Galileo and the Springtime of Science', *Proceedings of the Royal Institution*, 32 (1942): 310–25, 310.

34 For an account of this meeting see Christopher A.J. Chilvers, 'Something Wicked this Way Comes: The Russian Delegation at the 1931 Second International Congress of the History of Science and Technology' (DPhil thesis, University of Oxford, 2007).

35 Christopher A.J. Chilvers, 'Five Tourniquets and a Ship's Bell: The Special Session at the 1931 Congress', *Centaurus*, 57 (2015): 61–95.

36 W.T. Mitkewich, 'The Work of Faraday and Modern Developments in the Application of Electrical Industry' in *Science at the Cross Roads: Papers Presented to the International Congress of the History of Science and Technology held in London from June 29th to July 3rd, 1931 by the Delegates of the U.S.S.R.* (London: Kniga, [1931]), 105–12.

37 Boris Hessen, 'The Social and Economic Roots of Newton's "Principia"', in *Science at the Cross Roads*, 147–212. This was translated afresh in Gideon Freudenthal and Peter McLaughlin, *The Social and Economic Roots of the Scientific Revolution: Texts by Boris Hessen and Henryk Grossmann* (Dordrecht: Springer, 2009), 41–89. The 1931 quotation, given here, is on 176 and on 61 of the new translation.

For views of the background to and reasons for Hessen's paper see Gideon Freudenthal and Peter McLaughlin, 'Classical Marxist Historiography of Science: The Hessen-Grossmann-Thesis' in Freudenthal and McLaughlin, *Social and Economic Roots*, 138. See 32 for their discussion of recent responses including Loren R. Graham, 'The Socio-Political Roots of Boris Hessen: Soviet Marxism and the History of Science', *Social Studies of Science*, 15 (1985): 705–22 and H. Floris Cohen, *The Scientific Revolution: A Historiographical Inquiry*, (Chicago: University of Chicago Press, 1994), 331–32.

38 Chilvers, both in the title of his thesis 'Something Wicked' and in 'Five Tourniquets' repeated the incorrect addition of 'technology'.

39 For an account of the development of the usages of the terms see Steven Shapin, 'Discipline and Bounding: The History and Sociology of Science as seen through the Externalism-Internalism Debate', *History of Science*, 30 (1992): 333–69.

40 Douglas McKie, 'Memorandum on the Re-opening of the Department of History and Philosophy of Science, University College London' (10 January 1945), IC MS Dingle papers 1/6/1, 1.

41 G.N. Clark, *Science and Social Welfare in the Age of Newton* (Oxford: Oxford University Press, 1937), 61. For a discussion of Clark's critique see Freudenthal and McLaughlin, 'Classical Marxist Historiography', 30.

42 Robert Merton, 'Science, Technology and Society in Seventeenth Century England', *Osiris*, 4 (1938): 360–632, 565–66.

43 Robert Merton, 'Science and the Economy of Seventeenth Century England', *Science and Society*, 3 (1939): 3–27. See Freudenthal and McLaughlin, 'Classical Marxist Historiography', 31.

44 For example: A. Rupert Hall, 'Merton Revisited or "Science and Society in the Seventeenth Century"', *History of Science*, 2 (1963): 1–16 (note the omission of technology from his title); Roy Porter, 'The Scientific Revolution: A Spoke in the Wheel?' in Roy Porter and Mikuláš Teich eds., *Revolution in History* (Cambridge: Cambridge University Press, 1986), 290–316; Shapin, 'Discipline and Bounding'.

45 Alexandre Koyré, *Études Galiléennes*, (Paris: Hermann, 1939). For an example see James MacLachlan, 'Experimenting in

the History of Science', *ISIS*, 89 (1998): 90–92.

46 Joseph Singer, 'Historical Relations of Religion and Science' and William Inge, 'Conclusion' in Joseph Needham, ed., *Science, Religion and Reality* (London: The Sheldon Press, 1925) 85–148 and 345–89 respectively. Details of the editorial committee are given in the unpaginated front matter.

47 Bowler, *Reconciling*, 39.

48 Joseph Needham, *Chemical Embryology*, (3 volumes, Cambridge: Cambridge University Press, 1931), 1: 41–227.

49 Gary Werskey, *The Visible College* (London: Viking, 1978), 147, quoting Joseph Needham, *A History of Embryology* (Cambridge: Cambridge University Press, 1934), xvi.

50 Freudenthal and McLaughlin, 'Classical Marxist Historiography' seem confused about both Bernal and Needham whom they classify as 'eminent British historians' (29) although in 1931 they were both rising scientists, while on the previous page writing 'almost all the British participants were basically dilettantes'. Had they characterised Bernal and Needham as scientists rather than historians, this would have been (just about) a defensible position.

51 Bentley, *Butterfield*.

52 Thomas Mougey, 'Needham at the Crossroads: History, Politics and International Science in Wartime China (1942–1946)', *British Journal for the History of Science*, 50 (2017): 83–109.

53 Anna-K. Mayer, 'Setting up a Discipline, II: British History of Science and "The end of Ideology", 1931–1948', *Studies in the History and Philosophy of Science*, 35 (2004): 41–72, 55–56.

54 Frank A.J.L. James, 'Alfred Rupert Hall 1920–2009 Marie Boas Hall 1919–2009', *Biographical Memoirs of Fellows of the British Academy*, 11 (2012): 353–408, 361 and 363.

55 Shapin, 'Discipline and Bounding'.

56 Robert Young, 'The Historiographic and Ideological Contexts of the Nineteenth-Century Debate on Man's Place in Nature', in Mikuláš Teich and Robert Young, eds., *Changing Perspectives in the History of Science: Essays in Honour of Joseph Needham* (London: Heinemann Educational, 1973), 344–438.

57 Herbert Butterfield, *The Origins of Modern Science 1300–1800* (London: G. Bell and Sons Ltd., 1949), viii.

58   A. Rupert Hall, *The Scientific Revolution 1500–1800: The Formation of the Modern Scientific Attitude* (London: Longmans, Green and Company, 1954), 364.

59   Hall, 'Merton Revisited' (14) well illustrates the tangle that resulted from the paradox.

60   Charles Gillispie, *The Edge of Objectivity: An Essay in the History of Scientific Ideas*, (Princeton: Princeton University Press, 1960), 9.

61   Guy Ortolano, *The Two Cultures Controversy: Science, Literature and Cultural Politics in Postwar Britain* (Cambridge: Cambridge University Press, 2009).

62   Guy Ortolano, 'Breaking Ranks: C.P. Snow and the Crisis of Mid-Century Liberalism', *Interdisciplinary Science Reviews*, 41, no. 2–3 (2016): 118–132.

63   'History of Science and Technology at the Imperial College', *Nature*, 192 (1961): 1131.

64   A. Rupert Hall, 'Historical Relations of Science and Technology', *Nature*, 200 (1963): 1141–45.

65   Robert Merton, *Science, Technology & Society in Seventeenth Century England* (New York: Howard Fertig Pub., 1970), vii. This was a republication of his 1938 *Osiris* paper with an extended preface. Curiously and somewhat individualistically, Freudenthal and McLaughlin, *Social and Economic Roots* (32) suggested that Merton had ascribed some of this marginalisation to the 'idealistic proclivities of historians'.

66   However, vestiges of the arguments can still be seen in the recent discussions about the role of the Enlightenment in industrialisation. See J. Mokyr, *The Gifts of Athena: Historical Origins of the Knowledge Economy* (Princeton: Princeton University Press, 2002); M.C. Jacob, *Scientific Culture and the Making of the Industrial West* (Oxford: Oxford University Press, 1997); for a salutary corrective see Matthew Paskins, 'Sentimental Industry: The Society of Arts and the Encouragement of Public Useful Knowledge, 1754–1848' (PhD thesis, University College London, 2014).

7

'Come on you demented modernists, let's hear from you': science fans as literary critics in the 1930s

Charlotte Sleigh

Introduction

An archivist in the British Library is flicking through a stack of home-made, amateur magazines, made by science fans in the 1930s. He remarks on their value for the historian:

> You know, I like that word 'science-fiction', it so describes the minds of that time – you know, just on the verge of scientific discovery, half thrilled, half proud, sort of cocky about the possibility of their ultimate conquest over everything, and that phrase seems to illustrate how they grabbed this science and raced away with it in their imaginations as if they had known it all their lives.[1]

But these are not the words of a real archivist. They were written *in* the 1930s, placed in the mouth of a fictional archivist. They appeared *in* one of the amateur magazines in question, a fantasy about how such productions, and their makers and readers, would be seen with the benefit of future hindsight. The words are *themselves* a 'sort of cocky', half-wry assertion of the writer's significance, and the significance of his circles. They are, in one strange sense, right; by addressing themselves to a historian – by appearing in this very book that you are now reading – they became true.

This chapter takes as its topic the young generation of British science fans that gathered itself in the five or six years prior to the Second World

War. It discusses their writings about scientifiction,[2] published in their home-made magazines, and examines the ideological function of science to sustain self-identity and ambition. From within science fiction studies, fans' debates have traditionally been understood as arguments about how much science should go into literature, and how advanced or accurate that science should be. There was certainly a great deal of this. However, the thesis of this chapter is something more basic: that it is useful and revealing to understand the critical writings of fans as a sort of literary criticism performed *de novo*. The policing of science fiction, it is argued, was as much an act of demotic literary criticism as it was a flag planted for science in literature. The issue is thus turned on its head. Rather than asking what science they put in their stories, we can see how these young men constructed science ideologically. The surprising answer turns out to be that they constructed it through a literary-critical enterprise.

The men and their magazines

The British fan community has documented its own history rather well, though it is less well-known to academic historians.[3] One important starting point for its story concerns the publishing empire developed in the USA in the late 1920s by Hugo Gernsback.[4] Gernsback's pulp fiction magazines – *Amazing Stories, Wonder Stories* – were imported to the UK. Crucially, his decision to print reader letters, with full addresses, enabled fans to contact one another directly.[5] In 1930, a young journalist in Essex, Walter Gillings, set up the Ilford Science Literary Circle together with a fellow fan whom he had discovered – thanks to the letters pages – lived only four miles away from himself.[6] Three years later, Patrick Enever of Hayes began another science fiction collective, corresponding with the doyen of LA fans, Forrest Ackermann.[7] Many more groups grew up, some affiliating themselves as chapters of the US Science Fiction League.

These scattered groups of very young men had been born after the Great War, and had not felt the weight of the family breadwinner's obligation during the Great Depression. They did not, of course, know that another war was coming. In short, they existed in a brief historical moment of hope. They had also benefitted from the extension of the school-leaving age to 14 in the Education Act of 1918, and though from lower-middle- or working-class backgrounds, they were literate. At work in their late teens, the fans were often peripherally connected with the engineering industries – as apprentices, or clerks, and so forth. Even if not directly employed by such industries, they would have been aware of

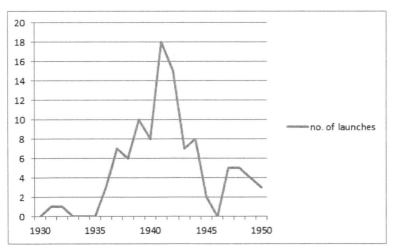

Fig. 7.1 Number of new fanmags launched per year, 1931–50.
Author's own graph.

the industrial dominance of their native towns and cities; the most active scientifiction groups were from Leeds, Liverpool and the Midlands. A Manchester group pursued rocketry in a similar scientifictional spirit.

It was the Nuneaton group that launched the first long-lasting fanmag, *Novae Terrae* (1936–39). When attendees at a 1937 meeting in Leeds formed the national Science Fiction Association, they adopted *Novae Terrae* as its official, nationwide publication. Figure 7.1 gives a sense of fannish activity in the period around the Second World War, showing the numbers of new titles launched each year.[8] Counting these titles is not an exact art; the cut-off from fantasy fanmags is hazy, and some titles were very slight one-offs. Nevertheless, it gives a sense of the growth in fandom.

The fanmags were a part of the phenomena of communication and mediatisation of science that were constituent of the modernist era. Science fans were inspired and obsessed by rockets and radio, but their raison d'être lay closer to home – the productions of the typewriter and the duplicator. These simple technologies made possible not only their articulation and sharing of ideas, but the creation of their personal and collective identities. The typewriter was not particularly new, but it was just becoming available to a wider demographic, whether at work or at home. Lesley Johnson, for example, was bought a typewriter by his mother to help him get a good job as a clerk – apparently typewriters were not provided. It cost £5 – the equivalent of six weeks' salary.[9] Such technologies caused the ambitions of being a scientist and being a writer to elide. When Arthur C. Clarke (see Figure 7.2) pontificated at the tender age of 17 that he was 'a scientist', what he was actually doing was writing for, and producing, the school magazine.[10]

The same went for the other fans; in producing their magazines they were playing at making the materiality of literature. On a basic level of time and resources the business of making and distributing paper publications outweighed everything else that the fans did. They justified the

Fig. 7.2 Arthur C. Clarke, aged 17, with his wireless equipment.
Rocket Publishing/Science & Society Picture Library.

text by eye, arranging it in columns or by page on the typewriter; they duplicated surreptitiously at work or on precious machines at home. They stitched the spines with their mothers' sewing machines; they posted copies around the country and collected dues and letters in return. Occasionally and temporarily professional production was possible; John Moores bankrolled the *Journal of the British Interplanetary Society* for a time, and Gillings approached financial sustainability with some of his efforts.

The fanmags (Figure 7.3) did not have large distributions. The most successful – *Novae Terrae*, the *Journal of the British Interplanetary Society* – peaked at 100–150 copies per issue, with perhaps twice that number of readers. Some of the writing is good, but not much of it; one or two of the contributors subsequently found fame as authors (Arthur C. Clarke, John Christopher), but not many. This poses an interesting challenge to the implicit values of scholarship. Usually historians and literary critics treat something as worthy of their attention if it is either a mass phenomenon, or if it is connected to a person or institution that is already regarded as important. The latter is self-confirming, resulting in ever deeper scholarly grooves being cut, in the modern period, for such figures as Woolf or Haldane. The fanmags were neither a mass phenomenon nor works of conspicuous quality or importance. One could perhaps make a better case for their being a mass phenomenon, if one takes into account the number of titles as well as the number of copies

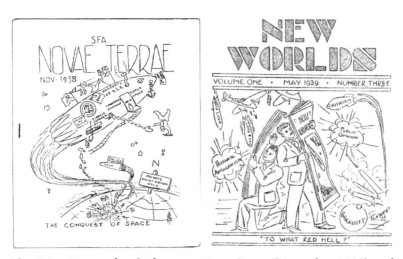

Fig. 7.3 Covers of early fanmags *Novae Terrae* (November 1938) and *New Worlds* (May 1939).

produced. However, one would find many of the same names cropping up again and again in the different titles. It would be better to say that the fanmags and their contributors are interesting precisely because they are not the 'usual suspects' of history, being ordinary working-class people rather than the better-known scientists of the period. As the historical tip of the science-fiction-reading iceberg, they may even be ancestors of that elusive beast of recent decades, science's public.

Canonisation and review as criticism

A lot of ink has been expended in trying to determine whether certain texts count as science fiction or not. Such accounts, as John Cheng points out, 'treat the genre transhistorically, locating texts chronologically to match its type and subject rather than examining the historical conditions of its emergence.'[11] The genre of science fiction is a constructed one, and a substantial part of the work of construction was done by the fans in their magazines. In later becoming a category employed transhistorically, science fiction has obscured some interesting literary and certainly historical questions one might ask: why did fans include the texts that they did; and what did they exclude, and why?

For one thing, it was the nature of professional scientifiction publishing (particularly the pulps) that stories and novels were promiscuously recycled and reserialised. As they moved from novel to serial form and back again, they were re-edited and transmuted. Thus the genre was created in part simply by virtue of selection and republication. More concretely still, fan organisations pooled their personal collections to form libraries, so that older and out-of-print texts could be borrowed by their members. This function of British fan clubs was present from the very earliest articulation of aims in 1933 and was still going strong just before the war.[12] A mail order system existed to facilitate the system across the country. Some fans – Lesley Johnson again – made a business out of sourcing and supplying new and old magazines and books. Thus there was a living and material process by which tales were republished, re-read and admitted to the living canon.

In addition to the book services of the clubs, fanmags published reviews of books and stories. These reviews were probably the most important element of the magazines, telling readers what should be read and what should not. The very first article in *Novae Terrae* launched straight off with 'three stories that most science fiction fans think finely written', and the theme of ranking continued throughout the magazine's life.[13] By and large reviews erred on the negative side, producing exasperation on the part of professional publishers. Hugo Gernsback blamed the demise of *Wonder Stories* (one of his many business failures) on the fans, complaining that they were 'bent on destroying scientifiction rather than building it up'. There was, he added, 'too much fault-finding and too little propaganda between scientifiction fans and the rest of the public'.[14] Eric Frank Russell, a successful writer sometimes at the receiving end of a sharp critical pen, distinguished between readers and fans, claiming that the latter ruined the trade by buying remaindered magazines instead of full-price ones, thus damaging their viability. Fans added insult to injury by proceeding to write letters 'telling the publisher how to publish, the printer how to print, the artist how to draw, the editor how to edit, and the authors how to write'.[15]

These were harsh judgements but right inasmuch as the definition of the fan included criticism. If nothing else, fans policed silliness and laziness in stories. Thus John F. Burke proposed, or presumed, a whole new club:

> Any of you reading this and agreeing are automatically enrolled as members of the SPISMDSWISTF – The Society for the Prevention of the Introduction of Sex, Mad Doctors and Space Warps Into Scientifiction. Come on, you demented modernists, let's hear from you.[16]

Burke's tongue-in-cheek criticism was just one small component of bigger critical activity, which occupied the majority of pages in the fanmags. D.R. Smith (1917–99), the so-called 'Sage of Nuneaton', apprenticed to machine tool manufacturer Alfred Herbert Ltd., provides a substantial case-study in point.[17] Scientifiction had been his gateway into what, compared with many fans, was a very wide range of reading, including classics, modern fiction and pornography. His regular column for *Novae Terrae*, 'Hymns of Hate', did for its literary targets exactly what the title suggested, resulting in at least one published apology. More than any other British fan, arguably, Smith attempted to position himself as something like an expert critic, and his writings ranged from a humorous taxonomy of fictional rays to more serious philosophical essays. Even meretricious texts had value to the fan-critic. 'The poor stories', he wrote, 'furnish the intellectual entertainment of analysing the defects'.[18]

However, the fans were not content to criticise for the sake of it. They were also keen to encourage what they saw as good-quality literature.

Literary criticism: debating the place of science in fiction

One measure of good scientifiction was the quantity and quality of science it contained. The fanmags led by example, many of them including articles and reporting (or advertising) lectures that were overtly scientific in nature – 'Why Bio-Chemistry?' and 'The Coal-Tar Cosmos', for example.[19] 'Why Bio-Chemistry?' represented a class of 'expert' science, apparently written by the well-known biochemist Jack Drummond. 'The Coal-Tar Cosmos', an exploration of organic chemistry, was written by one of the fanmag's own editors, Dennis A. Jacques; his hometown, Nuneaton, was a coal-mining centre, and many collieries had research laboratories attached (later unified in the 1950s when the National Coal Board moved its laboratories to a former factory site in the town). It is not known whether Jacques was employed in the coal industry, but certainly coal was a familiar aspect of his physical and social landscape, as it would have been for the other members of his group. As the fanmags continued and proliferated through the 1930s, such home-grown science came to predominate over the expert variety.

D.R. Smith's taxonomy of rays was an example of a broader genre of criticism, the humorous demolition of impossible science in pulps and novels. Writing in the very first issue of *Novae Terrae*, 'Space-Ray' noted that science, in stories, 'should be accurate in present day known details while deductions should be truly logical, and the scientific basis of the story original'.[20] In Glasgow, Donald G. MacRae concurred: 'unless a story is exceptionally well-written – Lovecraft, C.A. Smith etc. – it should be grounded on either a good basis of factual science … or, it should build up a logical science of its own'.[21] Along with other Glasgow fans he got up a petition to request more 'advanced' science in US scientifiction magazines.[22] Being able to out-science published authors on any of these scores (accuracy, plausibility, originality) was a source of enormous pride for these moderately educated young men. They became particularly excited by the prospect that such common-sense science might have real-world effects during the rocketry craze of 1936/7, in the course of which scientists' objections to the possibility of beyond-earth travel were painstakingly deconstructed and corrected by fans.[23] There was much earnest head-scratching about how 'advanced' the science should be in the professional scientifiction magazines when they – as the fans fervently hoped – eventually got off the ground in the UK. The consensus was that they would have to start slowly, in order to build up the readership from its current unsophisticated tastes.

But not all agreed. An early debate concerning the place of science in fiction erupted in the pages of *Novae Terrae* between Sam Youd (now better known by his pen-name John Christopher) and Arthur C. Clarke. Youd did not hesitate to name names: Jules Verne was overrated, as was the serial then on every fan's lips, *The Skylark of Space*. He claimed boldly:

> [T]he term 'scientific fiction' is in itself a literary impossibility. Story-writing is an art and as such cannot be mixed with science, the two are incompatible. In any kind of fiction the main thing is the style. Authors who sacrifice their story to scientific detail cannot, by any stretch of imagination, be called great.[24]

It is no surprise that Youd despised the Glasgow petitioners for more and better science.[25]

Clarke replied that cold-minded readers seeking only accuracy were a product of Youd's imagination.[26] He hastened to echo Youd's commendation of Lovecraft but claimed him for science fiction, managing to include him in the canon under the rubric of the commonly-expressed rule that whilst writers should not contravene known scientific laws, inventing new ones – or exploiting gaps in current knowledge – was fine.[27] Apparently – perhaps surprisingly – he thought Lovecraft did not bend the laws of *agreed* science. However, Clarke added an aesthetic criterion to the more usual realist rules about what constituted acceptable scientifiction. If there was one thing science should not be used for, it was decoration. Embroidering with science, Clarke wrote, was no substitute for style.

Style, then, seemed to be a point on which Youd and Clarke could agree. Indeed, this was a common position amongst the fans. They wanted literature that was elevated in some way: not trash. It was in fact a commonplace for scientifictionists to cite H.P. Lovecraft as an author who, though scientifically nonsensical, was worth reading on grounds of style alone. Clarke, however, was not so sure – at least in the context of this argument. With one breath, Clarke praised Lovecraft's stories on account of their 'wonderful style'; with the next, he dismissed them as 'bunk' at base.[28] Such a judgement appears to indicate that their much-touted style was cosmetic, at best – which, according to Clarke, was an oxymoron. Style that substituted for quality was no style at all, merely superficial gloss. A little later, Clarke went further, claiming that he could 'burlesque' Lovecraft's style – his 'creepy-creepy passages' – so easily that even this supposed quality of his writing was not all it was claimed to be.[29]

Not only did Clarke usurp Youd's critical criterion of style, but Youd also moved to adopt his antagonist's marker of plausibility. Youd criticised *Skylark* for the inaccuracy of its science; if Clarke must call Lovecraft 'bunk', then he must call Skylark 'bosh', as 'they [were] of the same order of probability'. Thus he demonstrated that he was better at exercising the very readerly quality he condemned in others.

It was a strange debate. As is often the case, the very proximity of its participants made the argument more intense. Sam Youd was shortly to publish (his first story appeared in 1941) in the very science-titled magazines whose contents and readers he professed to despise – as, of course, would Clarke. Both Youd and Clarke were aware of the frankly niche *Novae Terrae* and were in communication with its editors. To most purposes, they appear to be culturally indistinguishable. The two criteria on which the debate turned – accuracy and style – were unstable and inconsistently invoked. This instability perhaps indicates an uncertainty on behalf of the writers about whom it was that they addressed in their arguments. To insist upon accuracy was perhaps to proffer scientific skills, or perhaps to sentence oneself to life measuring a particular component on the factory floor. To have style – to be stylish – was a bold social manoeuvre that threatened to collapse or be called out as putting on airs at any moment. In adverting to one of D.R. Smith's targets '[John Russell] Fearn ... [at] the Blackpool Free Library' Clarke shied at book-based auto-didacticism, linking it with literary imposture.[30] Personal style was also an issue of concern for the fans. There was a typical look, and apparently Smith pulled it off best. But praise was often waspish. A fellow fan wrote:

> D.R. is perhaps the most typical fan I have ever encountered. Formerly the prize was divided between Arthur Clarke and Maurice Hanson, but D.R. is even more fannish than those two, which is saying something ... He wears spectacles and a preoccupied look. Affects unconventional clothes. His hair, a rich mouse in colour, dangles limply over his forehead ...[31]

More succinctly, but more caustically, William F. Temple said of his friend Clarke: 'He looks as if he hopes he looks like a scientist'.[32]

Clarke, at least, drew back from what, in style, had initially seemed a straightforward attribute. His boast that he could 'burlesque' Lovecraft was a particularly interesting move. It was a tinkerer's, an engineer's approach – 'look, I can rig up something just like it!' Pastiche in this instance lies somewhere between imitation-as-flattery and critique.

By troubling oneself to write a pastiche, one underlines the importance of the original text – but by showing that (any)one can do it, one demonstrates that the original was not, after all, a unique or perhaps even impressive achievement. Moreover, the act of pastiche part-conceals an invitation for the writer to be taken seriously; there is an element of his wanting to be told, 'what you've produced doesn't show that Lovecraft's bad; it shows you're *good*'. It is a low-risk invitation, because there is no way it can be turned back against the author: failing to write a good pastiche does not entail that one is a bad author. Pastiche, or burlesque, was another core element of many items in the fanmags: a peculiarly passive-aggressive way of interacting with professional authors by would-be professionals.

Readers, in one of the many questionnaires issued by *Novae Terrae*, tended to agree with Clarke in the debate. The editors reported: 'Regarding the Clarke-Youd controversy, 48% agreed with scientist Clarke, 20% with Youd, the remainder being non-committal'.[33] However imperfect the arguments were, Clarke better represented the image of literary engagement to which fans aspired. It is significant that the editors glossed him as 'scientist' in their report; this name went further than adverting to his desire to include science in story, but said something about his actual identity.

It was not just style and accuracy that were underdetermined: science, too, the supposed topic of the debate, remained undefined by Clarke and Youd. This was true for the fanmags more generally. The more they taxonomised scientifiction – which they did – the more fleeting their categories seemed to become. The prominence of 'real' science receded: such articles as those early ones on coal-tar and biochemistry quickly faded away. Indeed, some prominent members of scientifiction circles had very little interest in science at all. Such non-committal figures included Walter Gillings, one of the very earliest fans and the one who did most to advance scientific fiction as a professional activity in Great Britain. Once again, one is compelled to see the fanmags not so much as forays into science as into literary criticism.

The principles of fannish literary criticism

One does not have to look very far in the fanmags to see general formulations of literary criticism: first, making a good story, and second, using language appropriately and stylishly in its conveying. D.R. Smith wrote:

If the characters are interesting, natural and consistent in speech and action; if the style is free from stock phrases and correctly adjusted to the type of plot; if the story is developed smoothly and economically; then the elements of a good story are there. Add originality and a coherent plot to these qualifications and we have a standard that will be above that of most scientific fiction.[34]

It is an engineer's approach to fiction: making sure that the elements work well in themselves, and work together, harmoniously and economically, as a whole. Science, in the sense of engineering, is taken as a metaphor-in-practice for understanding literature. Stories are things to be constructed, tinkered with, polished: items of pride. In this sense, fan culture replicated the very pragmatic approach to texts taken by editors and publishers in the professional world. These advised, or required, their authors to re-write according to their commercial interests (or guesses). Editors, too, were ruthless in their comments: the check-box reasons for rejection received by would-be authors uncannily reflect the ruthless categorisation of their successful efforts by *Novae Terrae* as, for example, 'fair' or 'readable'.[35]

Stories were sometimes quite dramatically re-written for second or subsequent publication, particularly if transitioning from serial to novelistic form, or vice versa. There was little or no literary preciousness amongst authors. Texts were there to be tinkered with; they were a matter for great personal pride, but, like a gadget, they were not the product of disembodied genius but of hard work and hands-on experimentation.

And yet … the fans themselves were in love with words for the sake of words in their fanmags. They were trying out different genres of writing – journalism, essay, story, poems. They were interested in the nature of language itself. Several fiction writers were particularly intrigued by the evolutionary dimensions of language. Both L. Sprague de Camp and Lovecraft, for example, used ancient languages as cyphers to ancient times and forces in their stories.[36] Others speculated about the nature of future English in their forward-looking fantasies. In general, it was predicted to become Americanised in spelling, simplified in grammar. The two threads are brought together in, for example, *The Insect Men* (1930/6), which sees its heroes transported to a far-distant England which looks like 'New York gone mad' and features strange grammar and spellings such as Banes Airstn for Barnes Air-station.[37]

Forrest Ackerman, the Los Angeles fan who gave succour to the early group of fans in Hayes, was a keen exponent of Esperanto: a single language was needed for science, both to co-ordinate international rocketry efforts and to communicate on a rational, translatable basis with aliens.[38] British fans were divided in their responses to his proselytising efforts. Walt Gillings's Ilford Circle had predicted in the early thirties that the entire world would speak the same language by 1991, so appeared to be resigned, if not enthusiastic.[39] Other fans urged Ackermann to give up on Esperanto and learn Basic English instead.[40] For the British fans, ironically, Esperanto was not so much perceived as an international language as an American one, due to its connection with Ackermann. As such, it was associated with several other features of US scientifictional language: rational spelling, compulsive punning and portmanteau words (such as scientifiction itself). All of these provoked lively debate, with fans ranging themselves on both sides of the issues.

Rational language, and Basic English in particular, was dear to the heart of a new, professional, academic cadre of literary criticism that was emerging in parallel to science fiction fans, though almost certainly unknown to them. Its two leading figures were I.A. Richards and, less publicly, C.K. Ogden, both based – at least initially – at the University of Cambridge.[41] Literary criticism and Basic English represented for them the development of two trajectories in language. On the one hand there was emotive language, which a careful and proper critic could analyse in order to recover the emotional experience of a poet, for example. This emotive language was extremely dangerous, for if directed towards an ordinary person it could be highly effective as advertising, or worse, propaganda. Hence Ogden and Richards's second project: the development of a Basic English, whose workaday, unambiguous referentiality would mean that it could not be abused in such a way. It was perfect for science, too.

In the sense that they embraced intellect as the quality of their writing and criticism, fans echoed this high-culture parallel. For the fans, emotion was the wrong starting point. Intellect was the better tool, and science was a way of asserting the intellectual capacity of the common man. Critiquing science in stories was a way of showing that one could do science – even quite complex science, as for example in Clarke's objection to one story over its inaccurate treatment of the Lorenz-Fitzgerald contraction.[42] The splendidly named Festus Pragnell hinted boldly at the trajectory of such lay participation. 'Are such people as Einstein and Sir James Jeans so much more intelligent than the average man?' he asked. 'I doubt it. They are just specialists'.[43] In the anarchic world of fandom, science could be critiqued, tested and driven onward by ordinary persons. Such a philosophy was reflected in the Science Fiction Association's adoption of Archibald Low as its president – a non-establishment figure without formal scientific honours. The more idealistic fans held out for the application of common intellectual capacity, not only to literature, but also to the problems of the day. The Leeds group, for example, possessed a number of members who were self-identifying socialists and pacifists, explaining and expounding their position in scientifictional terms.

D.R. Smith too claimed that science fiction was 'greater and more durable' because it appealed to the 'head rather than the heart'.[44] Literature would eventually 'be weaned from emotion', he judged, but in the short term it needed also to entertain in order to win over readers to its greater mission. Such was the opinion of many fans, awaiting the day when 'advanced' science would be the topic of magazines in the USA, and even the UK.

By 1939, the next generation behind the fans was beginning to encounter something like Ogden and Richards's take on language in the classroom, via the successful book *The Control of Language*, written for school teachers, which ran through 11 impressions by 1957. The book's authors, Alec King and Martin Ketley, took the division of language into referential and emotive kinds further than Ogden and Richards had done, using it to accuse literary critics of sneaky subjectivity disguised as objective judgement. 'This is good', they noted, functioned as a rhetorical cover for the rather less impressive sentiment 'I like this'.[45] Unsurprisingly, perhaps, King and Ketley used 'scientific prose' as a term of unambiguous praise.[46] Moreover they were not sniffy about 'pure story-telling' which 'need be a narrative of nothing but external events, told in the most scientific prose'.[47] They could have been describing scientifiction.

C.S. Lewis was horrified by *The Control of Language*. Under the pseudonym of *The Green Book*, it was the subject of attack in a series of three lectures published as *The Abolition of Man*. 'It is an outrage that they should be … spoken of as Intellectuals', he raged.[48] The 'method of debunking', while working with the head, left none of the manly qualities behind, and cultivated none.[49] It would lead, he argued, to the production of 'men without chests', a metaphor which manages to summon the head-heavy figure of the alien (and future human) that had begun to emerge since the writings of Wells. In his third lecture, which shares its title with the book, Lewis proceeds from his attack on new-fangled literary criticism to the direction imposed upon humanity by science. Writing at times in a rather scientifictional mode, Lewis forecasts the steady diminution of humanity as 'debunked' values are set to one side and the pleasurable or utilitarian is pursued instead. It is an extraordinary claim to make – that faulty literary criticism can lead to the destruction of humanity. Lewis is quick to state that his argument is not opposed to science in general, but only a particular kind: it is the upstart scientific intellect of the uncultivated man. Lewis overtly disparaged 'the fiction of Engineers', which featured in his own taxonomy of science fiction as a very low form.[50] His description of the villain, Professor Weston, in *Perelandra* suggests a direct connection between the fans' magazines and the fate of humankind consequent upon debunked language:

> He was a man obsessed with the idea which is at this moment circulating all over our planet in obscure works of 'scientifiction', in little Interplanetary Societies and Rocketry Clubs, and between the covers of monstrous magazines, ignored or mocked by the intellectuals, but ready, if ever the power is put into its hands, to open a new chapter of misery for the universe.[51]

Although it would be implausible to propose that the fans were responding to emergent professional literary criticism in their writing, their potential to form a counter-narrative to elite criticism was perceived by Lewis. In his attacks on intellect-based, debunking criticism, and on 'the fiction of engineers' we see their use of language and their fiction linked as potent – and undesirable – social force. This chapter thus reinvigorates the 'two cultures' question. For one thing it pushes the question rather earlier than the timing of Snow's lecture might suggest, back to the 1930s. By positioning themselves as 'science fans', and yet commenting on literature, the historical actors in this chapter stepped on both terrains. The vitriol they attracted, and perhaps their wry self-awareness, reveal that this was a problematic combination even then. They also redefine the question, because rather than being a matter of science versus the humanities, it becomes a question of the moderately educated, technical classes versus the emergent academic elite. Science and literature, with a capital L, had more in common with one another than with the fans.

Science, for the fans, was a generalised toolkit that allowed one to have one's say, to apply general technical skills to any area of culture. Science was a statement about the fans' ownership of the text, and of particular democratic values of legibility and meaning inscribed within it. Science functioned as a manifesto of a particular class; in cashing out as a literary criticism it was finally crushed by the academy in the 1960s, which in decisively severing the authorial intent from the act of reading left no critical space for the authority of the reading-and-writing fan. Examining the history of the science fan suggests that the two cultures debate was about the defence of academic scholarship as authoritative voice in society.[52]

Let us close by way of returning to D.R. Smith. In his article 'The Future', Smith deals not, as one might expect, with some chronologically distant period of space travel, but with the prospects of narrative art. Smith dismisses contemporary high literature as too purely empirical: '[its authors] aim at portraying everything, from the sunset down, in terms of physical sensations. In this way they obtain absolute reality with little exercise of the reader's imagination, for reality in the absolute is for each of us what we see or hear or sense'.[53] Intriguingly, Smith seems to place modernism and realism in the same basket: 'new styles that … defy all rules of syntax … descriptive directness of the working class'. Today's literary critic would find this a crashing confusion of two different genres; but there it is. It indicates a very different way of dividing up the world. These categories function as antitheses to scientifiction, which is cerebral, imaginative, and inviting to the majority of readers.

By the end of his essay, Smith circles back from text to world, suggesting nothing less than that science fiction will engender future reality – not in the sense of rocket stories inspiring real rockets to be made, but in a more profoundly ontological sense. He proposes a cosmos that is spun out of language.

> The thought [of science fiction's success] has fearful implications. Carry it on to the end, when the last star has dissipated the last erg of energy to weary space, where nothing can move or be, but eternal intelligences, disembodied from matter and energy both, contemplating through eons of peace every atom in a motionless space. At length the long-dormant desire for amusement awakens again, and with their all-encompassing powers they prepare a book for their entertainment. Out of the ashes of the old a new universe is prepared. Weary of knowing every possible event in a stable universe they make the new such that the law of chance is its fundamental principle. And through the ever fruitful ages they read and comprehend every tiny wonderful detail of this book of their – Creation.[54]

Such a creation of reality out of words could be read as a statement of high modernism. Although this is an intriguing prospect, Smith is in some ways unrepresentative of the mass of fans. In the fanmags his voice emerges as a little more philosophical, and a little better expressed, than many. He is an intriguing stirp of intellectual history – a high modernist of science fiction that might have been. As it was, he was in his own words 'idle and unsocial', and after a dozen years or so – at the point of reaching maturity – he faded away from the fannish scene.[55] Taking that scene, and its magazines, as a whole, we see a modernist mixture that is hard to categorise in conventional scholarly terms. The engineer's approach to fiction-making combined with a passion for language, in an attempted participation in the materiality of intellectual culture. In the fanmags' pages, the industrial-modern tangled with the aesthetic-modern; the fan was a demented modernist, or, just possibly, a true one.

Notes

1　Eric C. Williams, 'Idle Chatter in the Vaults', *Novae Terrae* #28, 3, no. 4 (December 1938): 6–8.

2　'Scientifiction' was a commonly used portmanteau term to capture an interest in science pursued via fiction.

3　Rob Hansen, *Then* (© Rob Hansen, 1988–94. Unpaginated), vol. 1, chapter 1 'The 1930s: genesis'. www.ansible.co.uk/Then/then_1-1.html accessed June 2018.

4　John Cheng, *Astounding Wonder: Imagining Science and Science Fiction in Interwar America* (Philadelphia: University of Pennsylvania Press, 2012), 17–50.

5　The fans had an ambivalent relationship with American modernity; on the one hand, the USA – the New York cityscape – was regarded as the very embodiment of the future. On the other hand, there was a critical, jealous regard of the 'Yanks' and their supposed brashness.

6　See archival materials at www.fiawol.org.uk/fanstuff/THEN%20Archive/ISLC1.htm, accessed June 2018.

7　Rob Hansen, private archive.

8　Data from Rob Hansen, British Fanzine Bibliography 1931–1950. See www.fiawol.org.uk/fanstuff/biblio/thirt.htm, accessed June 2018.

9　Lesley Johnson, 'My Personal History of the British Interplanetary Society' (Liverpool, 1933–1937), A–F; MS at BIS archives, London.

10　A.C. Clarke, *Childhood Ends: The Earliest Writing of Arthur C. Clarke* (Rochester, Michigan: Portentous Press, 1996), 47.

11　Cheng, *Astounding Wonder*, 15.

12　Letter from John Elliott, President of the Science Fiction Club [Hayes], to Forrest Ackerman, 20 January 1933. Private collection (Rob Hansen).

13　'Space Ray', 'Science Fiction Ideals: No. 1 – The Perfect Science Fiction Story', *Novae Terrae* #1, 1, no. 1 (March 1936): 2–3; The stories were Abraham Merritt, 'The Moon Pool', Edward E. Smith [and Lee Hawkins Garby], 'The Skylark Of Space', and Richard Vaughan 'The Exile of the Quiet Sun' [The Exile of the Skies?].

14　E.J. Carnell, 'Americanisms', *Novae Terrae* #6, 1 no. 6 (August 1936): 13–15; 13.

15　Official Souvenir Report of The First British Science Fiction Conference Held in the Theosophical Hall, Leeds, January 1937 (Leeds: The Leeds Science Fiction League, 1937), 14pp; 9. Amateur publication. See www.fiawol.org.uk/fanstuff/then%20archive/1937conbooklet.htm, accessed June 2018.

16　John F. Burke, 'The New Cycle Needs Brakes', *Novae Terrae* #26, 3 no. 2 (Sept 1938): 6–7; This essay's expansion of 'modernism' to include the fans' activities finds a parallel in Paul March-Russell, *Modernism and Science Fiction* (Basingstoke, Hants: Palgrave Macmillan, 2015). March-Russell finds grounds for counting science fiction as a species of modernism; my aim here is to challenge and redefine modernism somewhat.

17　Rob Hansen, 'Forgotten Fans #6: D.R SMITH – The Sage of Nuneaton'. 2013. 8pp. Unpublished: by courtesy of the author.

18　D.R. Smith, 'The Eternal Dispute', *Novae Terrae* #26, 3 no. 2 (September 1938): 4–5; 5.

19　John C.H. Drummond, 'Why Bio-Chemistry?', *Novae Terrae* #18, 2, no. 6

(November 1937): 57; Denny Jacques, 'The Coal-Tar Cosmos' *Novae Terrae* #3, 1, no. 3 (May 1936): 6–7; 'The Coal-Tar Cosmos (Part 2)', *Novae Terrae* #4, 1, no. 4 (June 1936): 7 and 12; 'The Coal-Tar Cosmos (Conclusion)' *Novae Terrae* #5, 1, no. 5 (July 1936): 8.

20 'Space Ray', 'Science Fiction Ideals: No. 1': 2–3; 2.

21 Donald G. MacRae, 'Petition for Science', *Novae Terrae* #10, 1, no. 10 (February 1937): 4–5.

22 MacRae, 'Petition for Science', 4–5.

23 Charlotte Sleigh, 'Science as Heterotopia: The British Interplanetary Society before World War II', in *Scientific Governance in Britain, 1914–1979*, eds. Don Leggett and Charlotte Sleigh (Manchester University Press, 2016).

24 S. Youd, 'Fantasy v. Science' *Novae Terrae* #17, 2, no. 5 (October 1937): 11–13; 12.

25 S. Youd, 'Mr. Youd Replies [with response by Arthur C. Clarke]', *Novae Terrae* #20, 2, no. 8 (January 1938): 3–6; 3.Youd cites Glasgow petition as e.g. of fans valuing scientific accuracy above all else.

26 Arthur C. Clarke, 'Science Fiction v. Mr. Youd', *Novae Terrae* #18, 2, no. 6 (November 1937): 13–15.

27 Cf. 'Space Ray', 'Science Fiction Ideals: No. 1': 2–3; 2.

28 Clarke, 'Science Fiction v. Mr. Youd', 14.

29 Youd, 'Mr. Youd Replies', 6.

30 Clarke, 'Science Fiction v. Mr. Youd', 15.

31 Quoted in Hansen, 'Forgotten Fans #6', 3.

32 William F. Temple, 'The British Fan in his Natural Haunts: No. 3 Arthur C. Clarke', *Novae Terrae* #24, 2, no. 12 (June 1938): 14–17; 17.

33 'Novae Terrae' Panel of Critics, 'Report on Questionnaire No. 3 (January 1938 issue)', *Novae Terrae* #23, 2, no. 11 (May 1938): single sheet insert.

34 D.R. Smith, 'Concerning the Criticisms of Scientific Fiction', *Novae Terrae* #29, 3, no. 5 (January 1939): 5–7; 5.

35 Examples of rejection letters can be found in the archives of several authors, such as Sam Youd and John Wyndham, in the Science Fiction hub at the Special Collections & Archives, University of Liverpool.

36 L. Sprague de Camp, 'Language for Time Travelers' in *Years in the Making: The Time-Travel Stories of L. Sprague de Camp*, ed. Mark L. Olson (Framingham, MA: NESFA Press, 2005), 123–35; H.P. Lovecraft, 'The Rat in the Walls' in H.P. Lovecraft, *The Call of Cthulhu and Other Weird Stories*, ed. S. T. Joshi, new edition (Harmondsworth: Penguin, 2002), 89–108.

37 Alfred Edgar, *The Insect Men* (London: Amalgamated Press, 1936), 4–5 and 10–11. First appeared 3 May–21 June 1930 in *Modern Boy*.

38 Forrest J. Ackerman, 'Esperanto: Its Relation To Scientifiction', *Novae Terrae* #6, 1, no. 6 (August 1936): 3–6.

39 [Walter Gillings], 'England in 1991!' *Ilford Recorder* (31 July 1931). Page no. lost. See www.fiawol.org.uk/fanstuff/ THEN%20Archive/ISLC8.htm accessed June 2018.

40 J.B. Jepson, 'Answers to Forrest J.', *Novae Terrae* #10, 1, no. 10 (February 1937): 20.

41 See Charlotte Sleigh, *Six Legs Better: A Cultural History of Myrmecology* (Baltimore, Johns Hopkins University Press, 2007), 139–62.

42 S. Youd, 'Mr. Youd Replies [with response by Arthur C. Clarke]', *Novae Terrae* #20, 2, no. 8 (January 1938): 3–6; 6. See also, for example, D.R. Smith, 'Alas, Poor Einstein', *Novae Terrae* #2, 1, no. 2 (April 1936): 2–3 and 7.

43 Festus Pragnell, 'Originality in Science Fiction', *Novae Terrae* #13, 2 no. 1 (June 1937): 6–7; 7.

44 D.R. Smith, 'The Future', *Novae Terrae* #24, 2, no. 12 (June 1938): 11–13; 11.

45 Alec King and Martin Ketley, *The Control of Language: A Critical Approach to Reading and Writing* (London: Longmans, Green, 1957[1939]). See for example their brutal deconstruction of Coleridge on a waterfall, 20. See also their chapters on criticism, e.g. 158.

46 King and Ketley, *The Control of Language*, 30.

47 King and Ketley, *The Control of Language*, 176.

48 C.S. Lewis, *The Abolition of Man: Or, Reflections on Education with Special Reference to the Teaching of English in the Upper Forms of Schools* (London: G. Bles, 1946), 21.

49 Lewis's frequent use of the recent and slangy word 'debunk', particularly in the first lecture, is a semiotic hint at the downward social trajectory of literary criticism.

50 C.S. Lewis, 'On Science Fiction' [1955] in Lewis, *Of Other Worlds: Essays and Stories* (London: Harcourt, 1994), 59–73; 62.

51 C.S. Lewis, *Out of the Silent Planet; Perelandra* (London: HarperCollins, 2001), 251.

52 A similar conclusion is reached, via a very different route, by Guy Ortolano in his excellent book *The Two Cultures Controversy: Science, Literature and Cultural Politics in Postwar Britain* (Cambridge: Cambridge University Press, 2009).

53 Smith, 'The Future', 11.

54 Smith, 'The Future', 12–13. Smith echoes the cosmic outlook of Olaf Stapledon in this passage. Stapledon came closest to straddling the worlds of high and low culture: beloved by fans, but moving himself in cultured circles. Lewis did not know what to make of him; he confessed to finding his works a 'delight' (Lewis, *The Abolition of Man*, 30) but also observed that *Star Maker* ended 'in sheer devil worship'. Letter to Arthur C. Clarke (7 December 1943) in *The Collected Letters of C.S. Lewis*, ed. Walter Hooper, vol. 2 (New York: HarperCollins, 2004), 594.

55 Hansen, 'Forgotten Fans #6', 6.

Section 3
Mathematics and physics

8
Modern by numbers: modern mathematics as a model for literary modernism
Nina Engelhardt

Being modern = being mathematical(?)

> Let no one object that outside their field mathematicians have banal
> or silly minds, … but in their field they do what we ought to be doing
> in ours. Therein lies the significant lesson and model of their
> existence; they are an analogy for the intellectual of the future.[1]

In his humorous essay 'The Mathematical Man', published in German in
1913, the Austrian author Robert Musil puts forward the argument that
in order to negotiate the early twentieth century with its specific
challenges, people have to become 'mathematical men'. The 'we' in the
quotation refers to literary writers who are asked to take mathematicians
as a model for their work. Addressing mathematics, literature and the
question of how human beings ought to behave at the beginning of
the twentieth century, the opening citation brings together the central
concerns of this chapter: to examine how early twentieth-century writers
developed the idea that successfully 'being modern' – and 'writing
modernity' – calls for '*being* mathematical'.

In the following, I analyse Musil's essay as well as his novel *The
Confusions of Young Törless* (1906). The Russian writer Yevgeny
Zamyatin's essay 'On Literature, Revolution, Entropy, and Other Matters'
(1923) and his novel *We* (written in 1920, published in English in 1924)
provide material for comparison as they similarly develop, if in rather
different ways, the idea that for life and literature to be modern they
have to, in a sense, become mathematical. However, *We*, one of the first

dystopian novels, also illustrates the dangers of 'mathematising' society and literature. Inhumane regulation of lives and denial of personal happiness are among the possible results, as are some infelicitous effects on literature. These effects are evident, for example, in the poem 'Happiness,' which the protagonist of *We* values highly:

> I was enjoying a sonnet titled 'Happiness.' [...]:
> Forever enamoured are two plus two,
> Forever conjoined in blissful four.
> The hottest lovers in all the world:
> The permanent weld of two plus two ...[2]

We here taps into a widespread view of mathematics as the pinnacle of reason and as an example of exactitude and certainty that does not readily offer itself to be copied by life and literature. Yet, as the following discussion explores, Musil and Zamyatin's writings refer to a particular *modern* notion of mathematics, which questions formerly accepted mathematical positions. It is precisely this modern notion of mathematics that Musil and Zamyatin's texts present as a model for literature and for 'being modern'.

In the following discussion, I first introduce the modern understanding of mathematics and then analyse Musil and Zamyatin's engagements with it in light of new opportunities to negotiate the pressures of the early twentieth century and redefine the place and function of literature. The rapprochement between mathematical and literary positions that these examples indicate moreover demonstrates the need to reconsider the place of modern mathematics in relation to the division between the 'two cultures' of the natural sciences and the humanities.

Modern mathematics

In the early twentieth century, mathematics increasingly comes under attack as an instrument of streamlining, reduction and rationalisation. The sociologist Max Weber famously deplored the belief 'that one can, in principle, master all things by calculation'.[3] He invoked commonly understood characteristics of mathematics – clarity, simplicity, certainty – as reasons for its power and the influential role it plays in the 'rationalization and intellectualization and, above all, ... the *disenchantment of the world*'.[4] While Weber aimed to dispel the limited and limiting view of a calculable – mathematised – world, others endorsed the exemplary state of mathematics, celebrated its apparent rigour, reliability and certainty, and welcomed innovation and progress in the field. Mathematics had undoubtedly seen significant developments since the nineteenth century, and the modernisation of the discipline showed in the creation of new fields of research, new methodologies, sub-disciplines and mathematical entities. However, these developments went hand in hand with growing concerns about the nature of mathematical proof and truth. Gottlob Frege, a key figure in nineteenth-century mathematics and philosophy, summarised the situation: 'Mathematics should properly be a paradigm of logical rigour. In reality one can perhaps find in the writings of no science more crooked expressions and consequently more crooked thoughts, than in mathematics.'[5] Attempts to rectify the 'crooked expressions' failed and only revealed the seriousness of the challenges to mathematical exactitude and certainty. Thus, the process of modernisation in mathematics gave rise not only to hope for a professionalised and improved discipline, but also to a sense of crisis.[6] The emerging uncertainty existed next to, alongside and in competition with the image of mathematics as the pinnacle of certainty, rationality and absolute truth. The negotiation of these conflicting views took place among mathematicians themselves but also in a broader circle that felt a keen interest in the challenged role of mathematics, the resulting status of reason, and the possibility of attaining certainty and truth.

Frege's call for logical rigour was part of his programme to eliminate any references to intuition in mathematics and to demonstrate that arithmetic was reducible to logic. With this attempt, Frege founded the school of logicism – one of three schools that aimed to establish sound foundations for mathematics around 1900.[7] His project was not successful, however. Famously, just as Frege was about to publish the second volume of his *Grundgesetze der Arithmetik* (*The Basic Laws of Arithmetic*) in 1902, he received a letter from Bertrand Russell that called his entire project into question. The so-called 'Russell's paradox' shows that the axioms Frege drew on in his project were troubled by inconsistencies. Specifically, Russell detected a paradox in set theory in relation to the question of whether the 'set of all sets that are not members of themselves' is a member of itself: if it is a member of itself, then it is by definition not one of the sets that are not members of themselves; at the same time, if the 'set of all sets that are not members of themselves' is not a member of itself, then it logically is to be counted towards the sets that are not members of themselves. Due to the inconsistencies pointed out by Russell, Frege had to relinquish core elements of his views about logic and mathematics and accept that his work did not in fact establish a foundation for mathematics. Russell's discovery also was a major blow to his own hopes of putting mathematics on a foundation of logic: 'It was the discovery of one such contradiction, in the spring of 1901, that put an end to the logical honeymoon that I had been enjoying. I communicated the misfortune to Whitehead, who failed to console me by quoting, "never glad confident morning again".'[8]

Bertrand Russell is also the main protagonist in the graphic novel *Logicomix* (2009), by Apostolos Doxiadis, Christos Papadimitriou, Alecos Papadatos and Annie Di Donna. This graphic novel vividly, if at times with 'comical' licence,[9] illustrates the search for certainty and truth in mathematics and the mounting sense of anxiety about these issues in the late-nineteenth and early-twentieth centuries. I will draw on this graphic novel below, while recognising its limitations as a historical resource, and will introduce key issues in the foundational crisis of mathematics with the help of illustrations from it. I will support these ideas with a range of further evidence, before turning to my literary analyses.

Logicomix depicts Bertrand Russell's search for stable foundations for mathematics, from his first contact with numbers as a young boy to the publication of his major contribution to the discussion, namely the *Principia Mathematica,* written together with Alfred North Whitehead. Figure 8.1 shows Russell laying out his motivation: since science and knowledge greatly depend on mathematics, so the graphic Russell explains, it is vital to establish the rigour of mathematics itself.

A related picture emerges in fact from Russell's autobiography:

> I thought that certainty is more likely to be found in mathematics than elsewhere. But I discovered that many mathematical demonstrations ... were full of fallacies, and that, if certainty were indeed discoverable in mathematics, it would be in a new kind of mathematics, with more solid foundations than those that had hitherto been thought secure.[10]

Russell here echoes Frege's demand for rigour, and points to the logicist approach intended to clarify basic assumptions in mathematics. But *Logicomix* offers a striking graphic representation of Russell's sense of the failure of Frege's search for more secure foundations (Figure 8.2).

The illustration in Figure 8.2 suggests that Frege's supposedly foundational theory – illustrated by a tortoise in this picture – was faulty and that attempts to support it with a more comprehensive theory – a bigger tortoise – similarly failed. In (other) words: Russell's paradox can be resolved in a bigger foundational system, yet, this bigger system

Fig. 8.1 'Shaky Foundations' from Apostolos Doxiadis, Christos H. Papadimitriou, Alecos Papadatos and Annie Di Donna, *Logicomix* (London: Bloomsbury 2009), 112.

then includes paradoxes of its own and is itself in need of a more comprehensive 'stabilisation'. In his autobiography, Russell indeed uses the example of the elephant and the tortoise to describe this disconcerting situation in mathematics: 'I was continually reminded of the fable about the elephant and the tortoise. Having constructed an elephant upon which the mathematical world could rest, I found the elephant tottering, and proceeded to construct a tortoise to keep the elephant from falling. But the tortoise was no more secure than the elephant'.[11] *Logicomix* and Russell's autobiography clearly, and at times indeed graphically, communicate how the modern understanding of maths differs from earlier associations with rigour, certainty and truth, and why the impossibility of setting their field on secure grounds had reverberations beyond the circle of mathematicians immediately concerned with the

Fig. 8.2 'Turtles all the way down' from Apostolos Doxiadis, Christos H. Papadimitriou, Alecos Papadatos and Annie Di Donna, *Logicomix* (London: Bloomsbury 2009), 189.

problems. These reverberations certainly reached the writers Musil and Zamyatin, who both received a mathematical education at university level, and were aware of developments and concerns regarding mathematics. In the following discussion, I explore how their works engage with anxieties and opportunities tied up with the modern notion of mathematics and in what ways they introduce this as a model for literature and for 'being modern'.

Musil: 'The Mathematical Man'

Like *Logicomix* and Russell's writing, Musil's 1913 essay 'The Mathematical Man' emphasises the fundamental role of mathematics for other fields and for life at large. It introduces maths as the basis of the development of technology and machines, which thus means that it determines a large part of existence:

We may say that we live almost entirely from the results of mathematics … Thanks to mathematics we bake our bread, build our houses, and drive our vehicles … All the life that whirls about us, runs, and stops is not only dependent on mathematics for its comprehensibility, but has effectively come into being through it and depends on it for its existence[.][12]

Having argued that mathematics constitutes the underpinning of many aspects of life, the essay then introduces the 'foundational crisis' in mathematics:

suddenly, after everything had been brought into the most beautiful kind of existence, the mathematicians … came upon something wrong in the fundamentals of the whole thing that absolutely could not be put right. They actually looked all the way to the bottom and found that the whole building was standing in midair.[13]

Consequently, everything that depends on mathematics – all the technology and machines brought into 'existence' through it – lacks foundation. The feeling of the entirety of existence floating in midair probably resonated at the time of publication in the year before the outbreak of the First World War, yet, the humorous and exaggerated tone of the essay cautions against taking the assertion at face value. Moreover, the speaker in the essay cannot easily be equated with Musil but has to be understood as a fictional mouthpiece through which he develops a particular view in a hyperbolic manner. The concluding paragraph then spells out the serious concerns that underlie the self-professed 'playfulness'[14] of the previous lines, as it clarifies that the foundational questions in maths stand for a general crisis and 'lack of culture'[15] and that mathematicians' reactions to their problems can show literary writers and intellectuals how to respond to theirs.

Musil's essay celebrates mathematicians' pragmatism. The building of mathematics might stand in midair, but on the plus side, this does not stop mathematics from being extremely useful or life from being liveable: 'the whole building was standing in midair. But the machines worked! We must assume from this that our existence is a pale ghost; we live it, but actually only on the basis of an error without which it would not have arisen.'[16] The essay then praises the reaction of mathematicians who continue to believe in the power and value of maths. This constitutes a contrast to other fields and professions where fundamental questions are treated less pragmatically and criticism of reason leads to its renouncement: 'After the Enlightenment the rest of us lost our courage.'[17] Taking mathematicians as an example for dealing with crises, non-mathematicians too, so the essay suggests, should continue to build on reason, the Enlightenment tradition and its still valuable results – even if aspects of it might have become compromised.

Musil's essay then goes on to argue for the need for literary writers in particular to emulate mathematicians. The speaker complains that contemporary literature has turned away from the rational and has focussed on feeling and aesthetics, so much so that it injures understanding of life and has ruined 'our imaginative literature to such an extent that, whenever one reads two German novels in a row, one must solve an integral equation to grow lean again.'[18] The essay thus opposes early twentieth-century lamentations of rationalisation and excesses of the intellect and argues against limiting reason to science. Instead, it introduces the idea that successfully being modern requires a 'mathematical man' who holds on to reason despite its drawbacks, and it particularly calls for 'mathematical writers' who recognise the continued value of reason for their work. The following literary analysis of Musil's debut novel examines a fictional perspective on the suggestion that to live through and make sense of the crisis-ridden early twentieth century, people, most of all literary writers, have to become 'mathematical men'.

Musil: *The Confusions of Young Törless* (1906)

Törless, the protagonist of Musil's novel, is an adolescent in an elite boys' academy and he experiences the time of transition between childhood and adulthood as deeply unsettling. When his classmate Basini steals money and then submits to torture by his classmates, Törless becomes aware of a hidden side of society and of the disconcerting sexual attraction he feels towards Basini. But the confusion goes deeper; Basini triggers the realisation in Törless that '[t]here had always been something that his thoughts could not deal with.'[19] While partly a *Bildungsroman*, the novel goes beyond a focus on Törless's education and development. The philosophical and ideological positions the adolescent encounters open the novel to a broader picture of society in the early twentieth century which, like Törless, struggles with the loss of former certainties, revaluates the place of reason and rationality, and strives towards overcoming conflicting views.[20]

The novel signals the potentially leading role of mathematics when Törless, realising that what he calls 'the wooden ruler of rationalism'[21] cannot fathom certain aspects of reality, turns to mathematics to attempt to grasp the whole of confusing reality. As mathematics is commonly set at the apex of rationality this move might appear counter-intuitive, yet, Törless is prompted to turn to maths for help when he finds that several of its concepts seem to be contrary to reason: the infinite, for example, and irrational and imaginary numbers. These examples show, so Törless explains to his classmate, that mathematics 'sometimes runs so contrary to the understanding … The idea of the irrational, of the imaginary, of lines that are parallel and intersect in infinity – somewhere or other – excites me.'[22] Törless hopes that examining these curious cases in the most rational domain will help him come to terms with his wider confusions regarding the unintelligible elements of his teenage life and the special qualities of the early twentieth century.

Imaginary numbers prove the most fruitful example to examine the aspect of mathematics that resists rational understanding. Törless wonders about the very nature of an imaginary number: 'It doesn't exist … there can't be such a thing as a real number that's the square root of something negative.'[23] It here helps to remind ourselves of the definition of imaginary numbers: $i = \sqrt{-1}$. This is the same as: $i^2 = -1$. In the real number system, which is most commonly used in everyday calculations, square roots can only be positive. Törless's contention that imaginary numbers do not 'exist' echoes concerns by actual mathematicians who, well into the nineteenth century, considered imaginary numbers to be dubious. The very term 'imaginary number' suggests a problematical relation between $\sqrt{-1}$ and some kind of reality.[24] The problem for both Törless and nineteenth-century mathematicians is that imaginary numbers do not have a direct correspondence in nature; they do not relate to anything 'real'. They thus contradict realist notions, such as expressed in Galileo Galilei's famous phrase that mathematics is the language of the Book of Nature.[25] In this view, correspondence to nature ultimately guarantees the existence and meaning of mathematical objects. Mathematicians today happily calculate with imaginary numbers and accept them as 'real' or as 'unreal' as any other number.[26] In the early twentieth century, imaginary numbers were already widely accepted, so *Young Törless* primarily draws on historical and metaphorical questions around imaginary numbers.

In the novel, Törless concludes that, since there are no 'real' coun-terparts of imaginary numbers, mathematics does not have a direct referential relationship with nature. He reasons that it consequently has to be set on purely *mathematical* foundations and sets out to learn these. With this endeavour Törless mirrors turn-of-the-century research into the foundations of mathematics and, as the novel and Musil's non-fictional writing propose, into the fundamental structures of modern life itself. Unsurprisingly, teenage Törless does not acquire advanced mathematical knowledge, but his enquiries nevertheless result in an experience of the crisis in mathematics. Indeed, the novel points to the foundational debate that took place at the time of publication when Törless's friend doubts that even mathematicians themselves know the foundations of their field: 'These grown-ups and clever people have completely spun themselves into a web, one stitch supporting the next ... ; but no one knows where the first stitch is, the one that holds everything up.'[27] Törless himself maintains a pragmatist stance, arguing that the success of maths implies its truth. And when he moreover realises that the ambiguous, unexplained elements make mathematics more powerful, the novel foreshadows the argument in Musil's essay that uncertainty enables existence: 'we live it, but actually only on the basis of an error without which it would not have arisen.'[28] In *Törless*, the protagonist notices a similar process regarding the specific case of imaginary numbers, which allow one to arrive at a 'real' result that otherwise could not have been attained:

> in that kind of calculation you have very solid figures at the beginning, ... And there are real numbers at the end of the calculation as well. But they're connected to one another by something that doesn't exist. Isn't that like a bridge consisting only of the first and last pillars, and yet you walk over it as securely as though it was all there?[29]

The metaphor of imaginary numbers here illustrates the function of the not-fully-explainable to lead over stretches of missing ground and arrive at secure results. Methods to cover the gaps generated by lost certainties have wider significance for Törless, not least since he turns to mathematics in order to deal with the adolescent feeling of being in the air. When he recognises that uncertainties in the foundations of mathematics do not inhibit its usefulness, he can similarly accept confusions in his life, leave the critical bridging time of adolescence behind and approach adulthood, a once-again stable state. In other words, the way forward for Törless – and by implication for the young twentieth century and its search for foundations – is to emulate mathematics, to 'be mathematical'.

The model of mathematics helps Törless end the uncertainties of the in-between teenage years, but, as Musil's novel implies, he should have had it easier. Specifically, fictional literature should have provided a guide over the stormy waters of adolescence. Unlucky for Törless then that the school library is not well stocked: 'That illusion, that trick favouring personal development, was missing from the institute.'[30] This remark at the beginning of the novel alerts the reader to the fact that the imaginary domain is part of literature more prominently than of mathematics.[31] The fact that appropriate fiction is unavailable to Törless and he has to resort to mathematics is echoed in 'The Mathematical Man' and its lament that contemporary German literature is inadequate to the demands of its time. If Musil's essay then praises mathematicians' continued belief in reason as a model for literary writers, the novel illustrates how mathematics takes on a model function for Törless that literature should have fulfilled, and shows how its own novelistic engagement with maths negotiates the changing role of reason in the transitional age of the early twentieth century. Importantly, in both Musil's texts, mathematics has to be thought in its specifically modern form that puts into question its certainty and truth and involves the pragmatic use of imaginary domains.

Zamyatin: 'On Literature, Revolution, Entropy, and Other Matters'

Yevgeny Zamyatin also introduces what he calls the 'new mathematics'[32] as a model for a modern way of being and writing both in essayistic and novelistic form. Where Musil emphasises the continued value of maths despite uncertainties regarding its foundations, however, Zamyatin treats the absolute truth of mathematics and its applicability to the world as a given and establishes that problematical developments call for new understandings not of maths but of nature. In his view, the modern transformation of mathematics shows the necessity of a fundamentally changed understanding of reality and of a new literature appropriate to it.

'On Literature, Revolution, Entropy, and Other Matters' indicates some key concerns of Zamyatin's thinking even in its title, and the interrelation of literature, politics and science is also present in a text such as his dystopian novel *We* (1921), as are mathematical metaphors and examples, which are also used in the essay. In 'On Literature', Zamyatin explicitly contrasts traditional mathematics, exemplified by the immensely influential and long-lived geometry of Euclid, and a modernised new mathematics, associated with the work of Nikolai

Lobachevsky (1792–1856). Geometry, literally the measurement of the earth (from Ancient Greek: *gê-* 'earth' and *-metría* 'measurement'), had been largely determined by Euclid's work for over two thousand years. Zamyatin establishes geometry as a tool to survey and understand the world more broadly when he introduces an event that fundamentally changed the mathematical landscape in the nineteenth century: 'Lobachevsky cracks the walls of the millennia-old Euclidean world with a single book, opening a path to innumerable non-Euclidean spaces: this is revolution.'[33] Lobachevsky showed that Euclid's geometry is not the only possible geometry and that non-Euclidean geometries can be taken to better describe the world: Euclidean geometry is suitable to planes but not to curved surfaces, and since the earth is a sphere and therefore curved, it requires surveying with non-Euclidean geometry. When Zamyatin introduces Lobachevsky's work as a revolution, he establishes that with it, fundamental tools of thought and ways of understanding the world have to be reconsidered.

Having used maths as an example of how revolutionary changes institute new views of and approaches to the world, Zamyatin extends the argument to literature. He suggests that if non-Euclidean geometry reveals that measuring the earth with the long-trusted tools of Euclidean geometry is no longer adequate, so literary writers have to similarly realise that their traditional ways of describing the world produce overly simplistic representations of that world. Instead of adhering to realism and depicting the surface in conventional ways – 'clicking his Kodak (genre)'[34] – a literary writer has to take a similar step as mathematicians and establish new literary styles adequately to engage with reality. Zamyatin explicitly equates literary realism with Euclidean geometry and deduces the need for new literary forms from the advent of the new mathematics:

> Science and art both project the world along certain coordinates.... All realistic forms are projections along the fixed, plane coordinates of Euclid's world. These coordinates do not exist in nature. Nor does the finite, fixed world; this world is a convention, an abstraction, an unreality. And therefore Realism ... is unreal. Far closer to reality is projection along speeding, curved surfaces – as in the new mathematics and the new art.[35]

Zamyatin thus does not disavow the possibility of realist accounts but argues that when non-Euclidean geometry outstrips traditional mathematics, it implies the need for equally new literary styles to represent reality: 'Realism that is not primitive … consists in displace-ment, distortion, curvature, nonobjectivity.'[36] Only with such a deeper and more adequate understanding of reality can literature be useful to the 'new men'[37] of the twentieth century. Zamyatin thus urges literature to implement in its domain the revolution of the new mathematics – we might say: the modern mathematics – and its more truthful representa-tion of complex reality. In the next section, I analyse how Zamyatin establishes a similar call for a deeper understanding of reality and new literature in his novel *We*.

Zamyatin: *We*

Mathematics is omnipresent in *We*: the first-person narrator D-503 is a mathematician, citizens in OneState are not named but numbered, and the four basic rules of arithmetic regulate all aspects of life, including sexual relationships, the moral system and art. At first sight, *We* takes a critical stance towards mathematics as a means of instituting totalitarian control, and D-503's plan to celebrate the 'mathematically perfect life of OneState'[38] in writing appears to be misguided at best. The difference between D-503's rationally produced text – he begins by copying out a newspaper article and announces his intention to continue writing in this factual manner – and traditional literature is made explicit when D-503 discredits the literature of the readers' present: 'I wondered, that the ancients did not immediately see how completely idiotic their literature and poetry was.'[39] The sonnet 'Happiness' quoted above, however, alerts one to the fact that OneState's mathematised literature is not necessarily an improvement on the works of Shakespeare and Dostoevsky.

In his records D-503 describes people and relationships in mathe-matical terms, drawing on shapes common in Euclidean geometry: 'Her brows make a sharp mocking triangle';[40] 'she's got a simple round mind';[41] '[h]is forehead was a huge bald parabola';[42] 'we're a triangle, maybe not isosceles but still a triangle'.[43] Yet, when he meets the enigmatic woman I-330, the mathematical imagery changes: D-503 notices 'something about her eyes or brows, some kind of odd irritating X that I couldn't get at all'[44] and frets that '[t]his woman was just as irritating to me as an irrational term that accidentally creeps into an equation and can't be factored out'.[45] The mathematical terms here do not refer to the domain of certainty and rationality but to unknown and irrational values. Following the newly emerging emotion, confusion and sexual attraction expressed by these metaphors, D-503 begins a relationship with I-330 and encounters the Ancient House, a hideout with a secret tunnel to a world outside the bounds and rules of OneState. These experiences profoundly change his views, and he begins to measure the world along different coordinates: 'The coordinates of the whole business all begin, of course, with the Ancient House. The X, Y, and Z axes that recently began serving as the basis of my whole world all start there.'[46]

The terms describing D-503's changing views are remarkably similar to Zamyatin's essay, and if 'On Literature' then goes on to demand a transformation of literary coordinates, *We* does indeed exhibit a change in the style of D-503's records. Critics commonly comment on the fact that the rational style so consciously established at the outset of the novel is increasingly replaced by unordered, expressionistic diction. Gary Rosenshield describes this stylistic development as a 'transformation of the narrator from mathematician to poet'.[47] While Rosenshield's analysis very successfully traces changes in D-503's character and writing, the juxtaposition of mathematician and poet is misleading. Rather, shifts in D-503's writing style are precisely in concert with his changing mathematical views, which move from a traditional belief in its certainty to the acceptance of apparently non-rational aspects.[48] I consequently argue that the poet D-503 does not usurp the mathematician but that mathematics constitutes a crucial factor in producing D-503's poetic self.

Mathematics in *We* is not synonymous with reason; this becomes clear when concepts such as the unknown value X, irrational terms and imaginary numbers provide D-503 with vocabulary to describe his confusion and grasp experiences that take place outside of OneState's rational order. After the encounter with I-330, D-503 increasingly becomes aware of the fact that certain aspects of mathematics contradict OneState and its values. Notably, he remembers his first encounter with imaginary numbers and their troubling 'inexistence':

> Once Pliapa told us about irrational numbers – and I remember how I cried, I beat my fists on the table and bawled: 'I don't want $\sqrt{-1}$! Take it out of me, this $\sqrt{-1}$!' That irrational root grew in me like some alien thing, strange and terrifying, and it was eating me, and you couldn't make any sense of it or neutralize it because it was completely beyond *ratio*.[49]

Despite its apparently non-rational aspects, D-503 never doubts the truth of mathematics. Rather, he logically concludes that the existence of unknowns, irrationals and imaginaries in maths proves the fact that such elements similarly are part of life: 'For every equation, every formula in the superficial world, there is a corresponding curve or solid. For irrational formulas, for my $\sqrt{-1}$, we know of no corresponding solids … [I]f we don't see these solids in our surface world, there is for them, there inevitably must be, a whole immense world there, beneath the surface'.[50] Maths thus retains its status as ultimate truth and hence supports criticism of complete rationalisation. Indeed, mathematics also serves as a tool for the enemies of OneState. The revolutionary Mephi live outside the city-state and plan its downfall, and significantly, they possess advanced mathematical knowledge with which they motivate their revolution. If numbers go on to infinity, so their argument goes, then there also has to be an infinite number of revolutions. The often-repeated view that the Mephi are opposed to rationality is therefore only partly correct: when they justify their aims with the help of mathematics, they use 'the major language of reason'[51] against OneState, yet, this language itself includes irrational and imaginary elements.

D-503's realisation of the existence of irrational and imaginary aspects in mathematics and life also leads him to call for a new style of writing that takes account of these; or rather, he notices that they have been influencing his records all along: 'instead of the elegant and strict mathematical poem in honor of OneState, it's turning out to be some kind of fantastic adventure novel. Oh, if only this really were just a novel instead of my actual life, filled with x's, $\sqrt{-1}$, and degradations'.[52] Mathematical imagery thus also describes the style of D-503's records, which make up the novel *We*. They progressively deviate from the aim of objective and rational recording and instead become more and more subjective and disordered, increasingly including sudden shifts and fragmentary sentences. D-503's initially realist recording thus turns to a modernist style that reflects his confusions and the non-rational aspects of reality. With this 'non-rational' way of writing D-503 opposes the ideology of OneState, but this changes when he is forced to have an operation to remove a part of the brain and thus realign him with the totalitarian order. The surgery removes the 'imagination', a brain section that houses all irrational feeling and deviationist tendencies. After the operation, D-503 betrays I-330, negates the validity of irrational and imaginary perspectives, holding that 'reason has to win',[53] and returns to his rational writing style: 'It is day. Clear. Barometer at 760.'[54] 'Cured' of imagination, citizens are streamlined, completely happy and hardly human: 'not "men" – that isn't the word … Not men but some kind of tractors in human form.'[55] *We* thus establishes the imagination as the ultimate enemy of mechanisation and as at the core of what it means to be human. Not least, the imagination is a vital part of literary production, and D-503's resistance to totalitarian OneState manifests in a literary work – namely *We* itself. Yet, importantly, the imagination and its revolutionary potential also are part of mathematics. After the mass operation on the imagination, citizens of OneState will be unable to handle imaginary numbers and only have access to the 'rational' part of mathematics. Hope remains then that the revolutionary Mephi, who can draw on the entirety of this powerful tool to understand the world, might be victorious after all.

We distinguishes between a traditional notion of maths that lends itself to support rationalisation and modern mathematics, which implies a reconsideration of what counts as reality. As Zamyatin proposes in his essay, in *We* too, mathematics points to the fact that reality is not clearly ordered or wholly graspable, and the example of maths calls for a new way of writing that takes account of the 'mathematically proven' unknowns, irrational and imaginary parts of life. In Zamyatin's writing, modern mathematics thus works as a model for literature to abandon naïve realism and find new means of expression.

Being modern = being mathematical = being imaginary

Musil and Zamyatin's texts express the need for a new literature, not unlike Ezra Pound's famous modernist battle cry: 'Make it new!'. Pound, too, looks to science and mathematics as inspirations for the new literature he envisions. When asking literature to produce 'lasting and unassailable data',[56] he seems to imply that it can be true or false, just like scientific facts are commonly assumed to be true or false. Although the literary texts discussed here differ in the details of their engagements with the uncertainties of a specifically modern mathematics, they do all identify uncertainty and newness in mathematics itself and draw on precisely these aspects to indicate promising ways to modernise literature.[57]

Musil and Zamyatin's texts also show that nuanced engagement with mathematics does not necessarily place it at the heart of a threatening technology and inevitable rationalisation, but their modern notions allow them to connect literature and mathematical material in unforeseen ways. Musil's texts engage with concerns that maths loses its power as its foundations are discovered to be uncertain. His essay and novel propose that this does not inhibit the usefulness of maths but indeed leads to its status as a model of pragmatist use and continued value in changing times. In contrast to Musil's engagement with foundational questions and his emphasis on the enduring value of reason, Zamyatin does not challenge mathematics or its representational relationship to the world. Rather, he employs the example of maths to illustrate why writers cannot approximate twentieth-century life by representing surface reality and thereby champions greater attention to the irrational as the source of individuality and resistance to totalising movements. In this way, modern mathematics offers the writers different models to develop new literary forms with which to respond to the pressures of modern experience.

The fruitfulness of considering mathematics and literature together was and is not only noticed from the literary side. Indeed, since the 1990s historians of mathematics have paid increasing attention to aspects of crisis and anxiety in the process of modernisation; among them Herbert Mehrtens and Jeremy Gray most prominently consider the possibility of understanding the 'foundational crisis' of mathematics as part of the wider sense of crisis in the early twentieth century and to thus speak not only of a *modern* but of a *modernist* mathematics. In other words, while it is generally accepted that the nineteenth and early twentieth centuries saw modernisation and the consequent development of a modern mathematics, some historians of mathematics argue for regarding this transformation as part of more general movements and thus include mathematics in the notion of modernist culture.[58] Both Musil and Zamyatin's texts advocate considering mathematics as part of a broader understanding of culture when they connect its development with their analyses of the early twentieth century and with new trends in literature. The fact that both writers engage with mathematics in essayistic as well as novelistic work further indicates their challenges to boundaries between the factual and the fictional. Moreover, Musil plays with genre expectations when he adopts a fictional speaker in his essay, and Zamyatin uses a speech from the fictional character I-330 as an epigraph to 'On Literature' and takes up her argument in the main body of his essay. A further factor in their undermining of any clear distinction between factual and fictional writing is their emphasis on the importance of the imaginary for reality. Musil's texts show that 'inexistent' imaginary domains make possible transitions to once-more-stable states, and Zamyatin's *We* develops the idea that abandoning the imaginary means ceasing to be human. Both writers consequently establish imaginary numbers as a key part of modern mathematics and employ the concept as a central metaphor. 'Being modern' thus might partly be equated to 'being mathematical', but the 'equation' for Musil and Zamyatin also necessitates a third term: being modern '=' being mathematical '=' being imaginary.

Notes

1   Robert Musil, 'The Mathematical Man', in *Precision and Soul: Essays and Addresses*, eds. and trans. Burton Pike and David S. Luft (Chicago and London: University of Chicago Press, 1990 [1913]), 39–43, 42.
2   Yevgeny Zamyatin, *We*, trans. Clarence Brown (New York: Penguin, 1993), 65.
3   Max Weber, 'Science as a Vocation', in *From Max Weber: Essays in Sociology*, eds. and trans. H. Gerth and C. Wright Mills (London: Routledge, 1998), 129–156, 139.
4   Weber, 'Science as Vocation', 155.
5   Frege quoted in Jeremy J. Gray, 'Anxiety and Abstraction in Nineteenth-century Mathematics', *Science in Context* 17.1/2 (2004): 23–47, 26–27.
6   See Gray, 'Anxiety and Abstraction'.
7   The other schools are formalism and intuitionism. See Ernst Snapper, 'The Three Crises in Mathematics: Logicism, Intuitionism and Formalism', *Mathematics Magazine* 52.4 (1979): 207–216.
8   Bertrand Russell, *My Philosophical Development* (London and New York: Routledge, 1993), 58.
9   For an evaluation of accuracy, see Paolo Mancosu, Review of *Logicomix*, by Apostolos Doxiadis, Christos H. Papadimitriou, Alecos Papadatos and Annie di Donna', *Journal of Humanistic Mathematics* 1.1 (2011): 137–152.
10 Bertrand Russell, *Autobiography* (London and New York: Routledge, 1998), 725.
11 Russell, *Autobiography*, 725.
12 Musil, 'Mathematical Man', 41.
13 Musil, 'Mathematical Man', 42.
14 Musil, 'Mathematical Man', 42.
15 Musil, 'Mathematical Man', 42.
16 Musil, 'Mathematical Man', 42.
17 Musil, 'Mathematical Man', 42.
18 Musil, 'Mathematical Man', 42.
19 Robert Musil, *The Confusions of Young Törless*, trans. Shaun Whiteside (London: Penguin, 2001), 66.
20 Similar concerns also inform other writings by Musil, including his magnum opus *The Man without Qualities*. Alan Thiher provides an overview of Musil's

fictional work, his doctoral dissertation on the work of Ernst Mach, and his diaries, including a discussion of Musil's neologisms 'ratioïd' and 'non-ratioïd'. Allan Thiher, *Understanding Robert Musil* (Columbia: University of South Carolina Press, 2009), 19ff.
21 Musil, *Törless*, 9.
22 Musil, *Törless,* 90–91.
23 Musil, *Törless,* 81.
24 For the history of imaginary numbers, see Paul J. Nahin, *An Imaginary Tale: The Story of √–1* (Princeton: Princeton UP, 1998).
25 Galileo Galilei et al., *The Controversy on the Comets of 1618*, trans. Stillman Drake and C.D. O'Malley (Philadelphia: University of Philadelphia Press, 1960), 183–184.
26 The problem of the 'existence' of imaginary numbers vanishes when their applicability is bracketed and they are viewed in terms of their mathematical definition only: imaginary numbers are not part of – and thus do not 'exist' – in the real number system. However, in the complex number system both real and imaginary numbers are defined and 'exist'.
27 Musil, *Törless*, 91.
28 Musil, 'Mathematical Man', 42.
29 Musil, *Törless*, 82.
30 Musil, *Törless*, 10.
31 Language, too, encompasses imaginary elements for Törless when he understands metaphors to provide 'inexistent bridges' between feelings and words. The term 'metaphor' means 'to transfer, to carry over', and for Törless a metaphor indeed clarifies what literal language cannot express by carrying meaning over into a non-literal – imaginary – domain. For a discussion of language in *Törless*, see Matthias Luserke-Jaqui, '*Die Verwirrungen des Zöglings Törleß*: Adolescent Sexuality, the Authoritarian Mindset, and the Limits of Language', in *A Companion to the Works of Robert Musil*, eds. Philip Payne, Graham Bartram and Galin Tihanov (Rochester,

New York: Camden House, 2007), 151–173.

32 Yevgeny Zamyatin, 'On Literature, Revolution, Entropy, and Other Matters' in *A Soviet Heretic: Essays by Yevgeny Zamyatin*, ed. and trans. Mirra Ginsburg (Chicago and London: University of Chicago Press, 1970 [1923]), 107–112, 112.

33 Zamyatin, 'On Literature', 107.

34 Zamyatin, 'On Literature', 109.

35 Zamyatin, 'On Literature', 112.

36 Zamyatin, 'On Literature', 112.

37 Zamyatin, 'On Literature', 109.

38 Zamyatin, *We*, 4.

39 Zamyatin, *We*, 66.

40 Zamyatin, *We*, 168.

41 Zamyatin, *We*, 36.

42 Zamyatin, *We*, 222.

43 Zamyatin, *We*, 44.

44 Zamyatin, *We*, 8.

45 Zamyatin, *We*, 10.

46 Zamyatin, *We*, 90.

47 Gary Rosenshield, 'The Imagination and the "I" in Zamjatin's *We*', *The Slavic and East European Journal* 23.1 (1979): 51–62, 51.

48 Leighton Cooke notices that mathematicians mentioned in the novel point to the past and suggests that OneState is unable to creatively develop new mathematical concepts. I agree with T.R.N. Edwards's argument that the 'old-fashioned' mathematics at the ideological foundations of OneState is partly refuted by new mathematical theories. Leighton Brett Cooke, 'Ancient and Modern Mathematics in Zamyatin's *We*' in *Zamyatin's* We: *A Collection of Critical Essays*, ed. Gary Kern (Ann Arbor, Mich.: Ardis, 1988), 149–167.

49 Zamyatin, *We*, 39.

50 Zamyatin, *We*, 98.

51 Cooke, 'Ancient and Modern Mathematics', 165.

52 Zamyatin, *We*, 99.

53 Zamyatin, *We*, 225.

54 Zamyatin, *We*, 224.

55 Zamyatin, *We*, 182.

56 Ezra Pound, 'The Serious Artist', *Literary Essays of Ezra Pound*, ed. T.S. Eliot (New York: New Directions, 1968), 41–57, 42.

57 Musil and Zamyatin thus voice as a programme what Gillian Beer detects in 1930s literature: it is 'released from some of the constraints of mimesis' and becomes modernist, not least due to '[t]he questioning of substance in twentieth-century physics … It is harder to deny an "out there" that is undifferentiated, or irresolute'. Gillian Beer, 'Wave Theory and the Rise of Literary Modernism', in *Open Fields: Science in Cultural Encounter* (Oxford: Oxford University Press 2006), 295–318, 296 and 315.

58 See Leo Corry, 'How Useful Is the Term "Modernism" for Understanding the History of Early Twentieth-Century Mathematics?', in *Science as Cultural Practice: Modernism in the Sciences, ca. 1900–1940*, eds. Moritz Epple and Falk Müller (Berlin: Akademie Verlag, forthcoming); Leo Corry, 'The Development of the Idea of Proof up to 1900', in *The Princeton Companion to Mathematics*, ed. Tim Gowers (Princeton, Princeton University Press, 2008), 129–142; Moritz Epple, 'Kulturen der Forschung: Mathematik und Modernität am Beginn des 20. Jahrhunderts', in *Wissenskulturen: Über die Erzeugung und Weitergabe von Wissen*, eds. Johannes Fried and Michael Stolleis (Frankfurt am Main: Campus, 2009), 125–158; Moritz Epple, 'Styles of Argumentation in Late Nineteenth Century Geometry and the Structure of Mathematical Modernity', in *Analysis and Synthesis in Mathematics: History and Philosophy*, eds. Michael Otte and Marco Panza (Dordrecht and Boston: Kluwer, 1997), 177–198; Jeremy J. Gray, 'Anxiety and Abstraction in Nineteenth-century Mathematics', *Science in Context* 17.1/2 (2004): 23–47. Jeremy J. Gray, 'Modern Mathematics as a Cultural Phenomenon', in *The Architecture of Mathematics*, eds. José Ferreirós and Jeremy Gray (Oxford: Oxford UP, 2006), 371–396; Jeremy J. Gray, 'Modernism in Mathematics', in *The Oxford Handbook of the History of Mathematics*, eds. Eleanor Robson and Jacqueline Stedall (Oxford: Oxford UP, 2009), 663–683; Jeremy J. Gray, *Plato's Ghost: The Modernist Transformation of Mathematics* (Princeton and Oxford: Princeton University Press, 2008); Herbert Mehrtens, *Moderne Sprache Mathematiks: Eine Geschichte des Streits um die Grundlagen der Disziplin und des Subjekts formaler Systeme* (Frankfurt am Main: Suhrkamp, 1990).

9
Sculpture in the Belle Epoque: mathematics, art and apparitions in school and gallery

Lewis Pyenson

Felix Klein's models

The Belle Epoque saw an active trade in what are known as mathematical models – three-dimensional representations of complex mathematical surfaces constructed largely in plaster, but also in wood, metal and occasionally wire or string. Many hundreds of the models (around the time of its inauguration in 1928, the Institut Henri Poincaré in Paris acquired the collection of the Faculté des sciences, some 400 models) were produced in plaster and sold to universities and schools, each surface characterised by a particular equation. Mathematician Felix Klein – widely known today for having in 1882 described the non-orientable surface known as *Kleinsche Fläche* ('Klein Surface', subsequently known as *Kleinsche Flasche* or 'Klein Bottle' (see Figure 9.1) – was at the origin of the plaster-models trade. As a professor at the Munich Institute of Technology between 1875 and 1880, he collaborated with his colleague Alexander Brill and their students to formulate the objects, which were sold by Brill's brother Ludwig.[1] Brill then contracted other professors and their students to design additional models. Klein popularised a veritable zoo of models when he toured the United States in 1893, and following his visit American universities were quick to buy Brill's product. Around 1900, Martin Schilling took over Brill's enterprise.[2]

In the commentary he provided for the three-volume edition of his collected mathematical papers, published from 1921 to 1923, Felix Klein placed his interest in mathematical models in the context of a desire to

Fig. 9.1 *Kleinsche Flasche*, painted plaster, pre-1914 (approx. 25 cm long). Bibliothéque, Institut Henri Poincaré, Paris.

Image copyright © Lewis Pyenson, by permission.

facilitate an intuitive grasp of geometry, which animated him from his time as a student and assistant of Julius Plücker, the professor of mathematics and experimental physics at Bonn. Then, in 1870, before the Franco-Prussian War, he saw the collection of mathematical models at the Conservatoire des arts et métiers in Paris. Klein's publications from the early 1870s contain references to plaster models of surfaces and somewhat clumsy renderings of them in line drawings.

Klein's efforts fitted squarely into the activities of humanist colleagues who were collecting plaster copies of exemplary statues. In the last quarter of the nineteenth century and into the twentieth century, classicists hammered for plaster-cast 'laboratories of archaeology'. At the inauguration of the Cambridge Museum of Classical and General Archaeology in 1884, for example, the Keeper of Greek and Roman Antiquities in the British Museum, Charles Newton, contended that teaching archaeology without a cast museum was 'like trying to teach chemistry without a laboratory, or medicine without a hospital'. On the same occasion, the professor of archaeology at Strasbourg, Adolf Michaelis, reminded the gathering (in a letter) that the museum was 'as necessary a supplement to archaeological lectures, as a laboratory is to lectures on physics or chemistry'.[3] The promotion of casts in late nineteenth-century Dresden by Georg Treu was indeed, as historian Marcello Barbanera contends, 'the cohabitation between historical idealism present in the organisation of the Glyptothek, seen as a laboratory, and nearby artistic activity'.[4] Klein, attuned to this rhetoric, continually advocated for the importance of mathematical 'laboratories' with three-dimensional models and stereoscopic images of them.[5] At least with respect to the plaster models, he was successful. He reported in 1922: 'Today, no German university any longer lacks such a collection'.[6]

Nineteenth-century plaster casts

Plaster casts became readily available to the bourgeoisie in the eighteenth century.[7] By the middle of the nineteenth century, Ingeborg Kader emphasises:

> White plaster casts were … above all in Germany generally accepted and obligatory cultural symbols of a bourgeois elite. Their close tie to a philosophical content, documented by the color 'white', made it possible, 'in a rigorously moral and proper salon during the second half of the nineteenth century, for example, for the statue of an undressed Aphrodite or Idolino to be shown'.[8]

White statues in homes and museums were best seen against walls of contrasting colours.[9] Plaster mouldings came to doorways and ceilings in middle-class residences; ancestral life moulds surmounted doors and hung from walls. The bust of a regent broadcast a family's patriotism, just as phrenological heads adorned the office of quacks; an industry of moulders supplied this craving.[10] Rulers and their parliaments funded the vogue for collecting plaster casts of exemplary sculpture. In the middle of the nineteenth century, Henry Cole assembled a remarkable gallery of casts with international provenance; they are observed today at the Victoria and Albert Museum by students of art and the general public, just as they were in A Coruña, Spain, during Picasso's time as a precocious art student there.[11] With assistance from the Foreign Office, in 1867 Cole brought about an international convention for promoting reproductions of works of art, largely plaster copies.[12]

Producing copies of massive works in this way without damage to the originals was and is a demanding enterprise which developed sophisticated practitioners, notably at the Gipsformerei founded at Berlin in 1819.[13] It is possible that Klein and Brill were familiar with the extensive collection of plaster copies of sculpture in the Neues Museum on the Museuminsel in Berlin. The two mathematicians worked in the new building of the Munich Institute of Technology, an elegant Neo-Renaissance palace inspired by Gottfried Semper's teachings, and they undoubtedly knew about the collection in the nearby Glyptothek.[14] They may also have been aware of the labours of Heinrich Brunn, appointed professor of classical archaeology in 1865 at the University of Munich and busily acquiring plaster copies of classical sculpture, 379 in number by 1877, which were located in the former Jesuit College of St Michael close to the Institute of Technology.[15] As Klaus Borchard observes, with reference to the largest collection at Bonn, the nineteenth-century casts at institutions of higher learning made a literally spectacular impression on university students.[16] Felix Klein would not have been immune from the appeal, or at least the reputation, of the Bonn collection from his time there as a student of Plücker.

Three-dimensionality was ideally suited to Klein's pedagogical programme in mathematics. Unlike a painting on canvas or a print on paper, freestanding sculpture is designed to be seen from all perspectives. In conception and appreciation, it requires spatial intuition, and, if viewed under natural light, it changes subtly and continually in the course of a day. Galleries for displaying the nineteenth-century cast collections often featured skylights and clerestories. By the 1870s, newly constructed sculpture halls were fitted with gas lighting high above the works of art; this, on top of the soot produced by the gaseous flames, delivered a changing chiaroscuro to the figures and friezes, accentuating their form.[17]

In the absence of both gas and electrical lighting, a situation that would violate health and safety codes today, plaster sculptures were displayed clearly in the Niobe Room of the Neues Museum at Berlin (1906) and the Roman Room of the Glyptothek at Munich.[18] The unchanging glare of numerous early twentieth-century electric lights compromised sculpture even as it improved the view of paintings. Although there was a vogue for stereoscopic views of the statues, flat photographs, which students and scholars came to prefer over plaster copies by the end of the nineteenth century, fitted with celebrating two-dimensionality in painting.[19] Museum curators, following widespread concern about the industrial production of bronze casts, notably those of Auguste Rodin, were increasingly worried about authenticity; they observed that the plaster casts were only superficial representations.[20] By 1937 one German commentator could refer to the collections of plaster casts as 'ghostly chambers of horror'.[21] The qualities of seriality and eeriness made the plaster casts attractive to artists like Marcel Duchamp, Picasso, De Chirico and Carlo Carrà.

In both art and mathematics before 1914, plaster models cultivated the mind of the viewer by appealing to ideal forms. Although some sculptural copies improved on the original object, filling in craters and supplying missing appendages in the way that mounted fossil skeletons of dinosaurs do today, fragmentary works circulated widely. Observing a philosopher's face with a stove-in nose, an athlete minus his rod, or a goddess lacking arms, a viewer was called upon to look beyond the erosion of centuries to formulate a vision of ideal harmony. This act of imagination – necessary to appreciate originals like the Parthenon Sculptures at the British Museum – was also vital for an appreciation of the plaster mathematical models, since a number of them represent only a portion of a three-dimensional projection of a higher-dimensional surface. Both the sculptures and the mathematical surfaces were generally left white, thereby claiming an association with purity and, in exemplary works, the ideal essence of form.[22]

Klein's intuition

Felix Klein used the adjective *anschaulich*, or 'intuitive', to describe his approach to mathematics and to distinguish it from the formalist approach of Karl Weierstrass and Ernst Kummer in Berlin, who sought to refer all of mathematics to arithmetical procedures. Klein's approach closely resembled the cry of the Munich classical archaeologist Heinrich Brunn, in the 1870s and 1880s, for collecting and displaying three-dimensional images as a way of encouraging an *anschaulich* understanding of Classical Antiquity, which he contrasted to the dry grammar emphasised by philologists. It also fitted with other, aesthetic sides of late nineteenth-century Neo-Idealism represented by Friedrich Nietzsche and Julius Langbehn.[23]

The Neo-Idealist term *Anschauung*, late in the nineteenth century, differed from the psychological term *Vorstellung* or 'presentation'. *Vorstellung* is sometimes translated into English as 'idea', although the Platonic notion of idea also appears in nineteenth-century German works as a French borrowing, *idée*. In the nineteenth century, *Vorstellung* was popularised in his educational philosophy by Johann Friedrich Herbart (and later elaborated by Hermann Lotze). It has the double sense of mechanically describing a psychical phenomenon and also identifying the qualitative content of the phenomenon. *Anschauung*, in the view of Klein's contemporary, the Anglo-German chemist, philosopher and historian of science John Theodore Merz, 'implies something akin, though perhaps superior, to seeing or perceiving by means of the senses', in other words, 'intellectual sight'. By Klein's time, *Anschauung* 'acquired a meaning somewhat akin to the *amor intellectualis Dei* of Spinoza'.[24] It is best known in English from the compound word *Weltanschauung* – a grandiose, amorphous, all-encompassing 'worldview' or *vision du monde*.

Felix Klein championed *Anschauung* in his 1872 inaugural lecture as a 23-year-old professor at the University of Erlangen (for this reason it is known as the Erlangen Programme), the post he held before moving to Munich. The lecture set out an ambitious plan to organise geometrical knowledge as a coherent whole with the insight of group theory. At the centre of Klein's research programme was the search for invariants, spatial relationships in particular geometries that do not change under transformations (as they do under projective geometry where, in the example of the single-vanishing-point perspective of Western painting, conic sections – circles and ellipses – do change form).

Behind Klein's programme was great popular and specialist interest in Non-Euclidean geometries. Klein hoped to discipline this interest by moving the question of spatial *Anschauung* from mathematical theory into pedagogy, where it 'is to be valued very highly'. An independent matter was the physical truth of geometry, how one understands 'the full reality of the figures of space', interpreting ' – and this is the mathematical side of the question – the relations holding for them as evident results of the axioms of spatial *Anschauung*'. Physical models that one could see and touch were vital to achieve this end. Herbert Mehrtens has emphasised: 'The plaster model … represents mathematical reality. As an object of spatial intuition it is "evidence" for the mathematical theorems about surfaces'. The models made the realm of ideas visible.[25] The kingdom exhibited in concrete models derived, in the view of mathematician Slavik Vlado Jablan, from identifying 'the theory of symmetry as a universal approach to different geometries by registering manifolds, their groups of transformations and invariants of these groups'.[26]

Stated in this way Klein's proposal might be seen as part of the late nineteenth-century wave of sentiment not only for symmetries but also for lexicographic, historical and scientific cyclopedias – from the *Oxford English Dictionary* (James A.H. Murray) to the *Dictionary of National Biography* (Sir Leslie Stephen) to, in Germany, the *Allgemeine Deutsche Biographie* (Rochus, Freiherr von Liliencron), the *Realencyclopädie der classischen Altertumswissenschaft* (August Pauly and Georg Wissowa) and the *Flora Brasiliensis* (Carl Friedrich Philipp von Martius, August Wilhelm Eichler and Ignatz Urban). Klein, however, was interested not in providing a record of facts but rather in generalising from existing examples – in proposing mathematical theories. In his text, Klein distinguished between *Anschauung* and *Vorstellung* in the way indicated by Merz. Furthermore, he used the adjective 'abstract' to mark his proposal in the realm of mathematical theory.

French and Italian mathematicians enthusiastically embraced Klein's programme, resulting in a Continental culture of geometry and, eventually, algebraic topology.[27] Notably, they shared conundrums – what Thomas Kuhn identified as disciplinary anomalies. Principal among the conundrums were functions that lacked traditional attributes of differentiability and extension. These came to carry the name of 'monster'.[28] Into the first decade of the twentieth century, European mathematicians agonised over these unruly objects, which became part of a grand debate over the merits of the project to base all mathematics on arithmetical operations. Felix Klein resolutely opposed this reductionist approach, offering instead his algebraic programme – grounded in symmetry and harmony – as a way of conquering geometry.[29] The interest in Klein's programme extended from mathematics to philosophy. It formed the basis for the earliest Neo-Idealist philosophical reflections of Ernst Cassirer, who emphasised its importance until the end of his life.[30]

With widespread discussion of counter-intuitive mathematical 'monsters' during the 1890s, Klein softened his Erlangen view, in 1893, to hold that intuition has above all 'heuristic value' for 'pure mathematical science'.[31] By emphasising the notion of heurism, Klein placed himself firmly in the tradition of a generation of educational reformers, who advocated replacing the dry, formal methods of presenting science and mathematics in the schools with lively experiments and imaginative, visualisable lessons, based upon beautiful forms.[32] He followed the view of art historian Gustav Friedrich Waagen and architect Friedrich Wilhelm Schinkel (both prominent in the early decades of the Museuminsel at early nineteenth-century Berlin), who borrowed from Horace: 'First delight, then instruct'.[33] This approach to connecting fine arts (as part of technology) with history (as part of science, in the sense of *Wissenschaft*) motivates instruction in the nineteenth century. It is distinct from the early-modern notion of the *Wunderkammer*, where the inspiration of wonder united aesthetics and learning; it departs from the view in Modernity's twilight, that museums are places to provoke and shock.[34]

From the time that he was a young man, Klein was savvy enough to know that abstract programmes, if they are to attract attention, need a concrete reference in the human world. The model-industry he helped generate served to persuade students and the general public of the palpable forms behind what he explicitly identified as an abstract quest. Concretely, the models required display cases and a budget-line, bringing the professor of mathematics to emulate the professor of mechanical engineering, with his demonstration machines, and the professor of archaeology, with his collection of plaster copies.

Spread across Germany in the thousands, mathematician school-teachers were Felix Klein's natural constituency. Unlike either Greek or chemistry, mathematics figured as a subject at every kind of secondary school – classical as well as commercial. Trained in pure mathematics at universities, many mathematics teachers had completed a doctorate, and some of them continued to be active in research. At the end of the nineteenth century, they formed associations to defend their professional interests, and foremost among the interests was to secure the continuing presence of pure, mathematical reasoning in the school curriculum. They offered mathematics as a modern alternative to the Neohumanist classics, a proposition that found support from engineers. Klein's call for mathematical intuition to replace formalist rigor mortis, and his assemblage of plaster models of mathematical surfaces as an aid to intuition, fitted with the goals of many schoolteachers and engineers.[35]

On the occasion of an address honouring the eightieth birthday of Karl Weierstrass in 1895, Klein elaborated the call for intuition in his Erlangen Programme. He pointed to the 'great revolution' (*Umschwung*) in pedagogical texts produced by Weierstrass's formal mathematics. The formalist revolution, where equations had replaced diagrams and figures, had the regrettable effect of discounting intuition. 'Mathematics in no way can be discovered by way of logical deduction', he emphasised. Rather, 'next to the latter, intuition today retains its complete, specific importance'. Intuition was essential for arriving at the idealisation that characterises natural laws. In the natural sciences, 'logical consequences achieve their goal when intuition has accomplished its task of idealisation'. Mathematical intuition anticipated logical thought almost everywhere, and it covered 'at every point a wider compass' than logical thought. On the university level, 'intuition is frequently not only undervalued, but also whenever possible pushed aside'. It was an upside-down view of pedagogy, and a lop-sided vision of science. Enough, Klein affirmed. It was time to bring back intuition, which, along with formal presentation, constituted the roots and branches of the mathematical organism.[36]

Herbert Mehrtens has examined Klein's model industry in the context of modern mathematics. Mathematical models, he emphasises, were opposed to the dominant current of formal, arithmetically inspired mathematical research, and, although they inspired a circle of artists and sculptors later, they were outside the mainstream of modern art: 'Modern conceptual mathematics had no use for any visualisable "essence", while modernist purism in the arts introduced conceptuality with the help of the forms mathematical models provided as sculptural material'. Following Klein's comments in his collected mathematical papers, Mehrtens contends that Klein invoked intuitive mathematics for the ends of 'applied mathematics' – cyphers serving engineers and physicists, eventually schoolteachers. Even geometry, Klein's abiding interest, became 'mathematical theory about things that come into existence by axiomatic stipulation and have no existence or reality outside the theory'. In Mehrtens's view, 'The collections of models that had been gathered by the end of the century were presented as pedagogical means for the academic teaching of mathematics', above all the fostering of *Anschauung*.[37]

Bottles: Klein and Picasso

The surfaces of Klein's and Brill's plaster casts display poorly in light such as the fluorescent kind flooding today's classrooms. Neither are the models well adapted to the diffuse, northern light traditionally favoured by artists for their studios. They stand out dramatically, however, in the raking light of a single electric bulb. Presently, at the Institut Henri Poincaré in Paris, where several dozen are displayed, the ones in cases at the centre of a room, far from windows, are most clearly revealed. In the Belle Epoque, the models were kept in cabinets in mathematical seminar rooms and classrooms (preserving them from the soot produced by gas flames which would have enhanced their shape), providing distraction and contemplation for students whose gaze wandered from the balding head of a professor writing on a chalkboard or whose eyes tired from the spaghetti of symbolic forms on a printed page. A generation of mathematics students saw concrete representations of ideal entities and became habituated to non-figural abstraction more than a decade before Picasso and Braque painted their Cubist canvasses.[38]

One highly suggestive association exists between mathematical models and Picasso's Cubism. Picasso often used a small number of objects in his compositions from the years 1910–13: violin, guitar, tobacco pipe, wineglass and wine or spirits bottle and, finally, the human face.[39] Sound, taste, scent, and the desire to touch, all wrapped up in sight. It is fair to say that he, like Raymond Duchamp-Villon, wanted to transform these objects into abstract shapes by using representations of overlapping planes and straight lines.[40] What famously resulted were objects viewed from several perspectives at once, something not normally part of our field of vision. But over the preceding generation, one of Picasso's preferred, curved objects had already been represented as an 'impossible' sculptural structure. It is the 'Klein bottle', a vessel that can contain no liquid. In the collection of more than 300 plaster casts of mathematical surfaces at the Institut Henri Poincaré in Paris, the model of the Klein bottle is one of the very few that is painted. Its cousins figure in paintings by Picasso from this time.

Picasso's education as an artist in Spain involved drawing from models of Greek and Roman statuary, whether originals or more commonly plaster copies.[41] Large-limbed, prominent-nosed, and thin-lipped statuesque women appear in Picasso's Blue and Rose period paintings, a motif recovered in Neo-Classical paintings between the world wars. Indeed, one of the most expensive paintings ever sold at auction, Picasso's *Nude, Green Leaves and Bust* (1932), features a prominent Neo-Classical bust, whether of stone or plaster; his earlier painting, *Bust of a Woman, Arms Raised* (1922), shows a monochromatic head and torso as if it were a classical statue. 'The cubists?', asked the Impressionist painter Benoît Bénoni-Auran sympathetically in 1912. 'They invented nothing. They draw in accordance with the first principles taught in school', in the drawing classes at the Ecole des beaux-arts.[42]

Just as plaster statues provided examples of ideal human proportions, so, among nineteenth-century educators, mathematical models represented ideal, non-figural harmonies. It would have been odd if there had not been a collection of such models in one of the three schools of art where Picasso studied. The young Picasso, attentive to everything around him, could have seen a Klein bottle among the plaster, mathematical models displayed for students of design in Spain or at the Conservatoire des arts et métiers in Paris for the general public. Picasso's colleague when he was pioneering Cubism, Juan Gris, in 1902–04 had studied at the Escuela de artes e industrias in Madrid, a setting likely to have featured Klein's plaster models.[43]

Widespread familiarity with the Klein bottle may relate to Futurist Umberto Boccioni's sculpture, *Development of a Bottle in Space* (1913), where the inner and outer surfaces appear to interpenetrate (Figure 9.2).

The work was exhibited as two identical or very similar plaster models, one white and one painted red and orange, in 1913 at the First

Fig. 9.2 Umberto Boccioni, *Development of a Bottle in Space*, bronze cast (dated 1950) from a plaster or clay work of 1913 (39.4 × 60.3 × 39.4 cm).

Metropolitan Museum of Art, New York. Image copyright © The Metropolitan Museum of Art. Image source: Art Resource, NY.

Exhibit of Futurist Sculpture in Paris.[44] The significance of the sculpture (we do not know whether the several versions were cast or whether they were carved), reconstituted from scraps and cast into a definitive bronze well after Boccioni's death in 1916, has been affirmed by a number of scholars. Martin Kemp takes the sculpture as an example of how the Italian avant-garde artists 'immersed themselves in the aura of new space-time physics'. Christine Poggi comments on the sculpture in a sexual discussion of rods and cavities and centrifugal force. Majorie Perloff interprets *Development of a Bottle* as exploiting 'the collage principle of juxtaposition of disparate items without any explanation of their connection'.[45] But sex is no stranger to sculpture; space-time is a particular notion formulated by Hermann Minkowski which makes no explicit appearance in the writings of any of the avant-garde clan; and Boccioni's *Development of a Bottle*, far from an assembly of diverse objects, projects integral unity. Stephen Kern is closer to the mark when he sees the sculpture in terms of Boccioni's desire to transgress conventional boundaries, expressed in Boccioni's writing: 'Let's split open our figures and place the environment inside them. We declare that the environment must form part of the plastic whole, a world of its own, with its own laws'. The sculpture is an abstract exploration of interior space – an enterprise of Modernity.[46]

During the years leading up to the First World War, Picasso's most sustained involvement with bronze is recorded in spring 1914, in a Boccioni-like project: six casts of a Cubist absinthe glass. In the sculpture, Picasso includes a strainer holding a replica of a sugar cube (the noxious green spirit absinthe was usually consumed after having been strained through a sugar cube). Presumably Picasso or the foundry made a mould, which allowed the confection of six identical wax models, each of which could then be encased in a gated plaster shell, the usual procedure in the lost-wax method of casting. The commercially manufactured strainer – tin, as described for the version possessed by the Metropolitan Museum of Art, or silver-plated, according to some accounts – was placed on the bronze form after casting. Picasso decorated each of the bronze casts distinctly with paint and, on one, with sand.[47] The result is whimsical transformation of a popular vice, depicted in paintings during the Belle Epoque (some by Picasso, for example his Cubist *Glass of Absinthe* canvas of 1911). Viewed in one perspective, the statues portray a grotesque bust topped by a hat constructed of the spoon and sugar cube (Figure 9.3).[48] There is no doubt that Picasso's statues are glasses where inside and outside interpenetrate so that no liquid can be contained.

The contrast between Boccioni's and Picasso's exploded-bottle sculptures is striking. Boccioni captures the clean lines of force beloved of Futurists. His statue offers a complex landscape of bright patches and shadows, with both sweeping arcs and straight lines. Boccioni's *Development of a Bottle* is attached to a pedestal, which is fundamental to the piece. Boccioni coloured one of his plaster confections while Picasso finished his own sculptures so that the bronze medium is obscured. Boccioni's is solid plaster, while Picasso's is a collage, where he recovers the gritty surfaces of his pre-war Cubist paintings, adding in painted dots and daubs. In contrast to Boccioni's elaborate base, Picasso's sculpture balances on a mass that might be part of the spirit glass or might be an inverted cup.

Picasso likely saw or knew about Boccioni's *Development of a Bottle* and another of Boccioni's plaster statues, a bust of his mother titled *Antigrazioso*, also appearing in the 1913 Futurist exhibit. The *Antigrazioso* seems to be Boccioni's Futurist response to Picasso's 1909 bust of Fernande Olivier, *Head of a Woman*, known as the first Cubist sculpture and shown for a long time in Ambroise Vollard's gallery. Like Picasso,

Fig. 9.3 Pablo Picasso, *Glass of Absinthe*, 1914, painted bronze and tin spoon (22.5 × 12.7 × 6.4 cm).

Boccioni was a painter who had recently turned to sculpture. With the air of composing an impromptu, in his own absinthe sculpture, Picasso seems to reject the heroic qualities radiating from Boccioni's *Development of a Bottle*. If Boccioni is celebrating the mass-produced industrial bottle,

Picasso is treating glass and strainer with impish, perhaps even satirical delight. Picasso seems to say: The ordinary is best portrayed without fuss, and there are a great many ways to understand the essence of a glass of absinthe.[49]

Merging inside and outside is in itself no innovation. The Mediterranean tradition of anatomical drawing, beginning with Islamic treatises, passing through Leonardo and Vesalius and on to papier-maché, wax and plaster medical models in the eighteenth and nineteenth centuries, does just that.[50] So do exploded and cut-away technical illustrations from the time of Leonardo and Georgius Agricola, for everything from mechanical devices and mine excavations to botanical structure – traditional modes of instruction and explanation familiar to students of medicine, engineering and natural history in the Belle Epoque. The achievement of the Cubists lies in sanitising and de-industrialising these ways of seeing, permitting them to be displayed in a bourgeois dining room, salon or library. The Cubists shared with mathematician Felix Klein a search for ideal representations of interior essences.

Acknowledgements

The text is part of a longer work, *Motifs of Modern Art and Science*, © 2018 Koninklijke Brill NV, Leiden, Netherlands, and appears here by permission. For valuable discussions, help and comments, I am grateful to Brigitte Yvon-Deyme (Institut Henri Poincaré, Paris), Richard Staley (Cambridge University), Robert Marc Friedman (Universitetet i Oslo), Paul Greenhalgh (University of East Anglia, Norwich), and especially Liliane Beaulieu (Paris). My research could not have proceeded without the efficient collaboration of the Interlibrary Loan Service at Waldo Library, Western Michigan University. I acknowledge support from the Burnham-Macmillan Fund of the History Department, Western Michigan University.

Notes

1   David E. Rowe has recently studied the plaster models in relation to a particular class of mathematical objects: 'Mathematical Models as Artefacts for Research: Felix Klein and the Case of Kummer Surfaces', *Mathematische Semesterberichte*, 60 (2013): 1–24. He cites a letter of 1883 from Alexander Brill to Karl Weierstrass, 21–22, where Brill explains that his work on plaster models was aided by his friend Joseph Kreittmayr, a plaster modeller at the Bavarian National Museum. At the time, the Bavarian National Museum focussed on natural sciences, industry and crafts in a historical light; it was seen as a counterpart of the museums in South Kensington. *Encyclopaedia*

Britannica, ninth edition (1875–1889), s.v. Munich.

2 *Catalog mathematischer Modelle für den höheren mathematischen Unterricht, veröffentlicht durch die Verlagshandlung von Martin Schilling in Halle a.S.* (Halle: Martin Schilling, 1903); Peggy Kidwell, 'American Mathematics Viewed Objectively: The Case of Geometric Models', in *Vita Mathematica: Historical Research and Integration with Teaching*, ed. Ronald Callinger (Mathematical Association of America, 1996 [MAA Notes No. 40]), 197–208; William Mueller, 'Mathematical Wunderkammern', *Mathematical Association of America Monthly*, 108 (2001): 785–96, on 791. The most sophisticated study of the models is by Herbert Mehrtens, 'Mathematical Models', in *Models: The Third Dimension of Science*, eds. Soraya de Chadarevian and Nick Hopwood (Stanford: Stanford University Press, 2004), 276–306, here 294–99.

3 Mary Beard, 'Casts and Cast-Offs: The Origins of the Museum of Classical Archaeology, Cambridge', *Proceedings of the Cambridge Philological Society*, 39 (1994): 1–29, quotations on 3; also Mary Beard, 'Cast: Between Art and Science', in *Les moulages de sculptures antiques et l'histoire de l'archéologie*, eds. Henri Lavagne and François Queyrel (Geneva: Droz, 2000), 157–66, on 159–60, the term laboratory used by Cambridge classicists in 1882.

4 Marcello Barbanera, 'Les collections de moulages au XIXe siècle: Etapes d'un parcours', in *Moulages de sculptures*, 57–73, Treu on 69.

5 Robert E. Kohler, *Partners in Science: Foundations and Natural Scientists, 1900–1945* (Chicago: University of Chicago Press, 1991), 166 for Klein's efforts before 1920 to set up a mathematical laboratory at Göttingen.

6 Felix Klein, 'Vorbemerkungen zu den Arbeiten zur anschaulichen Geometrie', in *Felix Klein: Gesammelte mathematische Abhandlungen*, 2, eds. R. Fricke and H. Vermeil (Berlin: Springer, 1922): 36.

7 Francis Haskell and Nicholas Penny, *Taste and the Antique: The Lure of Classical Sculpture 1500–1900* (New Haven: Yale University Press, 1981), 79–91; Charlotte Schreiter, *Antike um jeden Preis: Gipsabgüsse und Kopien antiker Plastik am Ende des 18. Jahrhunderts* (Berlin: Walter de Gruyter, 2014).

8 Ingeborg Kader, 'Zur Rolle der Farbe in der metalen Repräsentation: Gipsabgüsse und die Farbe "Weiss"', *Moulages de sculptures*, 121–56, on 125, citing classical archaeologist Hanna Philipp in a publication of 1996.

9 Adrian von Buttlar and Bénédicte Savoy, 'Glyptothek and Alte Pinakothek, Munich: Museums as Public Monuments', in *The First Modern Museums of Art: The Birth of an Institution in Eighteenth and Early Nineteenth Century Europe*, ed. Carole Paul (Los Angeles: Getty Museum, 2012), 305–29, on 309, quoting from Leo von Klenze in his plans of 1830 for the Glyptothek in Munich.

10 Frank Matthias Kammel, 'Der Gipsabguss: Vom Medium der ästhetischen Norm zur toten Konserve der Kunstgeschichte', in *Ästhetische Probleme der Plastik im 19. und 20 Jahrhundert*, ed. Andrea M. Kluxen (Nuremberg: Aleph Verlag, 2001), 47–72.

11 Josep Palau i Fabre, *Picasso: The Early Years 1881–1907*, trans. Kenneth Lyons (New York: Rizzoli, 1981), 51, for Picasso's having registered for classes in copying from plaster and in figure drawing from plaster at the Instituto Da Guarda of A Coruña in 1894–95; Palau reproduces the page of a textbook of Picasso's from this time where Picasso draws geometrical figures in the margins, and also, on 43, the drawing of a face in profile defined by geometrical lines. Palau comments: 'We already find something of the future creator of Cubism, albeit in embryonic form'.

12 Malcolm Baker and Brenda Richardson, eds., *A Grand Design: The Art of the Victoria and Albert Museum* (New York: Harry N. Abrams, 1997), 132; David A. Scott and Donna Stevens, 'Electrotypes in Science and Art', *Studies in Conservation*, 58 (2013): 189–98, for a recent discussion of the nineteenth-century mania for copies, focussing on electroplating. The international convention took shape in the context of considerable uncertainty about the legal rights of an artist with regard to his work. Jacques de Caso, 'Serial Sculpture in Nineteenth-Century France', in *Metamorphoses in Nineteenth-Century Sculpture*, ed. Jeanne L. Wasserman (Cambridge, MA: Fogg Museum, 1975), 1–28, on 11.

13 Arthur Beale, 'A Technical View of Nineteenth-Century Sculpture', in *Metamorphoses in Nineteenth-Century Sculpture*, 29–56; Claudia Sedlarz,

'Incorporating Antiquity: The Berlin Academy of Arts' Plaster Cast Collection from 1786 until 1815: Acquisition, Use and Interpretation', in *Plaster Casts: Making, Collecting and Displaying from Classical Antiquity to the Present*, eds. Rune Frederiksen and Eckart Marchand (Berlin: Walter de Gruyter, 2010), 197–228.

14 Winfried Neerdinger and Katharina Blohm, eds., *Architekturschule München 1868–1993: 125 Jahre Technische Universität München* (Munich: Klinkhardt & Biermann, 1993), 49–52, for the palace inspired by Semper, with illustrations. It is doubtful that any institution of higher learning today has a principal structure erected over the past fifty years with such elaborate interior decoration. Eva Giloi, *Monarchy, Myth, and Material Culture in Germany 1750–1950* (Cambridge: Cambridge University Press, 2011), 147–54, especially 149, for Berlin; also for Berlin: Hans Georg Hiller von Gaertringen, *Masterpieces of the Gipsformerei: Art Manufactory of the Staatliche Museen zu Berlin since 1819*, trans. Wendy Wallis (Munich: Hirmer, 2012), 41–44; James J. Sheehan, *Museums in the German Art World: From the End of the Old Regime to the Rise of Modernism* (Oxford: Oxford University Press, 2000), 62–70 on the Glyptothek in Munich. Also Von Buttlar and Savoy, 'Glyptothek and Alte Pinakothek, Munich'.

15 According to the history of the collection, http://www.abgussmuseum.de/, accessed June 2018.

16 Klaus Borchard, 'Grusswort', in *Gips nicht mehr: Abgüsse als letzte Zeugen antiker Kunst*, eds. J. Bauer and W. Geominy (Bonn: Akademisches Kunstmuseum/ Antiken Sammlung der Universität Bonn, 2000), 7, on one of the leading collections known to Jacob Burckhardt, Karl Marx and Friedrich Nietzsche.

17 Many photographs of the nineteenth-century cast collections are found in Charlotte Schreiter, ed., *Gipsabgüsse und antike Skulpturen: Präsentation und Kontext* (Berlin: Dietrich Reimer, 2012), notably: Johannes Bauer, 'Gipsabgüsse zwischen Museum, Kunst und Wissenschaft: Wiener Abguss-Sammlungen im späten 19. Jahrhundert', 273–90, the Museum für Kunst und Industrie in 1871 on 276; Martina Dlugaiczyk, 'Gips im Getriebe: Abguss-Sammlungen an Technischen Hochschulen', 333–54, the Zurich Polytechnic circa 1880 on 335, the Delft Polytechnic around 1900 on 345; Marjorie Trusted, 'Reproduction as Spectacle, Education and Inspiration: The Cast Courts at the Victoria and Albert Museum, Past, Present and Future', 355–71, the V&A Cast Courts in the 1870s, with high gas fittings, on 367. Cf. the art display at the South Kensington Museum, circa 1857, a plaster cast of Michelangelo's *David* illuminated by a candle chandelier. Malcolm Baker, 'The Reproductive Continuum: Plaster Casts, Paper Mosaics and Photographs as Complementary Modes of Reproduction in the Nineteenth-Century Museum', in *Plaster Casts*, 485–500, on 487. An accessible and short survey in Charlotte Schreiter, trans. Catherine Framm, 'Competition, Exchange, Comparison: Nineteenth-Century Cast Museums in Transnational Perspective', in *The Museum is Open: Towards a Transnational History of Museums 1750–1940*, eds. Andreas Meyer and Bénédicte Savoy (Berlin: Walter de Gruyter, 2014), 31–43.

18 Suzanne L. Marchand, *Down from Olympus: Archaeology and Philhellenism in Germany, 1750–1970* (Princeton: Princeton University Press, 1996), 149, 366.

19 Trevor Fawcett, 'Plane Surfaces and Solid Bodies: Reproducing Three-Dimensional Art in the Nineteenth Century', *Visual Resources: An International Journal of Documentation*, 4 (1987): 1–23, 17–19 for Cole's enterprise and for the eclipse of plaster and electrotype copies by photographic images.

20 Rune Frederiksen and Eckard Marchand begin their comprehensive edited volume by addressing these points: 'Introduction', *Plaster Casts*, 1–20; also in *Plaster Casts*: Diane Bilbey and Marjorie Trusted, '"The Question of Casts": Collecting and Later Reassessment of the Cast Collections at South Kensington', 465–83, on 468–71. Kader, 'Rolle der Farbe', 125; Beard, 'Cast', 159–60; Penelope Curtis, *Sculpture 1900–1945* (Oxford: Oxford University Press, 1999), 73–79, on how emphasis on associating a work of art directly with the hand of the artist (and not his assistants or his duplicators) led to a declining focus on casting and a renewed interest in carving. The discussion, continuing with head-scratching about 'ready-made' art and mass-produced art, erupted into controversy at museums when Modernity

ended, in the 1980s. Alexandra Parigoris, 'Truth to Material: Bronze, on the Reproducibility of Truth,' in *Sculpture and its Reproductions*, eds. Anthony Hughes and Erich Ranfft (London: Reaktion, 1997), 131–51, on 131.

21 Dlugaiczyk, 'Gips im Getriebe', 351.

22 Kader, 'Rolle der Farbe', 133.

23 Marchand, *Down from Olympus*, 143–44, Brunn's rectoral address of 1885 titled 'Archäologie und Anschauung'; Nietzsche on 124–27; Langbehn on 150–55.

24 John Theodore Merz, *A History of European Thought in the Nineteenth Century, III, Part Two, Philosophical Thought* (New York: Dover, 1965), 205, 445, for a comparison of *Vorstellung* and *Anschauung*. Originally published 1912.

25 Felix Klein, trans. M. W. Haskell, 'A Comparative Review of Recent Researches in Geometry [1872]', *Bulletin of the New York Mathematical Society*, 2 (1892–1893): 215–249, on 244, Non-Euclidean geometry on 245; I have slightly modified Haskell's translation. Among other things, Haskell uses 'space-perceptions' and 'space-perception' for the German expressions 'räumlichen Vorstellungen', 'räumlichen Anschauung', and even 'Anschauung'. Klein, *Vergleichende Betrachtungen über neuere geometrische Forsuchungen* (Erlangen: A. Deichert, 1872). Fundamental is Thomas Hawkins, 'The *Erlanger Programm* of Felix Klein: Reflections on its Place in the History of Mathematics', *Historia Mathematica*, 11 (1984): 442–70. See also: Garrett Birkhoff and M.K. Bennett, 'Felix Klein and his "Erlanger Programm"', in *History and Philosophy of Modern Mathematics*, eds. William Aspray and Philip Kitcher (University of Minnesota Press, Minneapolis, 1988): 145–76. In my *Neohumanism and the Persistence of Pure Mathematics in Wilhelmian Germany* (Philadelphia: American Philosophical Society, 1983), I stated incorrectly on 55 that it was not published. As Hawkins writes, 'The *Erlanger Programm*', 444, although the 48-page address did appear in 1872 as part of the so-called programme literature (*Programm zum Eintritt in die Philosophische Facultät und den Senat der K. Friedrich-Alexanders-Universität zu Erlangen*), it was not widely available until reprinted many years later; it appears in Klein's collected mathematical papers. On the centrality of models in the Erlangen Programme,

Mehrtens, 'Mathematical Models', 288–89, quotation on 290.

26 Slavik Vlado Jablan, *Symmetry, Ornament and Modularity* (Singapore: World Scientific, 2002), 264; Dorothy K. Washburn and Donald W. Crowe, *Symmetries of Culture: Theory and Practice of Plane Pattern Analysis* (Seattle: University of Washington Press, 1988), 43–78, for an earlier synoptical discussion.

27 Erika Luciano and Clara Silvia Roero, 'From Turin to Göttingen: Dialogues and Correspondence (1879–1923)', *Bollettino di storia delle scienze matematiche*, 32 (2012): 9–220, for a comprehensive study of the fruitful interaction between Felix Klein and the mathematicians of Turin, Italy.

28 Solomon Feferman, 'Mathematical Intuition vs Mathematical Monsters', *Synthese*, 125 (2000): 317–332, cites Henri Poincaré's discussion of pathological functions in *Science et méthode*, and he observes that the term 'monster' was popularised later by Imre Lakatos in *Proofs and Refutations* (Cambridge: Cambridge University Press, 1976). Cf. Klaus Thomas Volkert, *Die Krise der Anschauung: Eine Studie zur formalen und heuristischen Verfahren in der Mathematik seit 1850* (Göttingen: Vandenhoeck & Ruprecht, 1986), 99, referring to Poincaré in *Science et méthode* and also to Poincaré in 1889 on the perversity of interest in exceptional functions.

29 Neo-Impressionist Max Liebermann's portrait of Klein in 1912 shows a confident man looking out across the universe of thought. In the painting, there is no suggestion of the asymmetries figuring in early twentieth-century art. Rüdiger Thiele, *Felix Klein in Leipzig* (Leipzig: Edition am Gutenbergplatz, 2011), 245–46 for the portrait, whose cost of 5000 RM was covered by Klein's friends.

30 Karl-Norbert Ihmig, 'Ernst Cassirer and the Structural Conception of Objects in Modern Science: The Importance of the "Erlanger Programm"', *Science in Context*, 12 (1999): 513–29, on 515–16, 523.

31 Mehrtens, 'Mathematical Models', 290.

32 Lewis Pyenson, *The Young Einstein: The Advent of Relativity* (Bristol and Boston: Adam Hilger, 1985), 178–83; Pyenson, *Neohumanism and the Persistence of Pure*

Mathematics in Wilhelmian Germany, 53–108.

33 Douglas Crimp, *On the Museum's Ruins* (Cambridge, MA: MIT Press, 1993), 301.

34 Falk Heinrich, *Performing Beauty in Participatory Art and Culture* (New York: Routledge, 2014), 63–65, where Heinrich points to the continuing tradition of 'wonder'; Anna Zakiewicz, 'To Inform or to Delight? The Vision of a Modern Museum in Changing World', in *Proceedings, International Conference on Social Sciences and Society, Shanghai, 14-15 October 2011*, ed. G. Lee, 264–67, see www.witkacy.hg.pl/opieka/teksty/ICSSS2011/404.pdf, accessed June 2018. We should remember that picture galleries were a feature of the seventeenth-century, both aristocratic and bourgeois. One prominent controversy highlighting the two views of museums is found in the *Enola Gay* exhibit at the National Air and Space Museum of the Smithsonian Institution in the 1990s. Lewis Pyenson, 'Western Wind: The Atomic Bomb in American Memory', *Historia Scientiarum*, 6, no. 3 (1997): 231–41. See Lewis Pyenson and Susan Sheets-Pyenson, *Servants of Nature: A History of Scientific Institutions, Enterprises, and Sensibilities* (London: HarperCollins/New York: W.W. Norton, 1999), 125–49, on museums.

35 Felix Klein's organisational work with the schoolteachers, discussed in terms of the international commission he helped create, in Gerd Schubring's analysis, 'The First Century of the International Commission on Mathematical Instruction (1908–2008)', eds. Fulvia Furinghetti and Livia Giacardi, see www.icmihistory.unito.it, accessed June 2018.

36 Felix Klein, 'Ueber Arithmetisirung der Mathematik', *Nachrichten von der Königlichen Gesellschaft der Wissenschaften zu Göttingen, Geschäftliche Mitteilungen*, 2 (1895): 82–91, on 83–84, 88–91; analysed in Pyenson, *Neohumanism*, 61, and in Volkert, *Krise der Anschauung*, 90. Hans Hahn, *Empiricism, Logic, and Mathematics: Philosophical Papers*, ed. Brian McGuinness (Dordrecht: Reidel, 1980), the introduction by Karl Menger. In this book is found Hahn's discussion from 1933 of counter-intuitive curves: 'The Crisis in Intuition', 73–102.

37 Mehrtens, 'Mathematical Models', 279–80, 290–91.

38 The point is made by Jack Burnham, *Beyond Modern Sculpture: The Effects of Science and Technology on the Sculpture of This Century* (New York: George Braziller, [1968]), 125. Eva Migirdicyan, *Les modèles mathématiques dans l'art du XXe siècle* (Doctoral thesis, Université de Paris Ouest Nanterre La Défence, 2011), 2 vols, for a long survey, focussing mostly on the period after 1918; on 223 for Naum Gabo's studies in mathematics and engineering at the Munich Institute of Technology in 1912; on 275 for Futurist Giacomo Balla's possible familiarity with the models in his studies at the Albertina School of Fine Arts, Turin, in 1894–95. Gerhard Fischer, *Mathematische Modelle* (Plates) and *Mathematical Models: From the Collections of Universities and Museums*, 2 vols (Braunschweig: Vieweg, 1986); Angela Vierling-Claassen, *Models of Surfaces and Abstract Art in the Early 20th Century*, see http://angelavc.files.wordpress.com/2011/06/bridges-paper-jan-2010.pdf, accessed June 2018.

39 Anne Umland, *Picasso Guitars 1912–1914* (New York: Museum of Modern Art, 2011), for a record of the paintings and sketches.

40 In a gibe directed against Art Nouveau, Duchamp-Villon claimed to seek 'a predominance of the straight line over the curved line, a predominance pushed to the extent of tyranny'. Marie-Noëlle Pradel, 'La Maison cubiste en 1912', *Art de France: Revue annuelle de l'art ancien et moderne*, no. 1 (1961): 177–186, on 183.

41 Almudena Negrete Plano, *La colección de vaciados de escultura que Antonio Rafael Mengs donó a Carlos III para la Real Academia de bellas artes de San Fernando* (Doctoral thesis, Universidad complutense de Madrid, 2012), for the extraordinary collection of classical, plaster models acquired by the Madrid academy where Picasso studied. During the nineteenth century, the Fine Arts Academy shared space with the Natural History Museum; the association dissolved around 1895, at the end of Picasso's time as a student there.

42 Mark Antliff and Patricia Leighten, eds., *A Cubism Reader: Documents and Criticism, 1906–1914*, trans. by Jane Marie Todd, Jason Gaiger, Lydia Cochrane, Lois Parkinson Zamora and Ivana Horacek (Chicago: University of Chicago Press, 2008), 240. Paul Greenhalgh, *The Modern Ideal: The Rise and Collapse of Idealism in*

the *Visual Arts* (London: V+A Publications, 2005), 222, for the Neo-Platonism of modernist artists in the years before 1914.

43 Antliff and Leighten, *A Cubism Reader*, 213.

44 Maria Elena Versari, '"Impressionism Solidified": Umberto Boccioni's Works in Plaster', in Frederiksen and Marchand, *Plaster Casts*, 331–350. Versari's discussion of Boccioni's *Development of a Bottle* is illuminating and careful. In El Lissitzky and Hans Arp, ed., *Die Kunstismen 1914-1924* (Erlenbach-Zürich: Eugen Rensch, 1925), 42, a photograph of Boccioni's sculpture is titled 'Bottle and Space' and dated 1911. It appears to be uncoloured plaster or clay, and it lacks the base of the bronze version.

45 Martin Kemp, *Art in History, 600 BC–2000 AD* (London: Profile Books, 2014), 167–68; Christine Poggi, *Inventing Futurism: The Art and Politics of Artificial Optimism* (Princeton: Princeton University Press, 2009), 168–69; Marjorie Perloff, *The Futurist Moment: Avant-Garde, Avant Guerre, and the Language of Rupture* (Chicago: University of Chicago Press, 1986), 55.

46 Stephen Kern, *Culture of Time and Space, 1880–1918* (Cambridge, MA: Harvard University Press, 1983), 197–98. Cf. Sigfried Giedion, *Space, Time and Architecture: The Growth of a New Tradition* (Cambridge, MA: Harvard University Press, 1941), 365–66, where Giedion sees the sculpture beginning, as Boccioni writes mystically, 'from the central nucleus of the object wanting to create itself, in order to discover those new forms which connect the object invisibly with the infinite of the apparent plasticity and the infinite of the inner plasticity'.

47 Doris Lanier, *Absinthe: The Cocaine of the Nineteenth Century* (Jefferson, NC: McFarland, 1995), 118–122.

48 The magisterial work on Picasso's sculpture: Werner Spies and Christine Piot, *Picasso: Das plastische Werk* (Stuttgart: Gerd Hatje, 1983), here 71–74, where the point is made about the grotesque head.

49 A point made in Jean Laude, *La peinture française (1905–1914) et 'l'art negre': (Contribution à l'étude des sources du fauvisme et du cubisme)* (Paris: Klincksieck, 1968), 1, 463, a work still pertinent for its general insight. Cf. other commentary: Perloff, *Futurist Movement*, 32, for Boccioni's sculpture in the context of nationalist rivalry; Jeffrey Weiss, *The Popular Culture of Modern Art: Picasso, Duchamp, and Avant-Gardism* (New Haven: Yale University Press, 1994), 239, where Picasso's absinthe sculptures are compared to illustrations on public sign-boards; Laude, *Peinture*, 1, 378–79, where the absinthe sculptures are said to bring 'transparency' to European sculpture, a trait held to derive from African art; Parigoris, 'Truth to Material', 143, where the absinthe sculptures are a slap in the face to bourgeois sensibility. It should be recalled that the addition to sculptures of material artefacts or their double, in the manner of Picasso's absinthe spoon, was no stranger to the Third Republic, from Edgar Degas's *Little Dancer, Fourteen Years Old* (1881) to Frédéric-Auguste Bartholdi's *Aeronauts and Homing-Pigeon Trainers of the Siege of Paris* (1906). Spies and Piot, *Picasso*, 72, make the point about Degas's little dancer; June E. Hargrove, 'The Public Monument', in *The Romantics to Rodin: French Nineteenth-Century Sculpture from North American Collections*, eds. Peter Fusco and H. W. Janson (Los Angeles/New York: Los Angeles County Museum/George Braziller, 1980), 21–35, a photograph of Bartholdi's destroyed bronze (his last major work), with its large balloon, on 28.

50 As currently displayed in a number of collections: Gretchen Worden, ed., *The Mütter Museum of the College of Physicians of Philadelphia* (New York: Blast, 2002); Gerard Tilles and Daniel Wallach, eds., *Le Musée des moulages de l'hôpital Saint Louis* (Paris: Doin, 1996); Walter Tega, ed., *Guide to the Museo di Palazzo Poggi, Science and Art* (Bologna: Museo Poggi, [2007]).

10
Architecture, science and purity

Judi Loach

Introduction

My concern here is architecture rather than construction – the role accorded to science in the aesthetic rather than technological aspect of buildings – in the avant garde around the Great War and beyond. It explains why science played a key role in some modern architectural theory, but how it only did so after being mediated through theories developed in other art forms, most notably painting but also music; and that it did so, at least in part, as a result of personal association with artists from other genres, through a broad range of scientific theorising, including psychological theories of sensual perception and its mental impact. It focusses on the work of the most influential theorist and designer of the period, one working across the fields of fine art and architecture: Le Corbusier (1887–1965).

My narrative tells how a young and hitherto unknown Swiss artist, Charles-Edouard Jeanneret (the future Le Corbusier), found, in certain Parisian avant-garde circles at the end of the Great War and the years that followed, a preoccupation with science – most overtly mathematics – which offered artists a way of vindicating the importance of their contribution to, and thus validating their own position within, a society that saw science as the inevitable means for 'becoming modern'. Seeing that respected figures embraced such ideas gave Jeanneret the confidence to further develop his own thinking along these lines, first within the two-dimensional sphere of painting, with the development of Purism, then with the translation of these theories into the three-dimensional forms of architecture, notably the design of villas. As he, together with his co-founder of Purism, Amédée Ozenfant, explained in their manifesto for it, they used 'the term "Purism" to express the characteristic of the modern spirit intelligibly within a single word',[1] implying that their new aesthetic movement would be that most appropriate for all the arts – eventually including architecture – in the modern age. They enunciated these theories in greatest detail with regard to painting, on the grounds that it was best suited to serve as the exemplar for all arts, since it was 'the least problematic art of the age'.[2] The implications of their theories would simply be assumed in their transfer to architecture, and consequently the bulk of this chapter deals with painting rather than architecture, in order to present certain somewhat hidden foundations of Le Corbusier's architectural theory. However, the usually overlooked presence of his older brother Albert, a musician turned eurhythmics practitioner, can be seen as playing a crucial role in introducing this Purist duo both to aesthetic theories derived from recent experimental science and to a wider belief in an intrinsic relationship between cosmic laws and human artistry, moreover one applicable to practitioners across all the arts.

Towards Purism

Charles-Edouard Jeanneret settled permanently in Paris in 1917 (seemingly oblivious of effectively moving to a war zone from a neutral country).[3] He had long been dissatisfied with his provincial home town of La Chaux-de-Fonds, then the world capital of watchmaking, dominated as it was by the commercial interests of bourgeois patrons.[4] He had been fixated on returning to Paris, the metropolis and milieu where he had spent over a year just a decade earlier,[5] effectively as a student, since, although employed by France's leading Modernist architect, Auguste Perret (1874–1954), whose technically advanced office pioneered concrete construction, he had spent half of each day in museums or in the Prints and Drawings department of the Bibliothèque Nationale.[6] Perret combined technical advances in structures with a sense of order and proportion intrinsic to the Classical tradition, which he had inherited from his own Beaux-Arts training, but whose forms he had much simplified, in large part due to working in this material: one poured into formwork around orthogonally arranged reinforcing bars. On his return visits to Paris, at least from 1913 onwards, Jeanneret had been accepted into the avant-garde Club artistique de Passy, a group founded in 1912 by the Symboliste poet Guillaume Apollinaire (1880–1918)[7] and Perret's brother-in-law, the writer Sebastian Voirol (1870–1930); meeting weekly in Perret's office, it brought together architects, artists, writers, actors, dancers and musicians committed to endorse and promote 'contemporary art and modern works by the present generation'.[8] It was evidently through them that he came to know Apollinaire personally,[9] and that he was already attending performances by the pre-eminent progressive theatre company of the day, Jacques Copeau, from the week that it opened in the newly renamed Théâtre du Vieux Colombier in autumn 1913.[10] On taking on the Athénée-Saint-Germain theatre, Copeau had commissioned the modern architect François Jourdain to remodel it, removing the proscenium arch and introducing a set of wide steps to either side of the stage so as to link it with the auditorium, thereby creating a space that could facilitate those developments that he sought in performance, thus producing an innovation of architectural as well as dramatic interest. Significantly, on settling in Paris in 1917 Jeanneret took long-term lodgings (in the event, 15 years) in the home not of any artist or architect but of one of the writers from the Club artistique de Passy, the Symboliste poet and playwright Charles Vildrac (1882–1971),[11] who had co-founded the artists' 'phalanstery' of the Abbaye de Créteil (1906–8) with his brother-in-law, the writer Georges Duhamel, the Cubist painter Albert Gleizes and others.[12] While then moving into the theatre, Vildrac was simultaneously involved in the modern art scene, running his own commercial gallery from 1910 to 1930, where he sold works by painters such as Seurat and Vlaminck; moreover, he covered the walls of his own flat with paintings that Jeanneret deemed 'the most beautiful and strongest of modern art', set off against walls painted in bright yellow or blue.[13] Jeanneret was constantly aware of the activities of Copeau's theatre company, as Vildrac was one of its principal playwrights and his son was also working in the company;[14] Jeanneret was evidently influenced by Vildrac's taste in contemporary art too, preferring the works of painters promoted by his gallery to those by well-known Cubists, notably Picasso and Metzinger.[15] Over the next couple of years, while setting up office as a promoter of concrete construction, Jeanneret would concentrate on becoming a great artist, soon working with a painter from Voirol's group, Amédée Ozenfant (1886–1966).[16]

In 1918 this duo was joined by Charles-Edouard's older brother Albert (1886–1973). The trio now worked, ate and holidayed together for several years, during which they simultaneously practised modern art and theorised on its translation into modern living, their plastic and corporeal explorations becoming the creative tools for testing and refining their aesthetic theories. Albert, however, was not a visual artist but a musician, and moreover one who had become an exponent of Emile Jaques-Dalcroze's *rythmique*.[17] This form of eurhythmics was initially developed in the light of contemporary physiology as a therapy for alleviating musicians' physiological problems engendered by excessive practising,[18] Albert himself abandoning a career as a solo violinist for this reason. *Rythmique* had subsequently developed as a type of (danced) performance in its own right, especially after Jaques-Dalcroze entered into a fruitful working relationship with the radical theatrical designer Adolphe Appia (in 1906), and then been invited by Wolf Dohrn (in 1910) to establish his institute in purpose-built facilities, designed by Heinrich Tessenow, at the Deutsche Werkstätte's garden city of Hellerau.[19] Consequently, Albert had been on stage there in 1913 for the legendary performance of Gluck's *Orphée*,[20] with ground-breaking, abstract sets by Adolphe Appia (1862–1928)[21] and lighting by Alexander Salzmann (1874–1934);[22] this was attended by Europe's leading dramatists, directors and stage designers, composers and writers, dancers and artists.[23] It is unclear whether Charles-Edouard was there (although he was certainly present at some rehearsals), as he had arranged to attend the performance in July, only to discover Claudel (whose *L'Annonciation faite à Marie* was to complete the programme) postponing it until autumn.[24] However, Charles-Edouard may have met Appia several years before, in 1910, while visiting Albert at Dalcroze's institute as soon as it moved from Geneva.[25] He is also likely to have known Salzmann, as Albert lodged with him, at least in 1912.[26] Given Charles-Edouard's subsequent visits to his brother at Hellerau, it seems likely that these relationships continued at least until the outbreak of the Great War.

Esprit Nouveau

In 1918 Jeanneret and Ozenfant, with the support of some avant-garde colleagues but in reaction to others, published *Après le Cubisme*,[27] effectively the first manifesto for their own artistic movement: Purism. And on the flyleaf they announced a series of subsequent publications, notably one on architecture that they claimed (as we shall see, impossibly) already in press; that treatise, *Vers une architecture* (misleadingly translated into English as 'Towards a new architecture'), would only appear in 1923, and then under Jeanneret's new name alone: Le Corbusier.[28] Meanwhile, these theories would be propagated through other publications,[29] above all through their Modernist magazine *L'Esprit Nouveau*, which they launched in 1920 with the support of the writer Paul Dermée (1886–1951),[30] already experienced in editing avant-garde magazines,[31] and the financial backing of subscribers encompassing practising artists and potential patrons. As the rubrics soon introduced for regular columns would indicate, the magazine would not only cover the arts – visual and performance arts, architecture and literature – but also science and (separately) 'applied science', *économique sociologique*, sport and aesthetics, the latter being considered as a philosophical – and even scientific – discipline.[32] Le Corbusier's famous architectural manifesto, *Vers une architecture*, in fact derived from the architectural articles in early issues of the magazine.

While *esprit nouveau* ('new spirit') was a common term at the time, it seems that they were referring to Apollinaire's usage, as it appeared in Paris soon after Jeanneret (who had already met the poet within Voirol's circle) settled there. Apollinaire's essay '"Parade" et l'esprit nouveau' appeared in a programme that Le Corbusier owned and – unusually for him – carefully preserved to the end of his life,[33] for Paris performances by Diaghilev's Ballets Russes in May 1917. In discussing the company's most avant-garde creation, 'Parade', which synthesised modern dance with sets by Picasso and music by Satie, it praised Picasso's sets and referred to Cubism's 'esprit nouveau', specifically because Apollinaire believed that Cubism had brought progress in art up to the level of that in the sciences and industry.[34]

Finally, Apollinaire's essay 'L'Esprit nouveau et les poètes' was 'performed' in a lecture-recital, the essay being 'illustrated' by poems read by actors, at the Théâtre du Vieux Colombier in late November that year,[35] and most probably attended by Jeanneret; it was published in *Mercure de France* the month after.[36] Here *esprit nouveau* was equated with 'the spirit of the time in which we live, a time fertile in surprises',[37] and was presented as 'above all, the enemy of aestheticism, formulae and any snobbishness' – evidently setting it up in opposition to the academies; indeed, this *esprit nouveau* had no intention to become a school but rather a great current encompassing all schools. 'It fights for the re-establishment of the spirit of initiative, for the clear understanding of its time and for opening up new visions of the universes both outside us and within, visions no less than those that scholars of all kinds discover each day and from which they draw out wonders'.[38] Its principal characteristics were its search for truth, in ethics as well as in imagination,[39] such that the artist's role in society was not only as creator but also as prophet.[40] While this 'new spirit' inherited the Romantics' sense of curiosity, exploring any fields that could help it to enrich all modes of life, it did not cut itself off from the past. Above all, it inherited from 'the classics', 'a solid common sense, an assured critical sense, holistic views of the universe and of the human soul, and a sense of duty' that restrained the display of feelings[41] (ideas with which Perret and his neo-Classically trained but Modernist-inclined colleagues would agree).[42] Apollinaire criticised the excesses of Italian and Russian avant gardes, specifically as lacking the necessary sense of order, which, along with the 'classical spirit', he saw as virtues intrinsic to French culture. It was in fact, overall, a rather patriotically French manifesto, from one of the many immigrants within the artistic avant garde in Paris, and evidently had a profound influence upon that milieu, including Ozenfant and the Swiss émigré Jeanneret, but perhaps most particularly Apollinaire's quasi-disciple, the Belgian émigré poet, Paul Dermée.

Purism vs Cubism

Less than a year later, in mid-October, the Ozenfant–Jeanneret duo's manifesto, *Après le Cubisme*,[43] presented their own new artistic movement, 'Purism', as a necessary advance on Cubism, which they portrayed as a movement of the pre-war period: a 'troubled art' reflecting a 'troubled time'.[44] By contrast, with the war now over, 'everything is getting organised, is becoming clearer, is getting purified … nothing is as it was before'[45] – except art, which now needed to root itself into this new reality if it was to survive (this last sentiment echoing Apollinaire's feelings about *esprit nouveau*). They claimed that laws – those of nature – govern human thought, and that scientific research was now, more than ever before, 'establishing' what these laws are, thus supplanting the Romantics' 'revelations'. To become appropriate for its time, art had to fit this new 'era of Science', whose spirit was 'a tendency towards rigour, precision, the best use of forces and materials … in short a tendency towards *purity*' (my emphasis).[46] Art needed neither to be machine-made nor to take machines as its subject, but needed, at a more fundamental level, to understand them, to absorb their spirit of 'rigorous facts, rigorous configurations, rigorous architectures of form, as *pure* and simple as those of machines' (my emphasis).[47] Ozenfant and Jeanneret would sub-sequently claim that 'pure science' and 'pure art' share a common 'spirit' – insofar as both are 'dependent upon number':[48] science seeks constants that express natural laws, while art seeks comparable 'invariants' (thus implying their contrast with Cubism, through their intrinsic connection with an ongoing aesthetic tradition, rather than radical break from that). In science and art alike, laws are thus human constructions, but ones that reflect the natural order. As human artefacts these 'laws' can be represented by numbers and lines, thereby replacing any mystical explanation of the universe, to the extent that that 'Nature is like a machine ... Geometry, physical and mathematical, defines the laws of forces, which are like main lines of order'.[49] The authors insisted that 'the laws of order are the laws of harmony', and that whereas defining 'the world's great ordering lines' led the scientist to formulations in number and lines (graphs), the artist's task was to express this universal, natural order through forms. Since scientist and artist alike would arrive at their visible but abstracted expressions through the same series of mental processes – induction, analysis, organisation and reconstitution – art can arrive at a 'certitude', equivalent to the 'probability' in scientific laws. And that 'certitude' is itself beauty.[50]

In all of this Ozenfant and Jeanneret were echoing Perret and Apollinaire, but they now went further, emphasising the parallel between scientific and aesthetic analysis, and thus justifying the utilisation of scientific methods for discovery of aesthetic laws.[51] In this age of science, painters needed to analyse their subject matter – nature – as science did, in both cases in order to understand its underlying order. Then, in a section significantly entitled 'Mechanism of emotion',[52] they defined the task of the artist, especially in this modern age, not as copying what they saw – the object provoking his or her emotions – as that would merely be equivalent to photography; instead they were to materialise that emotion provoked by the object that they sensed. The artist deserved this role due to his/her superior capacity as a 'resonator' (*résonateur*): to their sensitivity to the vibrations emitted by the emoting object that they sensed.[53]

This implied a model of human sense perception and cognition that would be spelled out in detail in subsequent publications, beginning in a little-noticed article just a couple of years later (1921).[54] Here, working from the premise of the universality of automatic physiological reactions to certain plastic forms, the picture would be defined as 'a machine for inciting emotions'.[55] Since the simplest geometrical forms produce the 'purest' (i.e. most direct) reactions in the human spirit, art must base itself on the universal language of geometry. Humans experience the greatest feeling of delight when perceiving the order underlying Nature, and in finding their own place within it; as a means of (re)creating such order, the work of art becomes a masterpiece of human creation, specifically by arousing in the spectator that sense of mathematical order that underlies all Nature. Whereas mathematics can only operate through symbols, and thus depends upon engaging the conscious intellect, art operates more directly, and not only upon the intellect but more broadly on the human spirit. Any artwork is thus an artifice with potential to put its spectator into the state desired by its creator. Obviously such a claim for the determinist potential of artworks has implications that deserve, even demand, teasing out, and since the authors seem to have preferred to leave these implicit, we will return to this later.

Ozenfant and Jeanneret's manifesto continued by stating that, since 'purified' forms are more effective in enabling any picture to realise its determinist potential, the artist needs to work with such 'purist' elements. Such a 'purist' form is not a copy of any natural form but an artist's creation, aiming to materialise the generality and invariability of any object. This concept of forms that capture the timeless essence of a thing through its carefully selected and simplified representation in material form has a dual resonance. First, it is most directly apparent in Ozenfant and Jeanneret's own 'Purist' paintings, where a single, most archetypal example of a bottle stands for all bottles, the most archetypal example of a glass for all glasses, and so on.[56] In Purist paintings these visual archetypes depicted *objets types* – the forms most commonly taken by such utensils because they had proved best adapted to their function – the bottles, glasses and so on used in one's local café. As such they had become the most widely used – and therefore produced – examples of their genre, demonstrating a kind of 'survival of the fittest', to transfer (as was frequently done in that period) a biological term to the realm of human production. Second, this concept of an ideal 'type' was a commonplace among industrial designers of the time, who generally believed that improvement of design quality depended above all upon being prepared to focus resources on the refining of an exemplary model before putting it into mass production. This was an approach with which Jeanneret would have been extremely familiar, due to working under Peter Behrens (1868–1940), a founder member of the Deutsche Werkbund, in Berlin for nearly half of 1910, and then again his happier contacts with the Deutsche Werkstätten at Hellerau, while visiting his brother there during 1910–4. This concept endowed the human inventor or maker with agency, becoming the active creator of the *objet type*, and thus paralleled by the role now accorded to a Purist painter, who was to identify such *objets types* and then create two-dimensional representations for them.

The ideas first articulated in *Après le Cubisme* were then reiterated, but invariably less explicitly, in the duo's subsequent articles. In their *L'Esprit Nouveau* article on 'Painting's destinies' (1923 – the same year as Jeanneret's major architectural tract, *Vers une architecture*, would appear),[57] they claimed that 'the distinctive quality of our age is to aim at perfect collaboration between sensation and emotion'.[58] They insisted that any painting could only impact upon the brain through physiological emotions transmitted by means of the visual sense, implying that such paintings therefore included not only those with narrative subjects but also, and at least equally, abstract ones. In this context their claim – that 'modern man knows that the only profound realities are those that move us directly'[59] – acquires a particular significance, and enables a shift to a broader argument; this again echoes one already announced in *Après le Cubisme*, justifying the shift from an aesthetics based in metaphysics to one on scientific foundations: '[Modern man] needs these idealised certainties which religion provided in earlier times; doubting religion and metaphysics, man is now taken back to himself, and the true world happens inside him.'[60] These ideas underlie, but are never made explicit, in the duo's final joint statement on Purism, in *La Peinture moderne* (1925),[61] but reappear more explicitly in early 1926, soon after their split.[62] In an article 'On Cubist and post-Cubist schools', written by Ozenfant alone, the advances of Cubism and its successors are presented in terms of emancipation from 'imitative' and 'symbolic' art, which depended upon naturalistic copying of 'scenes' in order to recreate the emotions originally evoked by those 'scenes'; instead the 'sensualist' art of these new schools acts directly upon the spectator's senses and emotions, through the *lyrical* effects of the artist's play of forms and colours.[63] In listing the reasons why such art appeals to 'modern man', Ozenfant virtually condensed much of the rhetoric reiterated throughout the duo's various Purist proclamations: he 'has habits of order, of clarity, of brightness, which make him want the same qualities in lyricism itself'.[64] But Ozenfant now goes further than before, in admitting an innate determinism (with Purist paintings as his implicit exemplar), by defining any work of art as 'an essential mechanism, intended to put today's subject into a physico-psychological state required by the artist'.[65] He concludes, returning to an idea first outlined in *Après le Cubisme*, by underlining the greater efficacy of non-narrative painting, in that its ability to move the spectator is independent of personal taste or ephemeral fashion.[66] As we shall see, the fact that this article was written specifically for an academic journal in psychology is not inconsequential; similarly, *L'Esprit Nouveau*'s initial subtitle – 'Revue internationale d'esthétique'[67] – would prove to be far from insignificant.

From science to art

In fact, the ideas underlying Purism were not original by now, but rather reflected those being advocated by a group of aestheticians who were just beginning to gain acceptance within a sector of the Paris establishment when Jeanneret first arrived in Paris. At the Sorbonne, the first Chair of Aesthetics – specifically a 'Chair of Aesthetics and Science of Art'– was established in 1920, and would be filled by a series of philosophers committed to psychophysics, while at the École Pratique des Hautes Études (EPHE), a Laboratoire de psycho-physiologie existed from 1889 and a Laboratoire de physiologie des sensations was established in 1897; both these laboratories, however, were located within the new Sorbonne building, thus encouraging dialogue between researchers in the two institutions. In setting out the remit of *L'Esprit Nouveau* within the preliminary pages of its first issue, the 'new spirit' was described as 'that which animates all scientific research', and the magazine's overall aim as the elaboration of 'an experimental aesthetic'.[68] The editors then claimed to belong to a group of aestheticians who believed that art 'has laws like [those of] physiology or physics',[69] contrasting themselves with others whose aesthetics was based on metaphysics, a position that they considered discredited by recent (and in fact not so recent) experience. In declaring their own commitment to an 'experimental aesthetics', they stated that it would exploit 'the same methods as experimental psychology'.[70]

In this they were aligning themselves with certain contemporary philosophers in Paris specialising in aesthetics, notably Charles Henry (1859–1926),[71] Victor Basch (1863–1944), Paul Souriau (1852–1926), Charles Lalo (1877–1953)[72] and Etienne Souriau (1892–1979),[73] all of whom sought to develop a new aesthetics, on scientific foundations;[74] indeed, it was to this end that Lalo and Etienne Souriau would eventually (in 1948) found the influential *Revue de l'Esthétique*. The very first two issues of *L'Esprit Nouveau* carried an article commissioned from Basch,[75] specifically on 'Aesthetics and the Science of Art',[76] just as he was taking up the newly established Chair of that name at the Sorbonne.[77] Basch argued that aesthetics should base itself on physics and physiology, as opposed to the older form of aesthetics, based on metaphysics; a related 'science of art' would analyse physical sensations so as to derive scientific laws about how artworks arouse pleasure in their viewers (or hearers). He therefore advocated exploiting psychological and physiological experiments, evidently drawing on the research carried out by various pioneers of experimental psychology, working in Germany from the mid-nineteenth century onwards,[78] most notably Gustav Fechner (1801–87), the founder of psychophysics,[79] but also Wilhelm Wundt (1832–1920), who had established the first laboratory for experimental psychology, at Leipzig in 1879.[80] In 1908 Fechner's work had been signalled within a significant

French-language publication on contemporary aesthetics (by none other than Lalo),[81] while in 1911 Wundt, as pioneer of experimental psychology, had been the subject of a French language monograph;[82] thus the work of both these German pioneers of the new field was brought to the attention of a relatively wide public in France in the years immediately preceding the Great War. Indeed, it is clear from Ozenfant's subsequent criticism of Fechner that he was aware of that German's aesthetic theories.[83] German developments in psychology, which had themselves informed the ideas of Fechner and Wundt, and which similarly adopted an anti-metaphysical approach based on physical experiments, had already been brought to a French readership, in 1879, through Théodule Ribot's *La Psychologie allemande contemporaine.*[84]

In turn, Charles Henry would contribute to *L'Esprit Nouveau* a long article on 'Lumière, couleur et forme', effectively the culmination of his written output; split over no less than four issues of the magazine, it was in fact a lecture delivered at the Sorbonne under the auspices of the avant-garde theatre group Art et Action[85] (itself closely related to the Cercle artistique de Passy). Henry, initially a historian of mathematics and a musicologist but progressively with an interest in psychophysics, was librarian at the Sorbonne until, in 1892, taken on by Haute Étude's Laboratoire de psychologie-physiologie, where he worked until the Laboratoire de physiologie des sensations was created there especially for him.[86] Reputed for his encyclopaedic knowledge and his simultaneous interest in contemporary science and arts, Henry moved in Symbolist and post-Impressionist circles, his scientific theories of colour exerting considerable influence on several painters, notably Paul Signac (who provided the illustrations for Henry's *Quelques aperçus sur l'esthétique des formes*, 1894–5)[87] and Georges Seurat (an artist promoted by Vildrac). Charles Lalo, who in 1933 would succeed Basch in the Chair of Aesthetics at the Sorbonne, would also contribute an article to the magazine, on 'Aesthetics without love', in other words, exemplifying an objective approach, as opposed to the then standard one, derived as that was from the Romantics.[88] It was in fact Lalo who, in 1908, had brought Fechner's psychophysics to the attention of the French, in invoking the German's research in support of his own thesis that visual perception psychologically conditions aesthetic sensation;[89] he had also been expected to contribute another article to the magazine, specifically on Fechner and experimental aesthetics[90] (the subject of his doctoral thesis),[91] but it never appeared. Paul Souriau, who held Chairs at the universities of Lille and then Nancy, was again a philosopher specialising in aesthetics who advocated the exploitation of psychophysics.[92] He too had much influenced painters, most notably Robert Delaunay, through his *L'Esthétique du mouvement* (1889);[93] moreover, it has been claimed that Le Corbusier, while still a student in his home town of La Chaux-de-Fonds, had probably become familiar with some of his works, most likely *La Beauté rationnelle* (1904) but possibly also his *L'Esthétique du mouvement*.[94] Paul's son Etienne, likewise a philosopher specialising in aesthetics (and again on a scientific, experimental basis),[95] held chairs at various universities, and from 1941 at the Sorbonne, where he succeeded Lalo in the Chair of Aesthetics. In 1948 Lalo and Etienne Souriau would play leading roles in the creation within this department of the Centre d'Etudes Philosophiques et Techniques du Théâtre, an institute that would invite Le Corbusier to present a paper at its inaugural symposium,[96] implying at least that the department felt his views to be sympathetic to their own, and probably that they had been in contact for some time; in turn, Le Corbusier appreciated this as arguably providing him with the most significant sign of acceptance by any French academic establishment that he would ever receive.

Golden Number and Golden Section

In *Après le Cubisme*, however, while tacitly retaining such deterministic concepts of how visual perception effects aesthetic appreciation, the emphasis shifts towards the ability of the human senses – common to geometrician and artist – to enable the human mind to apprehend 'universal harmony'. This harmony, being universal, enables the great classics of art, from the ancient Greeks or even before, to be presented as acceptable models for art in the modern world, insofar as they exemplify their accordance with the underlying laws of nature:

> The canons of antiquity, that are generally held to be artificial codes, or templates, were nothing other than founded on the correct understanding of the universality of natural laws that both govern the external world and determine the work of art. It was not codes, but rather correct yet mutable laws, that permit linking works by humans with those of nature.[97]

Ozenfant and Jeanneret claimed that 'the most ancient civilisations' (such as Egyptians, Greeks and Persians) had known these 'canons' – notably numerical relationships – as had 'the Goths', and that these had been rediscovered in the Renaissance but often misapplied from then on, being used as if just a set of strict regulations.[98] Likewise, today's artists should realise that science's discoveries are schematic, and should not be taken as rules to be followed literally: 'Science draws our attention to the orders that it discovers; science incites the artist to discover those new beauties defined by such order ... The scholar discovers harmony, source of beauty; the artist takes from that whatever is good for art.'[99] Progress in science and in beauty thus run in parallel; science and art collaborate mutually in enabling humankind's progress.

Despite the title and content of the Purist manifesto, this new movement evidently owed much to Cubism, especially in its conviction that mathematics had to play a fundamental role in art for the machine age. As would soon become apparent, Purism's rather vague references to geometry, 'harmony' and proportions were in fact principally to a series of constructions based on the Golden Section. This would be spelt out more clearly in the preliminary pages of the first issue of *L'Esprit Nouveau*, immediately after setting out the magazine's remit: that of elaborating an 'experimental aesthetics', exploiting 'the same methods as experimental psychology' so as to reveal 'laws' inherent in art equivalent to those of physiology or physics. That text continued, first, by citing, as 'an example of the true order at which we are aiming', the 'results of research [pursued by] one of us on the "Golden Section"', defined here as 'an extremely simple mathematical relationship'. It then proceeded to allege the Golden Section's ability to 'satisfy our aesthetic requirements', claiming the support of laboratory experiments for this, adding that the capacity of this proportional system to incite emotional responses had been proven invariable across social class and age of participants; this was presumably referring to Charles Henry's work at his laboratory investigating 'the physiology of sensations', but Fechner, whose psycho-physics led him to hypothesise the deterministic character of geometric forms and mathematical proportions, had investigated the effect of Golden Section earlier.

In fact, such implicit linkage of current French experimental psychology, inspired by German psychophysics, with the Golden Section proves to have been quite widespread in Parisian avant-garde circles in the early twentieth century. The most famous exponent there was Matila Ghyka (1881–1965), publishing on this from the late 1920s onwards,[100] and who referred explicitly to Fechner's psychophysical investigations into the Golden Number.[101] Le Corbusier has often been portrayed as a follower of Ghyka,[102] several of whose works he is known to have owned, and even annotated.[103] However, Ghyka was only publishing his own work in this area from 1927 onwards,[104] and Le Corbusier thus seems to have preceded Ghyka in adopting such ideas, having probably become aware of the Golden Section through Cubists whom he met through Parisian groups such as Perret's Cercle artistique de Passy, with which he had contact during the First World War if not even before. Certainly the relationship seems to have been less one of Le Corbusier taking Ghyka as his master, and instead rather one of mutual respect;[105] in 1948 Ghyka would write an article for none less than *Architectural Review* in praise of Le Corbusier's Modulor, his later development of a system of measurement integrating measurements from an idealised human body with Golden Number proportions;[106] Le Corbusier would only publish his own book expounding this system a couple of years later,[107] implying conversation between the two prior to that.

Second, the Purist manifesto claimed furthermore that this same Golden Section was equally 'one of the dependable measures passed on by artists of past times'. As such – on these two grounds – it enabled artists – implicitly, including those of today – to operate without resorting to the subjective and therefore variable criterion of taste; this was because, unlike the arbitrary *rules* laid down by the Académies des Beaux-Arts and their like, the Golden Section was no less than a *law* of aesthetics, since it was 'a physical and mathematical law, perceptible to our senses', and as such an element of the experimental aesthetics currently being developed through scientific experiments. Hence *L'Esprit Nouveau* committed itself to following 'laboratory work in aesthetics' alongside, and with equal interest as in, 'the experiments' being undertaken by modern artists, writers, painters, musicians and engineers'.[108]

In fact, the group of artists broadly described as Cubist prior to the Great War had effectively branded themselves as the 'Section d'or', most conspicuously through a group exhibition under that name, in 1912.[109] For Jeanneret, whose entrance into Parisian avant-garde circles from 1913 onwards had initially been through Voirol's Club artistique de Passy, Cubism would have seemed very different from how it is commonly perceived today – a movement largely defined by art dealers, and notably those with exclusive sales rights over Picasso and Braque. While these dealers presented those artists' works in their own galleries, and thus predominantly to a select audience of potential buyers, in the larger-scale, more public exhibitions Cubism was represented by a much broader range of tendencies; this encompassed artists such as Gris, Metzinger, Gleizes, Léger, Delaunay, Laurencin and Brancusi, but at times also involving certain Russian Constructivists (most notably Archipenko) and Italian Futurists – hence Apollinaire's criticism of their particular forms of avant garde – as well as members of the Dutch De Stijl movement, and future founders of Dada, such as Picabia and not least the Duchamps themselves (not to mention many more minor artists).[110] This larger, looser grouping of 'Cubists' was essentially a group of friends who met convivially each Sunday at the home and studios of the Duchamp brothers at Puteaux, or sometimes in the studio of Albert Gleizes at nearby Courbevoie. Naturally these artists shared those interests then current in Paris: African art, chrono-photography, non-Euclidean geometries and the visual representation of the fourth (temporal) dimension. In fact many, perhaps most, of these artists also attended the Passy group, meeting in Perret's office each Tuesday evening. Moreover, Apollinaire was arguably the leading spokesperson in both, and like the Passy group the Puteaux one encompassed writers and performers, including at times members of the previous Abbaye de Créteil, such as Jeanneret's future landlord Vildrac.

The Puteaux group seems to have adopted the Section d'or name for several reasons. This geometric construction served to illustrate their emphasis – one intrinsic to Apollinaire's concept of *esprit nouveau* – on how any art fit for the modern era should not break with the past but rather develop from the purest and most basic essence of Western artistic tradition, reaching back to antiquity and even beyond. In other words, art should now reassert certain universal and timeless principles, which endured because of their resonance with Nature itself – and thus, implicitly, with Man, as the most superior form of animal; as such, incorporation of the Golden Section could enable man to recover a sense of harmony with the cosmos that had been ruptured by unbridled industrialisation. At the same time, the 'laws' of 'Nature' were being revealed to a greater extent and degree than ever before, through scientific research. Initially one of the Duchamp brothers had been attracted to the Golden Section as a result of reading Leonardo da Vinci's treatise on painting in its 1910 translation, a less than neutral one as it was by the self-styled 'Sar', Joséphin Péladan, thereby ensuring that the Golden Section became endowed with somewhat esoteric significance. Perhaps the group's adoption of an actuary, Maurice Princet, as its 'mathematician' was intended to make them seem more objective, 'scientific'.[111] However, although some of these artists rigorously applied the Golden Section in their compositions (perhaps most notably Juan Gris and Jean Metzinger),[112] most did not.

By the time that Jeanneret and Ozenfant came to write their own manifesto, much of the supposedly scientific underpinnings of Cubism were being ridiculed, especially the possibility of representing the fourth dimension within a flat painted surface. One can therefore understand why their Purism now distanced itself from Cubism, and indeed overtly mocked its use of the fourth dimension, emphasising instead the need to portray the 'invariant', or universal. And hence also the somewhat covert nature of their references to the Golden Section – which they would soon promote in architectural form.

From painting to architecture: from Golden Section to 'regulating lines', and from looking to living

Given the interdisciplinary context of avant-garde groups in which the Ozenfant-Jeanneret trio operated – the Institut Jaques-Dalcroze, the Section d'or, the Club artistique de Passy, Copeau's Théâtre du Vieux Colombier and, not least, 20 rue Jacob,[113] where both brothers lived – it was inevitable that concepts developed within any art genre might be transferred to another. While *Après le Cubisme* mainly referred to 'art' in terms of painting, its subsection 'L'Esprit moderne' turned instead to architecture.[114] Here the recent constructions embodying this 'modern spirit' – those where harmony pervaded, because each element derived 'from a definite rigour, from the respect for and application of laws'[115] – were bridges, factories and dams, whose clarity of form enabled their viewers to 'recognise an underlying intent that had been clearly formulated'.[116] These artefacts – whose grandeur was claimed here as evoking Roman antiquity – were not designed by architects but by engineers and builders. Concrete was eulogised, here not so much as the latest material or technique but rather as the means that enabled 'the rigorous achievement of calculation; Number [with a capital 'N'], which is the foundation of all beauty, can from now on express itself.'[117] It was because machines were 'determined by number'[118] that they had evolved more rapidly than architecture, and now attained 'a remarkable *purification*' (my emphasis).[119] The 'intelligence' behind some machines makes them move humans, due to the 'proportions of their *organs* [sic] rigorously determined by calculation [and] the precise execution of their elements' like 'a projection of natural laws'. Underlying their theory is a conviction that the organic world of nature runs like a machine, according to laws as predictable as those discerned in physics. 'Order rules because nothing is left to fantasy'.[120] So far, engineers have brought us today's equivalent of Roman aqueducts, but soon architects will be able to create buildings comparable to the Greeks' Parthenon. Indeed, the methods and means of modern industry are going to enable us to fulfil an ideal of perfection of which the ancient Greeks could only dream.

When, in 1923, Le Corbusier – as he had now become – published *Vers une architecture* (claimed in the 1918 Purist manifesto to be in press but actually drawn from his subsequent articles in *L'Esprit Nouveau*) those ideas about architecture already expressed in the Purist manifesto five years earlier pertained. The term 'Purist' has since been applied to his ten villas built during this period of his career – between 1922 (the commission for the Villa Ker-Ka-Re at Vaucresson) and 1929 (the completion of the Villa Church at Ville d'Avray)[121] – due to their visual correspondences with Purist paintings: their limited range of clearly delineated forms, formal simplification, and rigorous compositions hinting at geometrical calculations on the architect's drawing board.[122]

In the case of the movement's hallmark villa, that commissioned by the wealthy Swiss banker Raoul La Roche (1923–5),[123] the implied fulfilment of Purist theory, hitherto only realised in two-dimensional paintings, was now made manifest not in three-dimensional architecture alone but simultaneously through painting and architecture together, understood here as complementary media.[124] This was underlined inside by the selection of artworks made by the architect for his client – dominated as it was by those of Jeanneret, Ozenfant and Léger. Furthermore, it was furnished with *types*, in the sense of objects which, regardless of age or provenance, represented exemplars of evolved design. Such objects fulfilled their practical functions while simultaneously serving as graceful forms, thus articulating the carefully proportioned spaces so as to fulfil an aesthetic function: bentwood Thonet chairs, like those habitually used in Parisian cafés; deep leather armchairs reminiscent of an English gentleman's club; traditional handwoven rugs, such as those from the Maghreb (widely imported into France at the time), laid on quarry-tiled floors typical in French vernacular buildings; bare light bulbs suspended from ceilings, the opaque white of their curvaceous forms highlighted by the contrast with the verticality of their thick black flexes; and even the mass-produced yet traditional table lamp common to public libraries, with its practical glass lampshade, diffusing light optimally for reading purposes. Any modern furniture introduced was of the simplest design, but elegantly proportioned, as was perhaps most notable in the tables – a simple rectangular slab of solid wood supported by a square of chrome standing on four chrome legs. All these were set against plain, pastel-painted walls and light (unbleached calico?) curtains. The effect was as if a two-dimensional Purist painting had been translated into three dimensions – and such that one could now enter into it. Moreover, as this was a villa, in which Raoul La Roche was to spend his time outside the bank where he worked, this was intended as a space for living in, in the sense of dwelling. It was devised quite literally to *frame* modern life so that it could be viewed clearly; or rather, it was to frame *a* modern life: that of an exemplary modern man, patron of the most modern painters and architect. And as such this building, called 'Villa' rather than 'Maison', was also a gallery into which fellow connoisseurs of modernity would be invited, yet not merely to view modern paintings but to experience a modern way of living.

In fact, this was not actually a single villa but rather a semi-detached pair of villas, albeit of unequal size, La Roche's being the larger of the two. Significantly the other villa was designed expressly for Albert Jeanneret, as he had now married a (relatively) wealthy Swedish singer, Lotte Raaf. In other words, the 'Villa Jeanneret' was designed as a space in which this disciple of Jaques-Dalcroze would work as composer of modern musical art, but equally where he and his family would incorporate *rythmique* into their daily regime. As such, this villa again becomes an exemplar not only of Purism in the sense of an art form to be viewed but as a lifestyle: by becoming a 'Purist' one would be enabled to act out modern life. Here one begins to understand how these buildings could be understood as implicitly translating into practice the determinist theory inherent within Purism – the theory most clearly enunciated in texts published after the initial manifesto *Après le Cubisme*, notably in Ozenfant and Jeanneret's 'Esthétique et purisme' or in Ozenfant's final essay, in a psychology journal, 'Sur les écoles Cubistes et post-Cubistes'; if a Purist painting was capable of producing those specific physiological, and subsequently intellectual, reactions in its viewer that the artist had built into their work devised purely for viewing, how much more effective would a three-dimensional, all-encompassing artwork be in enforcing the artist's desired responses upon anyone actually engaging with it – to the extent of physically entering into it and carrying on their everyday life inside it? And within this context of artwork as place of dwelling – home – should it be understood as a deliberately contrived mechanism for not merely enabling a human being to live a modern life but forcing one to do so?

In a 1921 article in *L'Esprit Nouveau*, Jeanneret and Ozenfant had gone further than the original Purist manifesto in another significant respect: in overtly promoting the Golden Section, but carefully avoiding that term with its then Cubist resonances, instead using the term *tracés régulateurs* (usually translated as 'regulating lines').[125] They thereby fulfilled, albeit silently, the promise hinted at in the preliminary section of the first issue of *L'Esprit Nouveau*, where the incorporation of the Golden Section was implicitly linked with recent scientific research into 'the physiology of sensations', with all that might imply for determining the reactions of those encountering modern artworks; this association would have been subtly reinforced by the fact that Lalo's article promoting an objective, scientifically based aesthetics began in the same issue,[126] while the first part of Henry's article, in the same vein, appeared in the following issue,[127] alongside the second part of Lalo's article. However, in their 1921 article they illustrated these *tracés régulateurs* as applied to buildings (including Le Corbusier's own Villa Schwob in his native La Chaux-de-Fonds)[128] rather than to their paintings;[129] as if to reiterate the point, in the following issue a review of this building, purportedly by another author, again used it to demonstrate this principle of *tracés régulateurs*.[130] Only in a further article, published in the magazine the next summer, do they impose such 'regulating lines' over a couple of their own paintings.[131] Le Corbusier's architectural manifesto *Vers une architecture* (1923, but actually a collection of articles published earlier in *L'Esprit Nouveau*) then devotes an entire chapter to the *tracés régulateurs*.[132] Nevertheless the introductory paragraph there introduces it in Purist language: 'The obligation of order. The *tracé régulateur* is an insurance against the arbitrary. It provides satisfaction for the spirit.'[133] A history of architecture from earliest times follows – his own approved canon of hypothetical primitivism, Greek antiquity, ancient Persian, French Gothic, Michelangelo in Rome and early French Classicism – case by case claiming consonance with this preferred geometric construction by overlaying such 'regulating lines' onto small photos or schematic diagrams.[134] For any critical reader the crudity of these means somewhat belies Purism's supposed rigour of method, and indeed that of this geometric construction. But Le Corbusier then went further, overlaying these *tracés* onto elevations his own buildings,[135] thereby implying his own use of such *tracés* as a design tool. Moreover, through such a presentation he incorporated it into the ongoing canon, thus accentuating his implicit assertion that his readers too should now employ it in buildings they design or commission. Yet by citing as exemplar a villa built in Switzerland in 1916, and thus prior to meeting Ozenfant let alone their joint invention of Purist theory, one suspects that this geometrical construction is being superimposed a posteriori. Furthermore, Tim Benton has shown that even in his 'Purist' villas Le Corbusier only rarely used *tracés régulateurs* for generating the design, instead more commonly applying them later to justify his intuitive sense of composition.[136] However, the visual appearance of *Vers une architecture*, whose almost collagesque composition and striking exploitation of typefaces recalls Apollinaire more than conventional architectural treatises, should have alerted readers to the fact that it is rather a propagandist tract, like so many avant-garde publications of the period. This chapter dictates compliance with a geometric construction that supposedly ordered the masterpieces of an enduring Western tradition rooted in classical antiquity, one ensuring their harmony because it embodied Nature's own ordering system.

This reflexive relationship between built dwelling and paper tract perhaps reaches its apogee in the Pavillon de l'Esprit Nouveau, erected within the 1925 Exposition des arts décoratifs to advertise the magazine of the same name. Le Corbusier's 'Immeubles Villas' ['Villa Blocks'] project of 1922 – in which instead of a block of single-storey flats he proposed stacking duplex villa-plus-garden units into a block – remained unbuilt. So he now presented one such unit in built form as the pavilion, suitably furnished as a modern home – with built-in furniture, office equipment … and Purist paintings – thus following the model set by the Villa La Roche, albeit at a more economical scale, so as to provide a prototype for Everyman. And he simultaneously published a complementary book with a photo of the built unit as its front cover,[137] while the pavilion itself plastered an enlarged version of the magazine cover over one facade. Although marginalised within the exhibition, it enabled the scientific, mathematical principles developed as the aesthetics of Purism within the plastic arts and then transferred to architecture to be viewed within a setting attracting at least as many visitors – and thus viewers – interested in those fine arts as in architecture. And to enable those visitors, if albeit fleetingly, to experience modern living.

Eurhythmic art and the rhythms of life

Evidently Jeanneret/Le Corbusier was remarkably keen to transfer these preoccupations with 'scientific' underpinnings for aesthetics – ones as much psychophysical as purely mathematical – into the realm of architecture. This exceptional dedication to exploring both disciplines – painting and architecture – simultaneously would remain a constant throughout his life, and creative production, as Eduard Sekler noted in reflecting on one of his last works, the Carpenter Centre for Harvard University:

> It is clear that with his commitment to painting, Le Corbusier among all architects of his generation was predestined to create

works that in architecture would display characteristics also applicable to paintings; after all he gave continuous architectural demonstrations of the 'synthesis of the major arts' – not in the sense of a superficial resemblance or coordination but in the sense of a fundamental identity of method and consequently of inner structure.[138]

Granted, Le Corbusier was unusual in operating in both architecture and plastic arts, where (as we have seen with the Section d'or group) certain preoccupations with geometrical constructions were already being pursued. But in his own case, and that of his collaborator Ozenfant, it seems too that he was taking seriously claims that the harmonious relations induced by integrating such geometry into buildings would affect more than just those buildings but also those viewing them, and would do so as automatically as would viewing a picture: 'a machine for inciting emotions'. In other words, what differentiated the Ozenfant–Jeanneret duo from their artist contemporaries was not merely their interest in, even integration of, the psychophysical theories propounded by certain scientist and aesthetician contemporaries, but their application of such theories to architecture. This combination of interests, taken together with Jeanneret/Le Corbusier's involvement in the design of spaces in which to carry out daily life, seems intended as a means of creating a new entity: a human being prepared for life in an industrialised and urban future yet simultaneously in harmony with the unchanging measures of nature, one who by thus regaining inner harmony of body and spirit would acquire greater energy.

The shadowy presence of the third member of this Modernist trio, Charles-Edouard's older brother, the musician and eurhythmics practitioner Albert, tends to be overlooked by art and architectural historians,[139] but most probably offers the explanation for this unique preoccupation. From his arrival in Paris in 1918 until he moved into the villa attached to La Roche's in 1923, Albert lived in the same building as his brother, and indeed shared most meals and holidays with the Purist duo – in other words, nothing less than their daily life. He was thus present when the original Purist manifesto was being written, through the first three years of *L'Esprit Nouveau* and up to the composition of *Vers une architecture*. Moreover, he contributed to virtually half the issues of the magazine, albeit mainly in the form of reviews of recent performances (ballets, operas and concerts).[140] His most extensive article, split over two issues (the second and third, no less) and entitled 'Rythmique' – summarising Jaques-Dalcroze's theory – was most revealing. Despite appearing under *L'Esprit Nouveau's* 'Music' rubric, this, like the developed form of *rythmique* itself, had a more than strictly musical relevance, dealing not so much with rhythm in music as with rhythm as the basis of all arts and even life itself. Within the context of industrialised society the human body is presented as a 'rhythmic machine', and the re-establishment of its natural rhythm becomes the means by which harmony will be established between body and spirit, thereby regaining energy, at once physical and mental (or even spiritual). *Rythmique* thus acquires a secular but quasi-metaphysical status, as a method to be followed by individuals but so as to thereby enable society to attain the form of organisation appropriate at once to both man's advanced evolutionary state and the conditions of modern (implicitly urban and industrial) life. The further advance of human society thus implicitly depends upon each individual's conscious attempt to rediscover one's own body's innate rhythms. This article won Albert considerable praise from the master himself, Jaques-Dalcroze, affirming that it encapsulated his own ideas very well.[141]

If Jaques-Dalcroze had not originally thought of his *Rythmique* in these broader terms, he had been led to do so by Wolf Dohrn, whose invitation to establish an institute at Hellerau stemmed from his own interpretation of *Rythmique* (in turn supported by Karl Schmidt, director of the Werkstätten) as an instrument of social reform, and indeed as source of the nation's politico-cultural renaissance; by establishing an institute for Dalcrozian *Rythmique* at its heart, the Werkstätte's garden city could become a 'German Olympia' – not merely an exemplary model for further garden cities but the national centre for cultural events that transcended the usual interdisciplinary boundaries among the various arts, between the arts and the sciences, and between body and spirit.[142] More immediately, Dohrn believed that the practice of *Rythmique* would heal the rupture caused within each individual by industrialisation's dehumanising mechanisation of work and life; it would thereby bring inner harmony to each of Hellerau's citizens, in turn inducing a comparable harmony within the new community. Dohrn's interpretation of *Rythmique*, and in particular his belief in its power to transform Hellerau's citizens into *Zukunftsmenschen* ('people of the future'), was informed by his reading of Karl Bücher's *Arbeit und Rhythmus* (1896),[143] in which that economist saw the 'derhythmatisation' of society and of each human within it as symptomatic of the rupture of a natural equilibrium, and which he blamed on modern civilisation and its fragmentation of life.

As Albert moved into a villa designed by his brother according to Purist theory, itself grounded in principles derived from current experimental science, so Charles-Edouard regularly attended Albert's *Rythmique* classes,[144] which derived from principles actually at least as rooted in contemporary scientific thinking, in this case about the human body, in other words, human physiology; yet at the same time *Rythmique* was equally indebted to a belief in its integral interrelationship with aesthetic sensibility and creativity, in other words psychology, thus echoing the psychophysical preoccupations of the Sorbonne aestheticians. It seems to me that Purism's particular incorporation of psychophysical theories is most likely due to Albert's presence, and any appreciation of its character here requires an understanding of the wider context of Dalcrozian *Rythmique*, and Albert's personal exposure to it. Jaques-Dalcroze had initially developed this body of theory and practice in response to the needs of his students, aspiring artists in musical practice, but he had not developed his ideas alone, within an exclusively musical milieu. Rather, these ideas derived from theories shared by an interdisciplinary group of artists to which he belonged, the Genevan Cercle indépendant, founded in 1887, which brought together all manner of artists: painters, writers of all sorts – poets, novelists and journalists – and musicians. It arguably constituted the leading avant-garde milieu in all Switzerland at this moment of nascent Modernism, its principal painter being none other than the nation's most internationally recognised of the day, Ferdinand Hodler (1853–1918),[145] while Jaques-Dalcroze was its most eminent representative for music. Its two key literary figures were Louis Duchosal and Mathias Morhardt, who while less well known to art historians today were at least as influential in their own day. The close interrelationship of its members is illustrated by Hodler's portraits of its members, most memorably Morhardt (1911),[146] and by Jaques-Dalcroze's settings of several poems by Duchosal – interdisciplinary collaborations.[147] In addition, all these established artists invited younger colleagues, including their own students, into the Cercle's meetings. Albert may have attended these informal gatherings, but even if he had not, as a disciple of Jaques-Dalcroze he can hardly have been unaware of their ideas.

Morhardt (1863–1939) was a Swiss journalist, literary critic and editor, who made his reputation in Paris whilst still in his early twenties, becoming editor of the Parisian daily *Le Temps* at the tender age of 25.[148] He wrote for art magazines sympathetic to the Symboliste movement, such as *La Vogue* and *Mercure de France* (the magazine that would publish Apollinaire's essay 'L'Esprit nouveau et les poètes'), becoming explicitly recognised as a member of the Symboliste group. Yet he is best remembered today as the contemporary promoter and biographer of the sculptress Camille Claudel (1864–1943), Rodin's one-time collaborator and lover.[149] Despite having moved to Paris early on he always retained close links with Geneva, and used his dual location to promote Swiss artists – most notably Hodler and Vallotton – by writing eulogistic reviews of their work in the Parisian press; as one of the leading defenders of Hodler's controversial painting *Night* when it was shown in Paris in 1891, the year after Geneva forbade its public exhibition, he was instrumental in securing Hodler's financial security, and subsequently his recognition by the French state, in an Officership of the Légion d'Honneur (in 1913). Morhardt was also a member of the Groupe néo-Malthusienne, meeting in Geneva from 1908–14 (but which had actually originated in Le Corbusier's home town of La Chaux-de-Fonds). Most importantly, however, he was an ardent Dreyfusard and an initiator of the Ligue des Droits de l'Homme, of which he was the second Secrétaire Général, from 1898–1911 (the year in which Hodler painted his portrait, seemingly to commemorate his period of office),[150] then serving on its Comité Central from 1913 to 1936. In other words, Morhardt was playing a major role internationally in political and intellectual circles. However, within our current context, it is even more significant that during the 1880s and early 1890s he participated in the Paralléliste group in Geneva, founding with them the literary magazine *Revue de Genève* in 1885; its contributors would include French writers such as the Catholic visionary Léon Bloy (1846–1917) and the Symboliste critic Charles Morice (1880–1919) – alongside Morhardt himself. With its promotion of Wagner's music, Verlaine's poetry and painters such as Hodler, it was the leading Symboliste periodical in Switzerland, demonstrating that a new cultural spirit had arrived in the Suisse Romande. Moreover, it attracted the attention of Symboliste magazines in Paris, notably *La Plume* (1889–1914); during this period Morhardt was also attending the artistic and literary soirées of the Genevan Cercle indépendant, held in Hodler's studio. Subsequently another *Revue de Genève* (1920–30) – which promoted international understanding especially between French and Germans, and regularly included a section devoted to the current activities of the Société des Nations – and in which Morhardt was again involved – sent a copy of each issue to *L'Esprit Nouveau*, where it also paid for full-page advertisements, thereby becoming one of its major supporters.[151]

Nevertheless, the member of those soirées who most determined the direction followed by its participants was a poet who never left Geneva and who is largely forgotten today, Louis Duchosal (1862–1901).[152] The son of working-class immigrants from Savoy – his father working mainly as a small-time, self-employed builder – Duchosal left school at thirteen for an apprenticeship with a lithographer. Ironically his lack of progress there, combined with the onset of the paralytic condition which would lead to his early death, enabled him to return to school and to indulge his passion for reading. At about sixteen he began writing poetry and attracted the attention of an established poet, Charles Bonifas, who introduced him to the artistic and literary discussions held each Sunday at the home of an older and better-known poet, Edouard Tavane. Simultaneously he benefitted from informal tutoring from the avant-garde poet Louis Tognetti, who led him to a more methodical choice of reading, and refined his taste. In 1882, together with Tognetti, he founded the short-lived satirical – and politically radical – magazine *Le réveil de Pippo*, in which he showed his opposition to Naturalism and affinity for Idealism, but also his appreciation of working- and artisan-class culture. The following year he continued to publish in this vein – as poet, short story writer and critic – in another magazine, *Le Foyer*, for which Bonifas also wrote. In 1885 Tognetti became the literary critic for *Le Genevois*, and secured Duchosal commissions there. Gradually this led to his writings being accepted by a large number of magazines throughout the Suisse Romande and beyond (in Lyons and Paris, and even in Rome). Most significantly, in 1889 he briefly edited the *Revue de Genève*, that momentous magazine which had been cofounded by Morhardt. Despite his physical disability, Duchosal became one of the very few writers in the Suisse Romande actually able to live off his writing (and indeed to support his parents from it as well). Yet his physical deterioration became marked from 1887, when Friedreich's ataxia was eventually diagnosed; the depressing knowledge of his forthcoming paralysis – of legs, then arms, larynx, throat and eyes – without loss of consciousness increased his introspection and related commitment to Symbolisme. Although already respected as a poet, Duchosal only began to publish his poems in book form in the last decade of his life, with two well received collections during his lifetime, two further collections being published posthumously;[153] he also wrote several plays,[154] but their success proved more ephemeral. His reputation spread throughout Francophone Europe, thanks largely to compatriots who had settled in Paris: Morhardt, Wagnon, William Ritter (whom Le Corbusier would seek out and indeed take as mentor), Louis Montchal and Léo Buchelin. However, he only enjoyed success briefly, as his advancing illness confined him first to a wheelchair, then to his flat, soon leaving him hardly able to speak.

Yet it was Duchosal who, through his own translation of German works, introduced the Cercle indépendant to Parallelism, a body of philosophical and aesthetic theory initially formulated by Fechner, in fact within psychophysics. This was based on a conviction in the parallel, and indissoluble interaction, between spirit and body, an idea first developed by Spinoza, and now effectively rediscovered as it was seen as pertinent within the context of industrialised society, which was widely perceived as responsible for shattering this natural relationship. Parallelism would in turn fundamentally change the direction of the work of all three principal colleagues – Hodler, Jaques-Dalcroze and Morhardt – thus affecting their three respective artistic disciplines: painting, music and literature. Parallelism generated theories in a number of fields of human enquiry, ranging from philosophy itself through biology and psychology to literature. For instance, in biology Parallelism described how developmental characteristics of groups of animals or plants can be expected to respond to similar environmental constraints. Although Parallelism had been implicitly present in ancient Greek philosophy, and subsequently in medieval philosophy, it had first been made explicit in Spinoza's *Ethics*. The concept then came to the fore due to Wundt's psychological experiments in the mid-nineteenth century, which stimulated the rise of psychophysical Parallelism. In transferring this concept from science to culture – indeed from biology to aesthetics – the Genevan Symbolistes promoted a quasi-metaphysical belief that the inherent order underlying all of nature, including man himself, was best expressed through repetitive patterns. Their poets suggested or even emphasised formal relationships inherent within their own creative work by use of repetitive patterns. Art historians have widely observed that the effects of Parallelism became evident in the paintings Hodler executed from the 1890s, in their idealisation of landscapes and a comparable tendency towards the iconic in his painting of human figures;[155] however, they have tended to recognise neither his translation into visual form of the underlying philosophical tenets of Parallelism in general nor indeed the Genevan *Paralléliste* group's particular understanding of Parallelism, despite Hodler's own writings providing evidence of his far more than superficial interest in this philosophy.[156] Nevertheless, the impact was probably most marked in the case of Jaques-Dalcroze, but in this case the debt to psychophysics, and especially to Parallelism, has been acknowledged in recent scholarship.[157]

It is significant that it was within this context of the Suisse Romande – between French and German cultures – that Duchosal translated Parallelist philosophy from German into French, and brought it to bear upon aesthetics, as early as the mid-1880s. This was over twenty years before its comparable translation would take place in Paris, since it would only be in 1908 that Charles Lalo would do so, in bringing Fechner's work in psychophysics to the notice of French audiences; for in his *L'Esthétique expérimentale contemporaine,* Lalo invoked Fechner in support of his own thesis that visual perception physiologically conditions aesthetic sensations.[158] Whilst Parallelism would also exert an influence on aesthetics in Parisian avant-garde circles, it would therefore not do so there until the twentieth century, so that Jaques-Dalcroze and his students, including Albert Jeanneret, were exposed to it, and indeed developing their own theories and practice from this basis, well before it reached Parisian contemporaries. Most significantly, at least within the context of this essay, many of the same characteristics are equally hallmarks of Parallelism and of Purism. It is most probably through this Genevan context, which motivated Albert to leave conventional professional practice as a solo violinist so as to fulfil avant-garde aspirations as practitioner of *rythmique*, that the Purists had become aware of psychophysics, and had done so specifically in the form of Parallelism.

Conclusion

Hitherto Le Corbusier's debt to scientific theory has focussed on mathematics, through his commitment to the Golden Section or Golden Number, leading to his promotion of *tracés régulateurs* and, in later (post-war) works, of his Modulor, both as mathematically based devices essential for creating harmonious designs.[159] Yet, as this chapter has demonstrated, his debt to new scientific thinking was not solely in the realm of mathematics but also, and even more fundamentally, in that of psychophysics. Moreover, his contact with, and understanding of these ideas probably derived at least in part to the transmission of early experimental psychology from Germany via Geneva, of which his musician brother is likely to have been aware well before such ideas circulated in Paris. These ideas can now be seen to have permeated Purism so that it had a far more determinist intent than has hitherto been appreciated, and in turn such determinism would underlie his subsequent architectural thinking and design. And it is this aspect that should now be taken into account in reappraising his intentions, and indeed the intended effect of his architectural designs: as mechanisms for bringing body and soul back into their natural state of harmony, and thereby transforming human beings into ones capable of attaining their full development within the modern world.

Notes

1 'Nous emploierons le terme "Purisme" pour exprimer en un mot intelligible la caractéristique de l'esprit moderne': Amédée Ozenfant and Charles-Edouard Jeanneret, *Après le Cubisme* (Paris: Edition des Commentaires, 1918), 53.

2 'la peinture étant l'art le moins mortifié de l'époque a servi d'exemple': Ozenfant and Jeanneret, *Après le Cubisme*, Introduction, n.p. [p. 7].

3 See his letters, e.g. recording sleep disturbed by warplanes, 18 August 1917: Le Corbusier, *Correspondance: Lettres à la famille, 1900–1925*, vol. 1, ed. Rémi Baudouï and Arnaud Dercelles (Gollion: Infolio, 2011), 412.

4 On Charles–Edouard Jeanneret in La Chaux-de-Fonds, see *La Chaux-de-Fonds et Jeanneret avant Le Corbusier*, eds. Mark Emery and Sylviane Ramseyer (La Chaux-de-Fonds: Musée des Beaux-Arts, 1987); Paul Turner, *La formation de Le Corbusier: idéalisme et mouvement moderne* (Paris: Macula, 1987); H. Allan Brooks, *Le Corbusier's Formative Years: Charles-Edouard Jeanneret at La Chaux-de-Fonds* (Chicago: University of Chicago Press, 1997).

5 Jeanneret was officially working for Perret from 1 July 1908–9 September 1909. See 'Introduction' to Le Corbusier, *Lettres à Lettres à ses maîtres: I: Auguste Perret*, ed. Marie-Jeanne Dumont (Paris: Editions du Linteau, 2002), 9 and 17.

6 e.g. Marie-Jeanne Dumont, 'Introduction' to *Lettres à ses maîtres: II: Charles L'Eplattenier*, ed. Marie-Jeanne Dumont (Paris: Editions du Linteau, 2006), 42–3.

7 On Apollinaire, see H. Allorge, 'Apollinaire (Guillaume)', in *Dictionnaire de la biographie française*, 3, eds. J. Balteau et al. (Paris: Letouzey and Ané, 1939), cols. 122–4; see also Victor Basch, *Etudes d'esthétique dramatique, 1: Le théâtre pendant les années de guerre … Guillaume Apollinaire* (Paris: J. Vrin, 1920).

8 Michel Corvin, *Le Théâtre de recherche entre les deux guerres: le laboratoire Art et Action* (Paris: L'Age d'homme, 1970), 67–8.

9 Witness, for example, Apollinaire's sending autographed copies to him of his *Le Poète assassiné* (Paris: Bibliothèque de curieux, 1916) and his surrealist drama, first performed in 1917, *Les Mamelles de Tirésias* (Paris: Sic, 1918), both preserved in Le Corbusier's library at the Fondation Le Corbusier, Paris.

10 *Le Corbusier: Correspondance croisée, 1910–1955: Le Corbusier, William Ritter*, ed. Marie-Jeanne Dumont (Paris: Editions du Linteau, 2014), 249, n. 2.

11 Le Corbusier most probably met the Vildracs via Auguste Perret: H. Allen Brooks, *Le Corbusier's Formative Years: Charles-Edouard Jeanneret at La Chaux-de-Fonds* (Chicago: Chicago University Press, 1997), 353. Marie-Jeanne Dumont states that Jeanneret first met Charles Vildrac in autumn 1913 (Le Corbusier, *Lettres à Auguste Perret*, 87); and that they met through Perret and Voirol (Le Corbusier, *Lettres à Auguste Perret*, 90), implying through the Club artistique de Passy, but without offering any sources for her statement. This seems to be confirmed by Jeanneret's letter to William Ritter of 3 November 1913, which confirms that they had met by then, and in an 'avant garde atmosphere' associated with Perret, so presumably the Club artistique de Passy: Le Corbusier, *Correspondance croisée*, 249.

12 On the Abbaye de Créteil, see Christian Sénéchal, *L'Abbaye de Créteil* (Paris: André Delpech, 1930).

13 'grands murs unis, qui jaunes, qui bleus, recouverts des plus belles et fortes peintures de l'art moderne', Jeanneret's letter to his parents, 22 November 1917, Le Corbusier, *Correspondance: Lettres à la famille*, vol. 1, 417.

14 Le Corbusier, *Correspondance: Lettres à la famille*, vol. 1, 417.

15 Jeanneret's letter to William Ritter: 23 December 1913, *Le Corbusier, Correspondance croisée*, 258.

16 In his own *Mémoires* Ozenfant claimed to have first met Jeanneret in May 1917: Amédée Ozenfant, *Mémoires: 1886–1962* (Paris: Seghers, 1968), 101. In *L'Art décoratif d'aujourd'hui* Le Corbusier claimed to have first met Ozenfant in 1918: Le Corbusier, *L'Art décoratif d'aujourd'hui* (Paris: G. Crès, 1925), 217; his diary entry of January 24, 1918, states that he had first met Ozenfant at a lunch with Art et Liberté (associated with the Club artistique de Passy) the previous day: Carol S. Eliel, 'Purism in Paris, 1918–1925', in *L'Esprit Nouveau: Purism in Paris, 1918–1928*, ed. Carol S. Eliel

(Los Angeles: LACMA, and New York:
Harry N. Abrams, 2001), 76, n. 20.
See also Turner, *La Formation de Le
Corbusier*, 154.

17 On Emile Jaques-Dalcroze and his
rythmique, see Regula Puskàs, 'Jaques-
Dalcroze, Emile', in *Dictionnaire historique
de la Suisse*; Emile Jaques-Dalcroze, *La
Rythmique* (Lausanne: Foetisch, two vols.
1903 and 1908); Wolf Dohrn's speech at
the laying of the foundation stone for the
Festspielhaus at Hellerau, 'Die Aufgabe
der Bildungsanstalt Jaques-Dalcroze', in
Der Rhythmus, Ein Jahrbuch, Vol. I (Jena:
1911), 2–19; Emile Jaques-Dalcroze,
Le rythme: la musique et l'éducation
(Paris: Fischbacher, 1920); Hélène Emma
Brunet-Lecomte, *Jaques-Dalcroze: sa vie,
son oeuvre* (Geneva and Paris: Jeheber,
1950); *Emile Jaques-Dalcroze: l'homme, le
compositeur, le créateur de la rythmique*,
ed. Frank Martin (Neuchâtel: Editions
de la Baconnière, 1965); Marie-Laure
Bachmann, *La Rythmique de Jaques-
Dalcroze: Une éducation par la musique
et pour la musique* (Neuchâtel: La
Baconnière, 1984); Alfred Berchtold,
'Emile Jaques-Dalcroze et son temps',
in *Emile Jaques-Dalcroze: l'homme, le
compositeur, le créateur de la rythmique*,
ed. Frank Martin (Neuchâtel: Editions
de la Baconnière, 1965), 27–157.
For an extremely succinct presentation
of Rythmique, see Emile Jaques-Dalcroze,
'Le rythme comme éducateur', in *Images
de la Suisse*, ed. Jean Ballard (Marseille:
Cahiers du Sud, 1943), 381–4.

18 See, for instance, Emile Jaques-Dalcroze,
*La respiration et l'innervation musculaire:
planches anatomiques en supplément à la
méthode de Gymnastique Rythmique*
(Paris, Neuchâtel and Leipzig: Sandoz,
Jobin & Cie, 1906).

19 Marco de Michelis, 'L'Institut Jaques-
Dalcroze à Hellerau', in *Adolphe Appia ou
le renouveau de l'esthétique théâtrale*, ed.
Jörg Zutter, (Lausanne: Payot, 1992),
21–47 (NB 30ff).

20 On this performance see Tamara Levitz,
'In the footsteps of Eurydice: Gluck's
Orpheus and Eurydice in Hellerau, 1913',
Echo, 3.2 (2001) [n.p.]; see also Davinia
Caddy, *The Ballets Russes and Beyond:
Music and Dance in Belle-Epoque Paris*
(Cambridge, Cambridge University Press,
2012), 209. The primary source is the
original programme: *Das Claudel-
Programmbuch* (Hellerau: Hellerau
Verlag, 1913); its title refers to the fact
that Paul Claudel's *L'Annonce faite à
Marie* was also performed.

21 On Adolphe Appia, see *Adolphe Appia,
ou le renouveau de l'esthétique théâtrale:
dessins et esquisses de décors*, ed.
Jörg Zutter (Lausanne: Payot, 1992);
Richard Beacham, *Adolphe Appia:
Artist and Visionary of the Modern
Theatre* (London: Routledge, 1994);
Edmund Stadler, 'Jaques-Dalcroze et
Adolphe Appia', in *Emile Jaques-Dalcroze:
l'homme, le compositeur, le créateur de la
rythmique*, ed. Frank Martin, 412–59
(Neuchâtel: Editions de la Baconnière,
1965).

22 On Salzmann, see Carla di Donato,
*Alexandre Salzmann e la scena del XX
secolo* (Rome: Carocci, 2015). With regard
to his lighting at Hellerau, see Salzmann's
essay 'Licht, Belichtung und Beleuchtung',
in *Das Claudel-Programmbuch*.

23 The audience included George
Bernard Shaw, Paul Claudel, Stefan
Zweig, Edward Craig, Constantin
Stanislavski, Max Reinhardt, Darius
Milhaud, Serge Rachmaninov, Anna
Pavlova, Rudolf Laban, Rabindrath
Tagore, Martin Buber, Rainer Maria
Rilke and Oskar Kokoshka.

24 Letter of 5 July 2013 from Charles-
Edouard Jeanneret to Perret (Le
Corbusier, *Lettres à Auguste Perret*, 85).

25 In a letter to his parents the young artist
refers to 'Appia' as *camarade* (as opposed
to 'Maître Dalcroze') after spending all
Christmas night drinking together, even
after Albert had retired to his own room:
Charles-Edouard Jeanneret to his parents,
1 January 1911, in Le Corbusier,
Correspondance: Lettres à la famille, vol. 1,
339. Marco de Michelis thinks that the
Jeanneret brothers' friend at this stage
was Theodore Appia, Adolphe's nephew
who, like Albert, was a student of
Jaques-Dalcroze and indeed became a
friend of both Jeanneret brothers: Marco
de Michelis, 'L'Institut Jaques-Dalcroze à
Hellerau', *Adolphe Appia ou le renouveau
de l'esthétique théâtrale*, ed. Jörg Zutter
(Lausanne: Payot, 1992), 43, n. 5. But
this does not fit easily with Charles-
Edouard's claim in the letter cited
(but which de Michelis seems not to
have seen) that his friend Appia was
'founder of great systems and yet-to-be-
patented inventions' ('fondateur de
grands systèmes et inventeur non encore
breveté'): Le Corbusier, *Correspondance:
Lettres à la famille*, vol. 1, 340).

26 De Michelis, 'L'Institut Jaques-Dalcroze à Hellerau', 22 and 43, n. 12, citing Albert's letters to his parents (18 February 1912 and undated: La Chaux-de-Fonds, Bibliothèque de la Ville, Dossier Jeanneret-Perret, XXI/51 and 53).

27 Amédée Ozenfant and Charles-Edouard Jeanneret, *Après le Cubisme* (Paris: Edition des Commentaires, 1918).

28 Le Corbusier, *Vers une architecture* (Paris: G. Crès, 1923); cf. *Towards a New Architecture*, trans. Frederick Etchells (London: Architectural Press, 1927).

29 e.g. 'Esthétique et purisme', *Le Promenoir* 4 (août 1921): 49–52, and 5 (s.d [novembre 1921]): 69–71 (the first part was accompanied by a drawing by Ozenfant, the second by a drawing by Jeanneret); Amédée Ozenfant, 'Sur les écoles Cubistes et post-Cubistes', *Journal de Psychologie normale et pathologique*, vol. 23, nos. 1–3 (15 January–15. March 1926): 290–302.

30 The Belgian Camille Janssen seems to have taken on this pen name before moving to Paris, in 1910. It was through his friendship with Apollinaire that he moved into the Cubist circle of the Section d'or group.

31 Having already launched a magazine, *Mosane,* in Belgium, in Paris he collaborated on the magazines *L'Action d'Art*, *SIC* (with Pierre-Albert Birot) and *Nord-Sud* (with the poet Pierre Reverdy), and would set up *Z* and *Interventions* as well as *L'Esprit Nouveau*, and after the latter's demise *Documents internationaux de l'Esprit nouveau*. Ozenfant already had experience in the publication of an avant-garde magazine, *L'Elan* (1915–7).

32 When rubrics were first introduced (*L'Esprit Nouveau*, 6) these included science and 'économique sociologique' alongside painting, performances (*spectacles*), literature, architecture, theatre and books; 'painting' was expanded to 'fine arts' (*Beaux-Arts*) from the next issue, and aesthetics added, with a separate category of 'engineer's aesthetics' in the following issue, and separate rubrics for exhibitions, music and news (*actualités*) from the issue after. Later on, separate rubrics were added for 'applied science', press, sport, music hall and circus.

33 *Les Ballets russes à Paris, Châtelet, mai 1917* (Paris: n.p., 1917). This text appears, simply under the title 'Parade', in Apollinaire, *Oeuvres en prose complètes*, 2, eds. Pierre Caizergues and Michel Décaudin (Paris: Pléiade/nrf/Gallimard, 1991), 865–7, with relevant notes, 1648–9. The programme is in Le Corbusier's library: Fondation Le Corbusier (henceforth FLC), Paris: V. 153. However, his notes and sketches in it show that he had not actually attended *Parade* but instead *L'Oiseau de feu* (with music by Stravinsky).

34 'à l'hauteur des progrès scientifiques et industriels': *Appolinaire, Oevres en prose complètes* 2, 865.

35 Guillaume Apollinaire, 'L'Esprit nouveau et les poètes', in Apollinaire, *Oeuvres en prose complètes*, 3, 941–54; notice, 1683–4; note, 1684–5; notes et variantes, 1685–8. The title page of the original manuscript reads: 'Guillaume Apollinaire, L'Esprit Nouveau. Conférence faite au Théâtre du Vieux-Colombier le 26 novembre 1917 accompagnée de Récitation de Rimbaud, Gide, Paul Fort, Fargue, Saint-Léger, Léger, Salmon, Divoire, Romains, Apollinaire, Reverdy, Jacob, Cendrars': Apollinaire, *Oeuvres en prose complètes*, 1685. Apollinaire's essay was read by Pierre Bertin, who together with Suzanne Méthivier and Henriette Sauret recited the poems: Apollinaire, *Oeuvres en prose complètes*, 1683.

36 *Mercure de France* (1 December 1918), *Oeuvres en prose complètes*, 3, 1685.

37 'L'esprit nouveau est celui du temps où nous vivons. Un temps fertile en surprises': Apollinaire, *Oeuvres en prose complètes*, 3, 954.

38 'Il lutte pour le rétablissement de l'esprit d'initiative, pour la claire compréhension de son temps et pour ouvrir des vues nouvelles sur l'univers extérieur et intérieur qui ne soient point inférieures à celles que les savants de toutes catégories découvrent chaque jour et dont ils tirent des merveilles': Apollinaire, *Oeuvres en prose complètes*, 3, 953–4.

39 Apollinaire, *Oeuvres en prose complètes*, 3, 943.

40 Apollinaire, *Oeuvres en prose complètes*, 3, 952.

41 'les classiques', 'un solide bon sens, un esprit critique assuré, des vues d'ensemble sur l'univers et dans l'âme humaine, et le sens du devoir': Apollinaire, *Oeuvres en prose complètes*, 3, 943.

42 See, for instance, Christopher Green and Jens M. Daehner, eds., *Une moderne Antiquité: Picasso, De Chirico, Léger, Picabia* (Paris: Hazan, 2012).

43 Ozenfant and Jeanneret's *Après le Cubisme* is dated 15 October 1918.

44 'cet art trouble d'une époque trouble': Ozenfant and Jeanneret, *Après le Cubisme*, 12.

45 'tout s'organise, tout se clarifie et s'épure … rien n'est déjà plus ce qu'il était avant la Guerre': Ozenfant and Jeanneret, *Après le Cubisme*, 11.

46 'une tendance à la rigueur, à la précision, à la meilleure utilisation des forces et des matières … en somme une tendance à la pureté': Ozenfant and Jeanneret, *Après le Cubisme*, 32–3.

47 'Des faits rigoureux, des figurations rigoureuses, des architectures rigoureuses, formelles, aussi purement et simplement que le sont les machines': Ozenfant and Jeanneret, *Après le Cubisme*, 34.

48 'L'art et la science dépendent du nombre': Ozenfant and Jeanneret, *Après le Cubisme*, 40.

49 'La nature agit à la manière d'une machine … La géométrie physique et mathématique définissent les lois des forces qui sont comme des axes d'ordonnance.' Ozenfant and Jeanneret, *Après le Cubisme*, 41.

50 'les lois de l'ordre qui sont celles de l'harmonie'; 'les grandes axes d'ordonnance du monde': Ozenfant and Jeanneret, *Après le Cubisme*, 43.

51 The titles of these sections are indicative: 'Parallèle entre les Méthodes d'analyse de la Science et celles de l'Art' and 'But de l'Analyse': Ozenfant and Jeanneret, *Après le Cubisme*, 42–3.

52 Ozenfant and Jeanneret, *Après le Cubisme*, 43–5.

53 Ozenfant and Jeanneret, *Après le Cubisme*, 43–4.

54 Ozenfant and Jeanneret, 'Esthétique et Purisme', *Le Promenoir*, 4 (août 1921): 49–52 ; and 5 (n.d. [novembre 1921]): 69–71. This hand-printed, avant-garde arts magazine was contemporary (1920–1) with *L'Esprit Nouveau*, with which it shared several authors (most notably the poet and future filmmaker, Jean Epstein, then a medical student in Lyons, but also the poet and literary translator Ivan Goll, Blaise Cendrars, Auguste Lumière, Fernand Léger and, posthumously, Apollinaire). But *Le Promenoir* was published in Lyons; its provincial production and short print runs (100 copies) due to hand printing explain how this article has been overlooked

hitherto. Nevertheless, *L'Esprit Nouveau* would praise it, as 'cette excellente petite revue', printing an extract from it as evidence of its quality (*L'Esprit Nouveau* 15, 1679–80). It is worth noting that its printer-publisher, Marius Audin, was a close friend of and occasional printer for a leading Parisian publisher of modern art books, Georges Crès, who would publish the series 'Collection de l'Esprit Nouveau', not least including Le Corbusier's *Vers une architecture* (1923) but also his *L'Art décoratif d'aujourd'hui* and *Almanach d'architecture moderne* (both in 1925) and Ozenfant and Jeanneret's *La Peinture moderne* (again 1925). On Audin's links with Crès, see Alan Marshall, 'Marius Audin: un imprimeur érudit d'entre-deux-guerres', in *Impressions de Marius Audin*, ed. Alan Marshall (Lyon, Musée de l'Imprimerie et de la Banque, 1995), 7–38, (8).

55 'une machine à émouvoir': 'Esthétique et Purisme', *Le Promenoir*, 4 (août 1921): 52.

56 See Ozenfant www.sfmoma.org/artwork/37.2991, accessed June 2018.

57 Ozenfant and Jeanneret, 'Destinées de la Peinture', *L'Esprit Nouveau*, 20 (n.d.), unpaginated. Later issues of *L'Esprit Nouveau* are undated, no doubt because financial pressures led to issues being published less regularly than had been promised to subscribers.

58 'La qualité distinctive de notre époque est de viser à la parfaite coopération de la sensation et de l'émotion': Ozenfant and Jeanneret, 'Destinées de la Peinture'.

59 'L'homme moderne sait qu'il n'est de réalités profondes que celles qui nous affectent directement': Ozenfant and Jeanneret, 'Destinées de la Peinture'.

60 'il a besoin de ces certitudes idéales qu'autrefois la religion donnait; doutant d'elle et des métaphysiques, il est ramené à lui-même et le monde vrai se passe au dedans de lui': 'Destinées de la Peinture'. See also Amédée Ozenfant and Charles-Edouard Jeanneret, 'La Formation de l'optique moderne', *L'Esprit Nouveau* 21 (1924), n.p.

61 Amédée Ozenfant and Charles-Edouard Jeanneret, *La Peinture moderne* (Paris : G. Crès, 1925), NB v.

62 Amédée Ozenfant, 'Sur les écoles Cubistes et post-Cubistes', *Journal de Psychologie normale et pathologique*, vol. 23, nos. 1–3 (15 Jan. – 15 March 1926): 290–302. The article is actually dated 1925: 'Sur les

écoles Cubistes et post-Cubistes', 290. Jeanneret and Ozenfant split over the hanging of paintings in the Pavillon de l'Esprit Nouveau, on July 25, 1925: *L'Esprit Nouveau: Purism in Paris, 1918–1925* ed. Carol S. Eliel (Los Angeles: LACMA and Harry N. Abrams, 2001), 65.

63   Eliel, *L'Esprit Nouveau: Purism in Paris*, 294–6.

64   'l'homme moderne a des habitudes d'ordre, de netteté, de clarté qui lui font désirer les mêmes qualités dans le lyrisme même': Eliel, *L'Esprit Nouveau: Purism in Paris*, 294.

65   'un dispositif impératif destiné à mettre le sujet contemporain dans un état physico psychologique voulu par l'artiste': Eliel, *L'Esprit Nouveau: Purism in Paris,* 298.

66   Eliel, *L'Esprit Nouveau: Purism in Paris*, 302.

67   This subtitle only lasted for the first three issues, after which it was replaced by 'Revue internationale illustrée de l'activité contemporaine'.

68   'celui qui anime toute recherche scientifique' and 'une esthétique expérimentale': *L'Esprit Nouveau* 1 (Octobre 1920), unpaginated section at beginning.

69   'l'art a des lois comme la physiologie ou la physique': *L'Esprit Nouveau*, 1.

70   'nous voulons appliquer à l'esthétique les méthodes mêmes de la psychologie expérimentale': *L'Esprit Nouveau*, 1.

71   On Charles Henry, see A. Fierro, 'Henry (Charles)', in *Dictionnaire de la biographie française*, 15, ed. M. Prevost et al. (Paris: Letouzet and Ané, 1985), cols. 979–80. See also: Eric Michaud, 'Esthétique scientifique de Charles Henry ou l'avènement du normal', in *Vers la science de l'art: l'esthétique scientifique en France, 1857–1937*, eds. Jacqueline Lichtenstein et al. (Paris: Presses de l'Université Paris-Sorbonne, 2013), 133–44; Jose A. Arguelles, *Charles Henry and the Formation of a Psychophysical Aesthetic* (Chicago: University of Chicago Press, 1972); Francis Warrain, *L'oeuvre psychobiophysique de Charles Henry* (Paris: Gallimard, 1931); J.F. Revel, 'Henry et la science des arts', *L'oeil*, Novembre 1964: 20–7. His own works relevant here are: 'Introduction à une esthétique scientifique', *La Revue contemporaine*, mai-août 1885, 441–69; *Cercle chromatique* (Paris: Verdin, 1888); 'Esthétique et psychophysique', *Revue philosophique de la France et de l'étranger*, 29 (1890), 332–6; *Harmonies de formes et de couleurs* (Paris: Librairie scientifique A. Hermann, 1891); *Quelques aperçus sur l'esthétique des formes* (Paris: Librairie Nony, 1895, originally published across four issues of *La Revue blanche*, 1894–5); and 'La lumière, la couleur et la forme', *L'Esprit Nouveau*, 6 (1921), 605–23; 7 (1921), 729–36; 8 (1921) 948–58; 9 (1921), 1068–75.

72   On Charles Lalo, see P. Pradet de Lamaize, 'Lalo (Charles)' in *Dictionnaire de la biographie française*, 19, ed. J. Balteau et al. (Paris: Letouzet and Ané, 2001), cols. 433–4.

73   On Etienne Souriau, see the issue of *Revue d'Esthétique* dedicated to him: *Revue d'Esthétique* 3–4 (1980).

74   On this, see Jacqueline Lichtenstein, 'Victor Basch et l'esthétique expérimentale: une histoire oubliée de l'esthétique française', in *Vers la science de l'art: l'esthétique scientifique en France, 1857–1937*, eds. Jacqueline Lichtenstein et al. (Paris: Presses de l'Université Paris-Sorbonne, 2013), 81–93.

75   The commissioning of the article is confirmed by Basch in his introductory paragraph: 'Vous m'avez demandé d'exposer aux lecteurs de *L'Esprit Nouveau*, Revue international d'Esthétique, la façon dont j'entendais et j'enseignais l'esthétique et la science de l'art': *L'Esprit Nouveau*, 1 (October 1920): 5.

76   Victor Basch, 'L'Esthétique et la Science de l'art', *L'Esprit Nouveau* 1 (octobre 1920): 5–12 ; and 2 (novembre 1920): 119–30. Basch seems to have thought highly of this article in that he reprinted it in his collected essays, and indeed as the opening essay: Victor Basch, *Essais d'esthétique, de philosophie et de littérature* (Paris: Félix Alcan, 1934), 4–34.

77   On Victor Basch (1863–1944) and his aesthetics, see Céline Trautmann-Waller, 'Victor Basch: l'esthétique entre la France et l'Allemagne', *Revue de métaphysique et de morale*, 34 (2002/2), 77–90.

78   See Victor Basch, 'Les grands courants de l'esthétique allemande contemporaine', in *La Philosophie allemande au XIXe siècle*, eds. Charles Andler et al. (Paris: Félix Alcan, 1912), 68–110.

79   On Fechner, see Isabelle Dupéron, *Gustav Théodor Fechner: Le parallélisme psychophysiologique* (Paris: Presses universitaires de France, 2000). His

works relevant to this context: Gustav
Fechner, *Zur experimentalen Ästhetik*
(Leipzig: S. Hirzel, 1871); *Vorschule der
Ästhetik* (Leipzig: Breitkopf & Härtel,
1876); *Sachen der Psychophysik* (Leipzig:
Breitkopf und Härtel, 1877); *Elemente der
Psychophysik* (Leipzig: Breitkopf und
Härtel, 1889).

80   On Wilhelm Wundt, see Georges
Dwelshauvers, *Wundt et la psychologie
expérimentale* (Brussels: M.
Weissenbruch, 1911); Serge Nicolas,
Psychologie de W. Wundt (1832–1920)
(Paris: L'Harmattan, 2003). His
publications relevant to the present
context: Hermann von Helmholtz and
Wilhelm Wundt, *Psychologie vom
naturwissenschaftlichen Standpunkt*
(1862); Wilhelm Wundt, *Beiträge zur
Theorie der Sinneswahrnehmung* (Leipzig
& Heidelberg: C.F. Winter, 1862);
Wilhelm Wundt, *Über die Physik der Zelle
in ihrer Beziehung zu den allgemeinen
Prinzipien der Naturforschung* (1867);
Wilhelm Wundt, *Grundzüge der
Physiologischen Psychologie* (Leipzig:
Wilhelm Engelmann, 1874); Wilhelm
Wundt, *Untersuchungen zur Mechanik der
Nerven und Nervencentren* (Erlangen:
Ferdinand Enke, 1876).

81   Charles Lalo, *L'Esthétique expérimentale
contemporaine* (Paris: Félix Alcan, 1908),
40–51.

82   Dwelshauvers, *Wundt et la psychologie
expérimentale*.

83   Amédée Ozenfant, 'Sur les écoles Cubistes
et post-Cubistes', *Journal de Psychologie
normale et pathologique*, vol. 23, nos. 1–3
(15.1.–15.3., 1926): 290–302 (301).

84   Théodule Ribot, *La Psychologie allemande
contemporaine: école expérimentale* (Paris:
Germer Baillière, 1879).

85   The lecture had been delivered on 7
November 1920. The article appeared in
L'Esprit Nouveau, 6 (1921), 605–23; 7
(1921), 729–36; 8 (1921) 948–58; 9
(1921), 1068–75. *L'Esprit Nouveau* 8
actually included a full-colour illustration
– the only one in the issue – of Henry's
'cercle chromatique'.

86   His major works are indicative of his
interests: *Introduction à une esthétique
scientifique* (Paris: La Revue
contemporaine, 1885); *Rapporteur
esthétique de Monsieur Charles Henry
permettant l'étude et la rectification
esthétique de toute forme* (Paris: G. Seguin,
1888); *Cercle chromatique: présentant tous
les compléments et toutes les harmonies de
couleurs avec une introduction sur la
théorie générale du contraste, du rythme et
de la mesure* (Paris: Charles Verdin,
1888); 'Esthétique et psychophysique',
Revue Philosophique, 29 (1890): 332–6;
Harmonies des formes et des couleurs
(Paris: A. Hermann, 1891); *Quelques
aperçus sur l'esthétique des formes*
(Paris: Nony, 1895).

87   Charles Henry, *Quelques aperçus sur
l'esthétique des formes* (Paris: Librairie
Nony, 1895); originally published across
four issues of *La Revue Blanche*, 1894–5.

88   Charles Lalo, 'L'esthétique sans amour',
L'Esprit Nouveau, 5 (1921): 491–9, and 6,
625–38.

89   Charles Lalo, *L'Esthétique expérimentale
contemporaine* (Paris: F. Alcan, 1908),
40–51.

90   'Fechner et l'esthétique expérimentale'
was promised for *L'Esprit Nouveau* 10.
See *L'Esprit Nouveau* 9: 1075.

91   Lalo, 'L'esthétique expérimentale de
Fechner'. See also Pradet de Lamaize, P.
'Lalo (Charles)', in *Dictionnaire de la
biographie française*, 19, cols. 433–4.

92   Indicative publications: *Théorie de
l'invention* (Paris: Hachette, 1881);
L'Esthétique du mouvement (Paris: Félix
Alcan, 1889); *La suggestion dans l'art*
(Paris: Félix Alcan, 1893); *L'imagination
de l'artiste* (Paris: Hachette, 1901); *La
beauté rationnelle* (Paris: Félix Alcan,
1904); *La rêverie esthétique* (Paris: Félix
Alcan, 1906); *Les conditions du bonheur*
(Paris: Armand Colin, 1908); *Traité de la
beauté rationnelle* (Paris: Félix Alcan,
1910); *L'esthétique de la lumière* (Paris:
Hachette, 1913).

93   Jacqueline Lichtenstein, 'Victor Basch
et l'esthétique expérimentale: une
histoire oubliée de l'esthétique française',
in *Vers la science de l'art: l'esthétique
scientifique en France, 1857–1937*, eds.
Jacqueline Lichtenstein et al. (Paris:
Presses de l'Université Paris-Sorbonne,
2013), 89.

94   See Jean-Louis Cohen, 'Introduction' to
Le Corbusier, *Toward an Architecture*,
trans. John Goodman (London: Francis
Lincoln, 2008), 14.

95   See notably, Etienne Souriau, *L'Avenir de
l'esthétique: essai sur l'objet d'une science
naissante* (Paris: Félix Alcan, 1929).

96   See Le Corbusier, 'Le théâtre spontané',
in *Architecture et Dramaturgie*, ed. André
Villiers (Paris: Flammarion, 1950),
149–58, with subsequent discussion,
Souriau, *L'Avenir de l'esthétique*, 158–68.

97 'Les canons antiques qu'on croit
   généralement être des codes artificiels,
   des calibres, n'étaient basés que sur la
   juste connaissance de l'universalité
   des lois naturelles qui gouvernent le
   monde extérieur et conditionnent
   l'œuvre d'art. C'était non des codes,
   mais des lois justes et souples qui
   permettaient de lier l'œuvre humaine à
   celle de la nature': Ozenfant and
   Jeanneret, Après le Cubisme, 47–8.
98 'Ces mêmes canons (triangles égyptiens,
   rapports numériques, etc.) étaient connus
   des plus anciennes civilisations. Les
   Egyptiens, les Assyriens, les Grecs, les
   Persans et les Gothiques les connurent; les
   Renaissants les retrouvèrent mais souvent
   les appliquèrent dans leur scolastique
   comme des règlements étroits': Ozenfant
   and Jeanneret, Après le Cubisme, 48.
99 'La science appelle notre attention sur
   les ordres qu'elle découvre; l'artiste est
   incité par elle à découvrir les beautés
   nouvelles que cet ordre définit … Le
   savant découvre l'harmonie, source de
   beauté; l'artiste y retient ce qui est bon
   pour l'art': Ozenfant and Jeanneret,
   Après le Cubisme, 49.
100 Notably: Matila Ghyka, Esthétique des
   proportions dans la nature et dans les arts
   (Paris: Gallimard, 1927); Le Nombre d'or:
   Rites et rythmes pythagoriciens dans le
   développement de la civilisation occidental,
   2 parts (Paris: Gallimard, 1931); Essai sur
   le rythme (Paris: Gallimard, 1938). For
   further detail of these works, see Judi
   Loach, 'Entre Valéry et Le Corbusier: la
   coquille et le Nombre d'or', in La Ville et la
   coquille: Huit essais d'emblématique, ed.
   Paulette Choné (Paris: Beauchesne,
   2016), 109–33 (NB 115–22).
101 Ghyka, Le Nombre d'or, 2, 155.
102 Most recently, in Cohen, 'Le Corbusier's
   Modulor and the debate on proportion
   in France'.
103 On his own library, see Turner, La
   Formation de Le Corbusier, 224–9, and
   Arnaud Dercelles, 'Présentation de la
   Bibliothèque personnelle de Le Corbusier',
   in Le Corbusier et le livre, eds. Arnauld
   Dercelles et al. (Barcelona: Collegi
   Arquitectes Catalunya, 2005), 6–78 (and
   on these books: 42). Le Corbusier owned
   the following books by Matila Ghyka:
   Esthétique des proportions dans la nature et
   dans les arts (Paris: Gallimard, 1927), FLC
   Z009, with Le Corbusier's annotations
   and containing a copy of his letter to
   Ghyka concerning tracés régulateurs;

Le Nombre d'or (Paris: Gallimard, 1931),
   FLC Z010, copy dedicated to Le Corbusier
   by Ghyka, and again with annotations by
   Le Corbusier; Essai sur le rythme (Paris:
   Gallimard, 1938).
104 Ghyka, Essai sur le rythme.
105 See, for instance, Ghyka's letter (no year
   given, but presumably 1931) responding
   to Le Corbusier's request for a copy of
   Esthétique des proportions, and in return
   asking Le Corbusier to write a review of
   his recently published second volume of
   Le Nombre d'or (Letter E2–3, Fondation Le
   Corbusier, Paris). I thank Tim Benton for
   having drawn this letter to my attention.
106 Matila Ghyka, 'Le Corbusier's Modulor
   and the Concept of the Golden Mean',
   Architectural Review, 614 (February
   1948), 39–42.
107 Le Corbusier, Le Modulor (Boulogne-sur-
   Seine: L'Architecture d'aujourd'hui,
   1950).
108 'Pour faire comprendre par un exemple
   d'ordre de vérité auquel on aboutit,
   nous citerons les résultats des recherches
   de l'un de nous sur la "Section d'Or".
   La "Section d'Or" est un rapport
   mathématique fort simple. L'expérience
   de laboratoire a montré qu'elle a cette
   particularité de satisfaire nos exigences
   esthétiques. On a fait diviser aux sujets
   une certaine longueur de façon à avoir
   les proportions les plus harmonieuses.
   La moyenne de centaines de résultats
   obtenus avec des personnes de tout rang
   social et de tout âge a donné la "Section
   d'Or". D'autre part, la "Section d'Or" est
   une des mesures que se transmettaient
   fidèlement les artistes d'autrefois. Sa
   connaissance dispense le créateur de
   recourir sans cesse au goût. Ce n'est
   plus une règle, ainsi que nous l'avons
   démontré; c'est une loi esthétique qui
   n'est d'ailleurs qu'une loi physique et
   mathématique perçue par notre
   sensibilité.
   Nous dégagerons de nombreuses lois
   semblables et, d'elle même, une
   esthétique expérimentale se constituera.
   C'est pourquoi nous suivrons les travaux
   d'esthétique de laboratoire avec autant de
   curiosité que les expériences librement
   instituées dans leurs œuvres par les
   artistes, littérateurs, peintres, musiciens,
   ingénieurs': L'Esprit Nouveau, 1 (October
   1920), unpaginated preliminary section.
109 On the Section d'or group, see ed.,
   La Section d'or, 1912–1920–1925, eds.
   Cécile Debray et al. (Paris: Éditions

Cercle d'Art, 2000), and in particular the essay within it by Cécile Debray, 'La Section d'or 1912 – 1920– 1925' (19–41).

110 See Ozenfant, 'Sur les écoles Cubistes et post-Cubistes', 291–4. Ozenfant confirms such broad membership of early Cubism – which he terms 'collective' or 'pure' Cubism, and which he sees as running from 1908 to 1912 (294) or 1914 (291) – in contrast with a more diversified and individualist Cubism, dominated by Picasso and Braque, from 1912 to 1918 (294).

111 On Maurice Princet's role in the Section d'or group, see Didier Ottinger, 'Le Secret de la Section d'Or: Maurice Princet au pays de la quatrième dimension', in *La Section d'Or: 1912–20–25*, ed. Cécile Debray (Paris: Cercle des arts, 2000), 63–8.

112 See, for instance, Gris's oil painting 'Man in café' (1912) together with its preparatory drawing (1911) in which the geometrical structure is clear (both now in the Philadelphia Museum of Art). For Metzinger, see, for instance 'Dancer in a café' (1912; now in the Albright Knox Art Gallery, Buffalo).

113 This address was more significant than was realised by Le Corbusier scholars, who remained unaware of its famous residents, at least until Jan Birksted's *Le Corbusier and the Occult* (Cambridge MA.: MIT, 2009), which drew attention to the residence there of the notorious American patron of the arts, Natalie Clifford Barney (124–6). However, she lived in the house at the back of the courtyard and used the temple (previously the Freemasonic Loge de l'Amitié) in the courtyard for her celebrated literary (and, to a lesser extent, musical) salons (Ibid., 124–5), while the Vildracs and their lodgers lived in the apartment block on the street front; I have yet to discover any evidence of Le Corbusier's participation in any of her salons.

114 Ozenfant and Jeanneret, *Après le Cubisme*, 26–8.

115 'd'une certaine rigueur, du respect et de l'application des lois': Ozenfant and Jeanneret, *Après le Cubisme*, 27.

116 'reconnaître l'intention nettement formulée': Ozenfant and Jeanneret, *Après le Cubisme*, 27.

117 'la réalisation rigoureuse du calcul; le Nombre, qui est la base de toute beauté, peut trouver désormais son expression': Ozenfant and Jeanneret, *Après le Cubisme*, 28.

118 'leur conditionnement par le nombre': Ozenfant and Jeanneret, *Après le Cubisme*, 28.

119 'un épurement remarquable': Ozenfant and Jeanneret, *Après le Cubisme*, 28.

120 'la proportion de leurs organes rigoureusement conditionnés par les calculs, devant la précision d'exécution de leurs éléments … une projection des lois naturels'. 'l'ordre règne parce que rien n'est laissé à la fantaisie': Ozenfant and Jeanneret, *Après le Cubisme*, 28.

121 Villa Ker-Ka-Re, Vaucresson (1922–3); Villa La Roche-Jeanneret, Auteuil (1923); Villa Lipchitz-Mestchanhoff, Boulogne-Billancourt (1924); Villa Ternaisien, Boulogne-Billancourt (1926); Villa Cook, Boulogne-Billancourt, and Maison Guiette, Antwerp (1926–7); Villa Stein-de Monzie, Garches (1927); Villa Baizeau, Carthage (1928); Villa Savoye, Poissy, and Villa Church, Ville d'Avray (1929). For further details, see Tim Benton, *The Villas of Le Corbusier and Pierre Jeanneret, 1920–1930*, rev. edn (Basel: Birkhauser, 2007).

122 Eduard Sekler, 'The Carpenter Center in Le Corbusier's Oeuvre: An Assessment', in *Le Corbusier at Work: The Genesis of the Carpenter Center for the Visual Arts*, eds. Eduard Sekler et al., 229–58. Cambridge: Harvard University Press, 1978 (231–2).

123 For Villa la Roche, see www.fondationl ecorbusier.fr/corbuweb/morpheus.aspx?s ysName=redirect150&sysLanguage =en-en&IrisObjectId=8286&sys ParentId=150, accessed June 2018.

124 See Christopher Green, 'The Villa La Roche: Painting in architecture', in *Le Corbusier: Architect of the century*, eds. Michael Raeburn and Victoria Wilson (London: Arts Council, 1987), 121–2.

125 Le Corbusier-Saugnier, 'Les tracés régulateurs', *L'Esprit Nouveau* 5 (March 1921), 563–72. ('Le Corbusier-Saugnier' was the name adopted for some of their jointly authored articles.)

126 Charles Lalo, 'L'esthétique sans amour', *L'Esprit Nouveau* 5 (Feb 1921), 491–9, and 6, 625–38.

127 Charles Henry, 'La lumière, la couleur et la forme', *L'Esprit Nouveau* 6, 600–5; 7, 729–36.

128 Lalo, *L'Esprit Nouveau*, 5 (Feb 1921), 572.

129 See http://miguelmartindesign.com/ blog/wp-content/uploads/2011/01/ Figure1.jpg, accessed June 2018.

130 Julien Caron, 'Une Villa de Le Corbusier. 1916', L'Esprit Nouveau, 6, 679–704. 'Julien Caron' was probably a pseudonym of Ozenfant.

131 'Réponse de Monsieur de Fayet', L'Esprit Nouveau 17 (June 1922), n.p.

132 Le Corbusier-Saugnier, Vers une architecture (Paris: G. Crès, 1923), 49–63.

133 'L'obligation de l'ordre. Le tracé régulateur est une assurance contre l'arbitraire. Il procure la satisfaction de l'esprit': Le Corbusier-Saugnier, Vers une architecture, unpaginated introductory section; repeated, 51.

134 'Temple primitive', plan (Le Corbusier-Saugnier, Vers une architecture, 54); small-scale sketch 'Façade de l'Arsenal de Pirée' (Ibid., 57); sketch 'Tracé des coupoles achéménides' (Le Corbusier-Saugnier, Vers une architecture, 58); 'Notre-Dame de Paris', photo of west elevation (Le Corbusier-Saugnier, Vers une architecture, 59); 'Le Capitole à Rome', photo of front elevation (Le Corbusier-Saugnier, Vers une architecture 60); 'Le Petit Trianon, Versailles', photo of front elevation (Ibid., 61).

135 Le Corbusier-Saugnier, Vers une architecture, 61–2. The first edition only illustrates the Villa Schwob; his note appended to p. 62 reads: 'Je m'excuse de citer ici un exemple de moi: mais malgré mes investigations, je n'ai pas encore eu le plaisir de rencontrer d'architectes contemporains qui se soient occupés de cette question; je n'ai, à ce sujet, que provoqué l'étonnement, ou rencontré l'opposition et le scepticisme.' Later editions of Vers une architecture would also include illustrations of two of Le Corbusier's Purist villas – Ozenfant's studio ('Le Corbusier et Pierre Jeanneret, 1923. Une maison.') and the Villa La Roche-Jeanneret ('Le Corbusier et Pierre Jeanneret, 1924. Deux maisons à Auteuil') – with tracés régulateurs similarly imposed.

136 Benton, The Villas of Le Corbusier.

137 Almanach d'architecture moderne (Paris: Crès: Collection de l'Esprit Nouveau, 1925). See http://fondationlecorbusier. fr/corbuweb/morpheus.aspx?sysId= 13&IrisObjectId=6447&sysLanguage= fr-fr&itemPos=2&itemSort=fr-fr_sort_ string1%20&itemCount=47&sysParentN ame=&sysParentId=25, accessed June 2018.

138 Sekler, 'The Carpenter Center in Le Corbusier's Oeuvre: An Assessment', 258.

139 Perhaps most notably, the catalogue for the exhibition 'L'Esprit Nouveau: Purism in Paris, 1918–1928' included an essay reassessing the role of Ozenfant yet failed to consider Albert's potential contribution at all: Françoise Ducros, 'Amédée Ozenfant, 'Purist Brother': An Essay on his Contribution,' L'Esprit Nouveau: Purism in Paris, 1918–1928, ed. Carol S. Eliel (Los Angeles: LACMA, and New York: Harry N. Abrams, 2001), 71–100.

140 Albert's articles for L'Esprit Nouveau were: 'La Rythmique', L'Esprit Nouveau 2–3 (1920); reviews of the two Ballets Russes commissions, Satie's Parade and Stravinsky's Sacre du Printemps, L'Esprit Nouveau, 4 (1921); 'L'Intelligence dans l'oeuvre musicale', L'Esprit Nouveau, 7 (1921); review of Satie's Socrate, L'Esprit Nouveau, 9; annual review, L'Esprit Nouveau, 11–2; review of the ensemble Léo Six, L'Esprit Nouveau, 13; review of the Concerts Wiener, L'Esprit Nouveau, 14; review of Mussorgsky's Boris Godunov, L'Esprit Nouveau, 17; review of Noces, L'Esprit Nouveau, 18; 'La Crépuscule des virtuoses', L'Esprit Nouveau, 19; 'Sur la musique moderne', L'Esprit Nouveau, 23.

141 Alfred Berchtold, 'Emile Jaques-Dalcroze et son temps', in Emile Jaques-Dalcroze: l'homme, le compositeur, le créateur de la rythmique, ed. Frank Martin (Neuchâtel: Editions de la Baconnière, 1965), 130.

142 De Michelis, 'L'Institut Jaques-Dalcroze à Hellerau', 31–32 and 28.

143 Karl Bücher, Arbeit end Rhythmus (Leipzig: Abhandlungen der Königlichen Sächsischen Gesellschaft der Wissenschaften, 1896).

144 Prior to Albert's arrival in Paris at the beginning of October 1919 (Letters to parents, 2 October 1919) Le Corbusier had been using Müller's gymnastics (Letters to parents, 9 May 1919. See Le Corbusier, Correspondance. Lettres à la famille. Vol. 1. See also Jørgen Peter Müller, Mein System (first edition: Copenhagen: Holger Tillge, 1904); these were fashionable in avant-garde groups in the early twentieth century, and usually linked with hygienism, frequent bathing and fresh air.

145 On Hodler, see Beatrice Meier, 'Hodler, Ferdinand', in Dictionnaire historique de la Suisse www.hls-dhs-dss.ch/textes/f/ F19084.php, accessed June 2018; Oskar Bätschmann, 'Hodler, Ferdinand', in SIKART: Lexicon zur Kunst in der Schweiz. This is published by the Schweizerisches Institut für Kunstwissenschaft/Swiss

Institute for Art Research (SIAR) and derives from the institute's earlier, hard copy *Biographical Lexicon of Swiss Art* (1998). See also *Ferdinand Hodler*, ed. Ulf Küster, with texts by Oskar Bätschmann, Sharon Hirsh, Ulf Küster, Jill Lloyd, Paul Müller, Peter Pfrunder and Noëlle Pia (Berlin: Hatje Cantz, 2013); Oskar Bätschmann et al., *Ferdinand Hodler (1853–1918): Catalogue raisonné der Gemälde*, four vols. (Zurich: Scheidegger and Speiss, 1998–2018).

146 Now in the (private) Sammlung Thomas Schmidheiny, Jona, Switzerland. See also his portrait of c.1913, now in the Dübi-Müller Foundation collection, at the Musée cantonal des beaux-arts, Soleure, Switzerland. On Hodler's portraits, see Oskar Bätschmann et al., *Ferdinand Hodler (1853–1918). Catalogue raisonné der Gemälde, 2: Die Bildnisse* (Zurich: Scheidegger and Speiss, 2012).

147 Jaques-Dalcroze's cantata *Paysage sentimental* (c.1900) sets Duchosal's *Impression Lyrique* for soprano and string quartet or quintet (or, alternatively, for soprano, violin and piano); in c.1905 he set Duchosal's *Trois Mélodies* for voice, in 1905 *Mon pauvre coeur* for mezzo or baritone, and in 1914 *Le Cavalier bleu* for mid-range voice, all with piano accompaniment: Henri Gagnebin, 'Jaques-Dalcroze compositeur', in Martin, *Emile Jaques-Dalcroze*, 270. See also Tibor Dénes, 'Catalogue complet', in Martin, *Emile Jaques-Dalcroze*, 461–572.

148 On Mathias Morhardt see Laurent Langer, "Morhardt, Mathias", in *Dictionnaire historique de la Suisse*, see www.hls-dhs-dss.ch/textes/f/F27750.php, accessed June 2018; also *Journal de Genève*, 12 avril 1939, and *La Patrie Humaine*, 28 avril 1939.

149 His lengthy article 'Mlle. Camille Claudel', published in the Symboliste magazine *Mercure de France* in March 1898 (709–55), would remain the standard reference until the 1980s.

150 Marcel Baumgartner, *Ferdinand Hodler: Sammlung Thomas Schmidheny* (Zurich: Schweizeriches Institut für Kunstwissenschaft, 1998), 76, n. 11.

151 The 'Revues reçues' section in *L'Esprit Nouveau* regularly acknowledged receipt of copies of the *Revue de Genève*; full-page advertisements for the *Revue de Genève* appear in issues 1–12 and 17 of *L'Esprit Nouveau*.

152 On Duchosal, see Daniel Maggetti, 'Duchosal, Louis', in *Dictionnaire historique de la Suisse*, see www.hls-dhs-dss.ch/textes/f/F15939.php, accessed June 2018; Jean Violette, 'Notes sur la vie de Louis Duchosal', in the posthumous collection of his earlier but hitherto unpublished poems, *Posthuma* (Lausanne: Payot, 1910), 5–29; François Vincent, *Le poète genevois Louis Duchosal dans l'intimité 1887–1901* (1910); and Vaché Godel, 'Louis Duchosal, poète de l'agonie', *Journal de Genève*, 29 September 1962.

153 *Le Livre de Thulé* (Lausanne: F. Payot/ Paris, Grassart, 1891; second edition Lausanne: F. Payot, 1912); although evidently Symboliste, this won the poetry prize from the relatively conservative Société des amis des lettres de la Suisse Romande. *Le Rameau d'or, 1890–94* (Geneva: Charles Eggimann, 1894); *Derniers vers* (with preface by Philippe Mounier, Geneva: A. Jullien, 1905); *Posthuma* (Lausanne: Payot, 1910). *Le Livre de Thulé* and *Le Rameau d'or* were published together as a single volume in 1912 (Lausanne: Payot, 1912). In addition several poems were published separately in the 1890s.

154 First, the one-act comedy, *Marquise de vos beaux yeux me font mourir d'amour* (Geneva: Stapelmuhr, 1889; second edition Geneva: Aubert-Schuchardt, 1894); his best known, *Polichinelle et Compagnie* (Geneva: Pages Littéraires, 1897); another one-act play, *Le Guet* (Geneva: Stapelmuhr, 1898).

155 For example, Marcel Baumgartner, *Ferdinand Hodler: Sammlung Thomas Schmidheny* (Zurich: Schweizerisches Institut fur Kunstwissenschaft, 1998).

156 See his essay, 'Le Parallélisme' (c.1908), first published in German under the title 'Uber das Kunstwerk', trans. Ewald Bender, in the Berlin newspaper *Morgen* (1 January 1909): 23–6. It seems to have first been published in French posthumously, by Hodler's biographer, Carl Albert Loosli: *Ferdinand Hodler* (Zurich: Rascher, 1919: three vols.), vol. 1, 63ff. See also Niklaus Manuel Güdel's edition of Ferdinand Hodler, *La Mission de l'artiste* (Geneva: Notari, 2014), 75–158. The first attempt to seriously consider the philosophical implications – as opposed to simply their aesthetic consequences – seems to be a couple of pages in the catalogue

for a major exhibition of Swiss Symbolist art: Valentina Anker, *Le symbolisme suisse: destins croisés avec l'art européen* (Bern: Benteli, 2009), 156–7.

157 Roxana Vicovanu, 'De la "grammaire du geste" au "grammaire de l'art": la contribution d'Adolphe Appia et d'Emile Jaques-Dalcroze à une définition du rythme et du mouvement vivants', in *Vers la science de l'art*, eds. Lichtenstein et al., 211–28.

158 Charles Lalo, *L'Esthétique expérimentale contemporaine* (Paris: Félix Alcan, 1908), 40–51.

159 Roger Fischler, 'The Early Relationship of Le Corbusier to the "Golden Number"', *Environment and Planning Bulletin*, 8 (1979): 95–103 ; Judi Loach, 'Le Corbusier and the Creative Use of Mathematics', *British Journal for the History of Science* (special edition, ed. J.V. Field & F.A.J.L. James, on 'Science and the Visual'), vol. 31, 2, no. 109 (June 1998): 185–216; Jean-Louis Cohen, 'Le Corbusier's Modulor and the Debate on Proportion in France', *Architectural Histories*, 2.1: 23, 1–14 http://dx.doi.org/10.5334/ah.by.

11

A Portrait of the Scientist as a Young Ham: wireless, modernity and interwar nuclear physics

Jeff Hughes

Addressing the Liverpool meeting of the British Association for the Advancement of Science from its Presidential chair on 12 September 1923, Sir Ernest Rutherford lauded progress in physics since the Association had last met in the city in 1896. Following the discovery of X-rays and radioactivity, physical science had experienced 'a period of intense activity when discoveries of fundamental importance have followed one another with ... bewildering rapidity', marking 'the beginning of what has been aptly termed the heroic age of Physical Science'. As Director of the Cavendish Laboratory, Cambridge, and one of Britain's leading public scientists, Rutherford gave his audience an overview of recent and contemporary developments in his own field – subatomic physics, and particularly the new physics of the nucleus which he himself had pioneered at Manchester and Cambridge. Though the period had seen 'bold ideas in theory, as the Quantum Theory and the Theory of Relativity so well illustrate', for the most part, he stressed, 'the epoch under consideration has been an age of experiment, where the experimenter has been the pioneer in the attack on new problems'. Praising government support for science in the form of the Department of Scientific and Industrial Research (DSIR), and the role of Dominion scientists (of whom he himself was the leading representative) in this great 'tide of advance' in knowledge, Rutherford felt it 'a great privilege to have witnessed this period, which may almost be termed the Renaissance of Physics'.[1]

The Liverpool meeting was one of the most successful that participants could remember. Over 3,000 BAAS members attended, and there were 7,000 paying visitors to the Scientific Exhibition at the Central Technical School and 15,000 attendees at the free public lectures.[2] Rutherford's address was innovative – for the BAAS, at least – in being illustrated by lantern slides that were duplicated simultaneously in an overflow hall elsewhere in the city, making the lecture 'much less formal in delivery than has been customary in recent years'.[3] Peter Chalmers Mitchell – perhaps the nearest thing to a scientific correspondent in 1923[4] – observed that Rutherford and his colleagues 'were in a state of intense intellectual excitement … bringing together and discussing the various aspects of these new views of Nature on which they are engaged'. This excitement was driven by the fact that '[i]n the last quarter of a century the ideas about the nature of matter and of the forces of which matter is an expression have been completely changed'. Indeed, '[t]here has been what may be called a revolution of thought, to be compared with the enormous change made in thought by Darwin more than half a century ago'.[5] While some in the Philharmonic Hall audience were disappointed at Rutherford's dismissal of the promise of readily available subatomic energy, this recognition of a significant shift in the grounds of knowledge over the previous two decades was widely shared.[6] Indeed, Rutherford's theme was taken up enthusiastically – and largely uncritically – by the press. Two *Times* leaders on consecutive days eulogised the 'Heroic Age of Physics' and 'Truth and Simplicity'; *The Manchester Guardian* reported 'Imaginative Research' and a 'Quickening Tide of Scientific Advance'; and the local *Liverpool Daily Post and Mercury* waxed lyrical on 'The New Physical Synthesis'.[7]

Rutherford's 1923 address has often been seen as an unproblematic, confident and self-evidently meaningful statement of the historical and contemporary characteristics of emerging nuclear science. In part such a perception can be attributed to the enormous impact that nuclear physics made later in the twentieth century and thus its founders were given a disproportionate voice in forming its history. In a hagiographic 1967 essay on Rutherford, for example, C.P Snow claimed that at the 1923 BAAS meeting Rutherford 'announced, at the top of his enormous voice: "We are living in the heroic age of physics"', and 'went on saying the same thing, loudly and exuberantly, until he died fourteen years later'.[8] For most of Rutherford's audience, however, his address was significant for an entirely different reason. A contemporaneous account of its delivery is revealing:[9]

> Sir Ernest Rutherford had before him, on a table some feet away, a microphone, which looked like a lozenge about six inches across. As his speech was picked up by this little instrument, it was converted

into feeble electric currents, and these currents, after magnification, were relayed forty miles by telephone wire from Liverpool to Manchester. At Manchester, part of the speech current was used to operate the Manchester wireless station, and the remainder passed on to London to operate the London station.

At London, the speech current was connected, again strengthened, and sent by trunk line to Glasgow, Newcastle, Cardiff and Birmingham, and out by wireless from each of these stations, with the result that every word uttered by Sir Ernest – and even the coughs of members of his large audience – were heard simultaneously all over the United Kingdom.

Broadcast live by the nascent British Broadcasting Company to wireless listeners all over Britain using an elaborate system of inter-city telephone lines made available by the Post Office, and the BBC's transmitters around the country, Rutherford's 1923 address was the first simultaneous national broadcast of a major public event.

Trumpeting this 'Miracle of Broadcasting' as 'The B.B.C.'s Biggest Experiment', the inaugural issue of the BBC's magazine *Radio Times* on 28 September described for 'listeners' (the novel use of the term requiring the use of quote marks) the technical and organisational achievements that made it possible. Despite some criticisms of the length of the lecture – Rutherford spoke for an hour and a half, severely testing the endurance of some in his extended audience – the BBC nevertheless reported itself 'overwhelmed with letters of congratulation', with the 'gratifying percentage of fifty-five to one in favour of speeches of this description'.[10] To many 'listeners-in' and contemporary commentators, this astonishing simultaneity, underpinned by a well-promoted conjunction of physics and engineering, a nation-wide communications infrastructure linking the systems of two state institutions (the Post Office and the BBC) and increasingly domesticated but very material wireless technology, seemed like the very height of modernity and progress.

Historians of modernity have become increasingly interested in media and communications technologies and their roles in reframing and reforming the lived experience of modern social and cultural life. The new media technologies of modernity – film, the mass press and communications technologies, from telegraphy through telephony to wireless and organised broadcasting – had profound implications for the public sphere, national identity and governmentality.[11] Wireless was one of the most significant socio-technical phenomena of the interwar years. As Laura Beers and Geraint Thomas point out, in this period the development of the media 'was arguably the principal vehicle driving the creation of a more truly national culture in the 1920s and 1930s'.[12] As well as transforming communication and creating a new form of the 'mass' public sphere and new ideas of citizenship,[13] wireless and the 'new aurality'[14] had diverse and widespread impacts: social (through new listening habits, hobbyism and the age, gender and spatial reprogramming of domestic and social life around broadcasting,[15] and the novel problems of noise pollution which came with a new sonic culture[16]); material (through the creation of extensive physical infrastructure,[17] and the production of new, increasingly design-driven consumer technologies[18]); industrial and economic (broadcasting spawned a large new light industrial manufacturing sector, with implications for gender distribution in the workforce[19]); governmental (not least in policing,[20] military,[21] intelligence and espionage activities[22] and jurisdictional and regulatory issues[23]); and political and ideological (with state,[24] national,[25] international, imperial and broader geopolitical dimensions[26]).

Wireless also helped redefine the form and content of intellectual culture. Historians have begun to map both the ways in which organised broadcasting was incorporated into existing cultural forms (e.g. their representation and signification in contemporary literature and art[27]) and, reciprocally, the ways in which wireless broadcasting allowed those who controlled it to promote various modernisms in literature, music, architecture and other cultural forms.[28] As far as the relationship between interwar wireless and *science* is concerned, however, historians have tended to treat early radio broadcasting simply as a vehicle for science popularisation.[29] Gillian Beer has explored the relationships between interwar radio, popular physics and cultural modernism, and nicely points out that as broadcasting became domesticated and familiar in the late 1920s, wireless became both a useful simile by which to explain some of the more abstract concepts of 'modern' physics and an increasingly important vehicle for that discourse (at least for those prepared to listen).[30] The BBC certainly became an important medium for broadcasting science in the interwar years, both in an expository way and through the airing of debate about the social meaning and values of science; it was a fertile meeting-place for ideas from the worlds of scientific and humanistic culture before C.P. Snow artificially separated them.[31] It may even be that through the choices of its producers, the BBC helped to promote not just scientific values but 'modern' science, just as it promoted 'modern' avant-garde music, for example.[32] But as the contemporary performance and reception of Rutherford's 1923 address affirm, the new medium was not merely a platform for the exposition of modern science: it can be seen at another level as an *exemplar* of it. Even more than this, however, wireless technology and culture fed back into the *practice* of science in ways that have not generally been recognised. In what follows, I want to explore how wireless shaped physics – Rutherford's physics.

Not least because of its characterisation by many contemporaries as 'modern' (or any of numerous synonyms) and as distinct from an antecedent 'classical' physics, the new early-twentieth-century science of radiations, quanta, atoms and nuclei is often taken alongside relativity as an emblematic achievement of modernity, particularly when cast as the origin of such an earth-shatteringly 'modern' outcome as the atomic bomb.[33] This fixed star in the historiography of modern physics has undoubtedly led many historians to assume that these predecessor sciences were epistemologically and ontologically stable. In that sense the development of relativity, quantum physics and nuclear physics are often held as paradigmatic achievements of scientific modernity. Much analysis of cultural modernism and the sciences therefore tends to essentialise this 'modern' physics and to see it as a given – a grounded, self-evident domain which indeed could be appropriated, reflected and refracted by other contemporary cultural forms as they sought to create, sustain or modify their own senses of modernity.[34] Yet the 'modern' physical sciences were *not* epistemologically uncontentious or ontologically fixed.[35] Thanks to the work of Richard Staley and Imogen Clarke, among others, we are now coming to understand that 'modern' physics was a contingent accomplishment, an assemblage whose content and boundaries were negotiated and negotiable, rather than fixed, and whose projected disciplinary identity was an actors' construction, pitched against an antecedent (and equally constructed) 'classical' physics, rather than a self-evident reflection of natural categories.[36]

Despite the hubristic public claims of Rutherford, broadcast in his 1923 BAAS address and promulgated widely elsewhere in the 1920s, the knowledge-making practices of what would come to be called 'nuclear physics' were in fact fragile, and drew on material and other resources from the contemporary surroundings. In particular, during the 1920s, as wireless technology and the practical skills to manipulate electronic equipment became more widespread, Cavendish physicists appropriated and deployed them to construct new forms of instrumentation which were widely reproducible, which created new forms of relatively stable experimental practice and which allowed the effective multiplication of contexts in which nuclear science could be carried out. Nuclear science, we are beginning to understand, was itself fundamentally shaped by the burgeoning interwar wireless industry. In what follows, I explore some of the ways in which cutting-edge wireless practices were shared, developed and disseminated through an institutional and social form characteristically associated with modern broadcasting: the wireless society.

Through an exploration of the activities of the Cambridge University Wireless Society (CUWS), I aim to show the extent and depth of modern wireless culture in interwar Cambridge University, and to demonstrate some of the ways in which young researchers, expert in wireless technique from constructing and adapting their own wireless sets in the context of broadcasting, were able to use the social and technical spaces created by the CUWS to develop links with industrial, governmental and military researchers in ways which helped channel their skills into physics research. In addition, a small but significant overlap of membership between the CUWS and the Signals Section of the Cambridge University Officers Training Corps (CUOTC) illustrates an important and hitherto unexamined intersection between military interest in modern wireless techniques for communications and researchers' interest in the same techniques for use in the university science laboratory. I conclude by asking whether the result, a reflexively modern techno-scientific enterprise, constituted a *scientific* modernism comparable to the literary and other cultural modernisms with which we are more familiar. We begin by returning to Rutherford and the Cavendish Laboratory in the early 1920s.

Experimentation at the Cavendish

When Rutherford arrived at the Cavendish Laboratory in 1919, he brought with him a programme of research into radioactivity and the structure of the nucleus, the intensely charged central core of the atom that he himself had identified during his time at Manchester. He gradually built up a research group using various techniques and technologies to explore atomic constitution.[37] In keeping with the dominance of 'bomb historiography', we tend to think of the interwar Cavendish as being mainly about nuclear physics. However, throughout the period the Cavendish also had a substantial research group using wireless to explore the ionosphere. Headed until 1924 by Edward Appleton, like Rutherford a former J.J. Thomson student, this group had strong connections with industry, the BBC, departments of state and the military, for which Appleton acted as a consultant.[38] When Appleton himself went to a professorship at King's College London in October 1924, his own student Jack Ratcliffe took charge of the ongoing radio-ionospheric work at the Cavendish.[39] Aitor Anduaga and Chen-Pang Yeang have given us detailed accounts of wireless and ionospheric research at the Cavendish

Laboratory (and elsewhere) in the 1920s, emphasising the links between academic, industrial and military laboratories, as well as the National Physical Laboratory, the DSIR's Radio Research Board laboratory, the BBC and the General Post Office, in the wider context of imperial tele-communications projects. Working at a variety of locations in Cambridge, this small but extraordinarily well-connected research group operated alongside Rutherford's, and placed the Cavendish directly between the state and the emerging construct of the ionosphere.[40]

Against the success of the applied radio and ionospheric research at the Cavendish, Rutherford's own programme of nuclear research was in the doldrums by the mid-1920s. Soon after the up-beat message of his 1923 BAAS Presidential Address, his research became mired in controversy when a small group led by Hans Pettersson at the Institut für Radiumforschung in Vienna began its own disintegration experiments using similar techniques to those deployed at the Cavendish. It found very different results from those obtained in Cambridge, leading to increasingly acrimonious exchanges in print about the reliability of the experimental results in each laboratory.[41] The controversy was excruciat-ingly embarrassing for the Cavendish, not only because of Rutherford's hubristic public pronouncements and the presumed international dominance of the Cavendish in nuclear science, but because public controversy threatened to undermine both the epistemological certitude the physicists sought to project and the putative value of scientific inter-nationalism through which they sought to gain moral and intellectual authority in a world whose values had been shaken by war.

The controversy came to a crux in late 1927, when Rutherford's deputy James Chadwick visited the Vienna laboratory in person, and took charge of the experiments. He ran a series of control trials in which he was able to demonstrate to his own satisfaction, and to that of Stefan Meyer, Director of the Vienna laboratory, that the optical particle-counting techniques employed there were unreliable – that is, did not correspond to those developed and deployed at Cambridge over several years – and that the Viennese experimental results therefore could not be trusted.[42] One of the most significant consequences of Chadwick's visit to Vienna and his uncovering of the discrepancies between the experimental protocols used in Vienna and Cambridge was the stark realisation – in both places – of the fragility of existing particle-counting methods, and the difficulty of establishing standardised, replicable and robust methods for achieving reliable and reproducible results. The response to these epistemological troubles was threefold: a new receptivity by experimentalists to the kinds of mathematical theory deprecated by Rutherford in his 1923 BAAS address, particularly the new wave mechanics associated with Heisenberg, Gamow and others, which seemed to offer new interpretative possibilities;[43] a turn to machines – the particle accelerators that would come to define a new regime of ever-larger and ever more expansive and expensive 'Big Science' that continues to the present;[44] and the use of electronic counting methods to create more stable and reproducible detectors than the optical techniques hitherto deployed.[45]

The large accelerators – cyclotrons, linear accelerators, Van de Graaf generators – that increasingly came to replace radioactive sources of subatomic projectiles relied on the development of the electrical industry and materials capable of withstanding (and techniques for manipulating) high voltages. At the Cavendish, John Cockcroft was able to draw on his electrical engineering background and his association with the Manchester electrical engineering company Metropolitan-Vickers to develop particle accelerators.[46] As Peter Galison has well demonstrated, electronic techniques took rapid hold in nuclear science labs in the late 1920s and early 1930s, complementing photographic techniques for capturing and representing sub-atomic phenomena.[47] These electronic particle detectors, too, drew on industry – the burgeoning radio industry that mushroomed alongside organised broadcasting in the 1920s. But how were the technologies and practices of broadcasting and modern 'listening-in' parlayed into state-of-the-art instrumentation in sub-atomic physics?

The answer lies in a culture of young wireless devotees whose well-honed skills with valves and circuits allowed them to manipulate electronic devices in such a way that they could be tailored specifically for use in nuclear experiments. They were part of a much broader culture of wireless constructors and enthusiasts in the 1920s, ranging from professional physicists and electrical engineers, through semi-professional hobbyists to amateurs with sophisticated technical skills. Collectively they constituted a new cadre of technical adepts at the cutting edge of wireless technology. During the 1920s a wide range of magazines were published to cater to constructors, experimenters and listeners in wireless, including *Popular Wireless, Amateur Wireless, Wireless Constructor, Practical Wireless* and *Wireless World,* several of them publishing articles of a demanding technical standard.[48] Asa Briggs notes that by about 1921 there was 'usually at least one "wireless enthusiast" in every village', some of whom 'passed into the radio business either as salesmen or manufacturers'.[49] At the same time, wireless clubs sprang up around the country as wireless amateurs proliferated. Early examples (e.g. Derby, founded 1911) sometimes acted as institutional bases for amateur wireless transmissions before regulated broadcasting came into existence in the early 1920s. By November 1922, when public service broadcasting began with the inauguration of the British Broadcasting Company's 2LO station, there were over a hundred wireless societies in towns up and down Britain.[50]

Wireless clubs and societies connected enthusiasts. They provided a space in which lectures and technical demonstrations could further members' knowledge, skills and materials (including, sometimes, broadcasting equipment) could be shared. And, through the Wireless Society of London, to which 63 wireless societies across Britain had become affiliated by 1921, they acted as a focus for lobbying on regulation, licensing and interaction between the enthusiast community and the Post Office (as the state's regulatory agency for wireless), and were particularly effective in agitating for the relaxation of wartime Defence of the Realm Act (DORA) restrictions on wireless use.[51] When some measure of decontrol came in 1924, one of the foremost public authorities on wireless, W.H. Eccles (a former assistant to Marconi, and from 1919 Vice-Chairman of the Imperial Wireless Committee) observed that:

> The Radio Society and its affiliated Societies comprise every kind of amateur – the home constructor who has fallen a victim to the fascination of making or improving tuners and amplifiers, the ripe enthusiast who welcomes morse as much as music and constructs reflex circuits and other modern marvels, and that matured transmitter who fishes in the oceans of space ....[52]

This distinctly modern techno-scientific social space of the wireless society was to be found in Cambridge too. The Cambridge University Wireless Society was established in 1920. Its origins lay in the pre-war work of J.J. Thomson's Cavendish Laboratory, a reservoir of skilled practice in gas discharges and vacua – the technologies essentially underlying valves. During the war, a number of Cavendish physicists had gone to Marconi's experimental establishment at Brooklands, and were soon seconded to the Royal Flying Corps, where they continued their work on valve development and wireless technique. Richard Whiddington, one of J.J. Thomson's students, joined a cell of other former Cavendish students – Owen Richardson, Frank Horton, Gilbert Stead and George Bryan – at Imperial College, London, to work with valve manufacturer S.R. Mullard in developing the high vacuum R5 receiving valve which was introduced into military service towards the end of the war.[53] By 1918, academic physicists had proved themselves invaluable as operatives, innovators and researchers at various levels, and particularly in wireless.[54] Eccles noted that 'when the war came, the amateurs penetrated in their hosts into the armies, and turned their wireless experience and their talents to the design, construction, operation and improvement of apparatus for use in war.'[55] After the Armistice, many of those who had served in signals during the war returned to their universities with their technical knowledge and ambitions considerably expanded and an extended sense of confidence with wireless technique. At the same time, with support from the popular press, Marconi and other manufacturers began peacetime experiments in broadcasting, though the Post Office (the regulatory body for wireless) sought to control the extent of such development so as to protect the airwaves for military and imperial communications.[56]

It was as these researchers returned to the university, joined by a new generation of wireless-attuned freshmen undergraduates, and against a background of vigorous public debate over the possibilities of broadcasting, that the Cambridge University Wireless Society was founded on 13 October 1920. Dominated by physicists and engineers, the Society quickly affiliated with the Wireless Society of London, and the Society's officers agreed that 'Colonel Stratton and Sir Ernest Rutherford be asked to become Vice-Presidents of the Society', and that Edward Appleton should be asked to become an honorary member.[57] Astrophysicist Frederick Stratton had been involved with the University's Rifle Volunteers since his arrival in Cambridge in 1901 and the Cambridge University Officers Training Corps from its formation in 1908, had helped introduce wireless to the CUOTC in 1910, and served with distinction in the Royal Corps of Signals during the war.[58] Rutherford had worked on wireless as a research student at the Cavendish in the 1890s, and on acoustic devices for submarine detection during the war.[59] Appleton had become involved in military wireless through his pre-war membership of the CUOTC, and had acted as a wireless instructor during the war.[60] This connection with the CUOTC would continue to be significant, as we shall see, and clearly the fledgling CUWS drew heavily on wartime experience. At the first formal meeting of the CUWS on 31 October 1920, for example, Stratton lectured on the development of wireless in the British Expeditionary Force. This military ethos dominated its early existence.

With a membership subscription fixed initially at 2/- per term, the Society met roughly monthly in term-time thereafter, each meeting being organised around a talk given either by a member of the Society or by a visiting speaker. In November 1920, Mr Wynn read a paper on 'Duplex Telephony in Aircraft' and '[t]he President described a 4 valve receiver of his own manufacture, following which a general discussion took place'.[61] In January 1921, '[a]t the conclusion of private business, Mr. Beale described & demonstrated a three valve amplifier, a universal timer and other apparatus of his own design and manufacture'. In February, one of the very first invited speakers from outside the Society's own membership was Admiral Sir Henry Jackson, a pioneer of military wireless and now Chairman of the new DSIR Radio Research Board, who gave a paper on 'The Educational Value of Wireless'.[62] Several of the early 'academic' speakers, too, were closely connected with the military applications of wireless. In February 1922 the Society's guest was Richard Whiddington, the former Cavendish student who had worked on valves and wireless signalling during the war, now Professor of Physics at Leeds, who spoke on 'Wireless Circuits in Minute Physical Measurements'.[63] In January 1923, Gilbert Stead of the Cavendish, one of Whiddington's wartime colleagues, spoke on the 'Development of the Thermionic Valve During the War'. He had originally served at HMS *Vernon*, the Admiralty Signal School, and 'gave some account of the various difficulties encountered during the early days of triode development & some specimens of early valves produced at the Signal School were passed round for inspection'.[64]

As these examples indicate, the Society drew on the wartime experiences and contacts of its members to maintain strong connections with the military scientific-technical establishment. Among the other military personnel invited to speak were Flt. Lt. De Burgh of the RAF (6 November 1922), G.W.N. Cobbold of the Signals Experimental Establishment, Woolwich (15 November 1922) and Flt. Lt. Hugh Leedham who gave a talk entitled 'Wireless in the Air Force Overseas' on his experiences using wireless in aircraft operations in Iraq (7 November 1923); and N. Hecht of the RAF Wireless Department who spoke on 'Aerials' (9 February 1925). In addition to these links with the military at the national level, the Society maintained a close connection with the CUOTC, with several prominent members belonging to both groups, and Stratton both a Vice-President of the CUWS and Commanding Officer of the OTC signals section. This connection assumed material form in 1921, when the Society obtained a room in the Engineering School building and discussed keeping its own permanent set of apparatus and applying for a receiving licence.[65] Later that year it agreed that the CUOTC 'be allowed use of the Society's room & aerial on Sunday afternoons'.[66]

As well as its close military ties inside and outside the university, the CUWS acted as an intermediary between Cambridge wireless enthusiasts and world of radio in industrial and government laboratories. Among those who came to speak were Peter Eckersley of the Marconi Company (20 November 1921, before his move to the BBC); E.H. Shaughnessy of the Engineer-in-Chief's Office of the General Post Office (6 March 1922); J. Hollingworth of the National Physical Laboratory (20 November 1922, 17 November 1924); Eckersley again, this time as Chief Engineer of the BBC (22 October 1923); and R.L. Smith-Rose and F.M. Colebrook of the National Physical Laboratory Radio Department (19 November 1923 and 23 February 1925 respectively). Renowned valve designer Captain C.F. Trippe, formerly of the Royal Engineers and now of the Osram Valve Company (3 December 1923) 'described in detail the manufacture of a typical receiving valve ... [and] ... modern methods of obtaining high vacua. He then outlined the manufacture of other types such as "dull emitters" and transmitting valves, pointing out where the technique differed from that of the common receiving valve'.[67] The vote of thanks to Trippe was given by Rutherford, now himself a minor radio celebrity, and clearly a regular attender at CUWS meetings.

In addition to inviting speakers, the Society organised a visit to the works of Marconi at Chelmsford (1 March 1922), bringing its members into close contact with industrial researchers and industrial radio practice. This was characteristic of a shift in the activities of the CUWS through the mid-1920s. With the formalisation of organised broadcasting and the expansion and diversification of the commercial wireless industry, the Society's members were increasingly interested in the fast-developing technological landscape of wireless, so that speakers from the new and rapidly expanding BBC and the radio industry were much in demand.[68] As well as Eckersley, for example, H.L. Kirke of the BBC Chief Engineer's Department came to speak on 'Valve Transmitters' (3 November 1924) and 'Single Wavelength Working' (26 November 1928); and A. Cameron of the engineering staff lectured on 'Broadcasting Stations' (19 October 1925). Researchers from the National Physical Laboratory and the Post Office Research Department at Dollis Hill were also frequent visitors, emphasising the links developed between the CUWS and state scientific and technical institutions.

This shift was in significant part demographic. With the rapid development of organised broadcasting, the number of wireless amateurs and enthusiasts in Britain increased substantially from the mid-1920s. A new generation of boys and young men, probably already technically minded and mechanically dextrous from construction toys like Meccano, grew up constructing cat's whisker crystal sets and tuning in to whatever broadcasts they could in the still sparsely populated ether of the early 1920s. Numerous children's books fostered interest in and enthusiasm for wireless-mindedness specifically among young boys – in *Tony D'Alton's Wireless* by Arthur Russell, for example, the exemplary young hero, '"Sparks" D'Alton, the eminent young amateur experimenter of Wilbermere College!' embarks on a series of wireless-based japes and pranks in a private educational establishment, indicating the book's likely readership.[69] For more advanced readers, Oliver Lodge (one of the fathers of wireless and the paterfamilias of British physics) and many others produced popular books on wireless and ether.[70] Indeed, in February 1923, Lodge drew a large crowd at the CUWS for his talk on 'The Roots of Wireless' – for which the vote of thanks was again given by Rutherford.[71]

By the late 1920s, proprietary wireless equipment was becoming affordable thanks to the mass production of valves, kit sets for home construction and so on, and growing numbers of youngsters immersed themselves in the world of valves and circuits.[72] Successive cohorts of students then arrived at university skilled to various degrees in the arts of wireless circuitry and valvesmanship, sometimes fostered and developed at their (often public) schools. In this context, the Society's military connections became less important than they had been immediately after the war when many of the CUWS's first members had military experience. Informal meetings now assumed much more importance in the Society's activities. Typically held in someone's college rooms (often Ratcliffe's at Sidney Sussex), such meetings would include presentation of an informal paper or commentary on a topical issue as a prelude to a broad discussion. With the Society's new atmosphere of informality and a field developing at breathtaking speed both technologically and commercially, the CUWS also offered a space for these adepts to exchange practical information at an informal level and, increasingly in the 1930s, materials and techniques. It increasingly emphasised actual radio lab and workshop practice (witnessed by the increasing number of 'show-and-tell' presentations), and the Committee announced in 1930 that 'discounts were available to members of the Society with a number of electrical firms' in the town.[73] With regular sales of 'junk' (redundant) equipment, the art of 'scrounging' electronic hardware became one which could be shared with like-minded enthusiasts on social occasions. Following the trend established in the mid-'20s, visiting speakers from industry and the world of broadcasting predominated in the late '20s and early '30s. M.C. Hoxie of RCA Photophone Inc., S.A. Stevens of the Westinghouse Co., J. Hollingworth, R.H. Barfield and R.L. Smith-Rose of the DSIR Radio Research Board and Noel Ashbridge, the new Chief Engineer of the BBC, all appeared on the programme in 1929 and 1930.

Physics research students from the Cavendish Laboratory continued to dominate the Society's Committee. From the radio side, Ratcliffe was a

Committee member for several years, and his student Eric White was elected Secretary in 1929. White was succeeded in 1930 by Shirley Falloon, a Queen's College undergraduate physicist, dedicated radio enthusiast and inveterate gadgeteer with 'green fingers' for electronic equipment.[74] Virtually all of Appleton's and, subsequently, Ratcliffe's students in the Cavendish radio group were CUWS members – but so were a significant proportion of other Cavendish research students – among them the ex-Metropolitan-Vickers apprentice John Cockcroft, who was a Committee member.[75] We shall return to him shortly. In October 1931, Cavendish research student Wilfrid Bennett Lewis was elected President of the Society. Lewis's career in modern wireless is exemplary. As a boy, Lewis 'acquired a large collection of Meccano but his real passion was for electrical apparatus and wireless'. A bench in his bedroom was 'covered with a succession of crystal sets ... progressing to thermionic valves and short-wave radios'. This interest continued through his time at Haileybury School, where he also joined the Officers' Training Corps.[76] When he arrived at Cambridge to read Physics it was inevitable that he should become involved with the CUWS, and his first performance at an informal meeting was in October 1929, with a paper on 'The Detector and the Problem of Selectivity without Reproduction of Side-bands'.[77]

As CUWS President, re-elected year after year in the 1930s, Lewis continued the practice of structuring the Society's programme around numerous speakers from industry and the civil establishments. Visitors from the Mazda Valve Company, the Telegraph Condenser Company, Murphy Radio, the General Electric Company and others graced the Society's meetings, often to large audiences. H.M. Dowsett, Research Manager of the Marconi Wireless Telegraph Company, increasingly involved in the development of an imperial wireless communication network, had an audience of 400–500 enthusiasts for his talk in November 1933.[78] Lewis also increased the frequency of visits to industrial labs, with the Mazda Valve Works at Enfield and the Post Office transmitter at Rugby featuring on the term-card in 1932–1933.[79] In keeping with his reputation as the enthusiast's enthusiast informed on every aspect of wireless practice, Lewis significantly expanded the informal and cooperative aspects of the Society's activities. In February 1933 the informal meeting consisted of a 'junk sale', with coffee and cigarettes provided to lubricate commerce. In 1934, Lewis began a 'spot-the-errors' competition, in which talks by CUWS members contained numerous deliberate errors which the others were invited to detect. The winner, C.H. Westcott – yet another Cavendish research student – 'spotted no less than 7 errors and was accordingly awarded the prize of free membership of the Society for one year'.[80] At the Society's Annual General Meeting in June 1935, Lewis 'opened a discussion on scrounging which became too informal to record coherently, but the following items may be noted. A cheaper substitute for Plymax was said to be Metaplex sold by Peto Scott for 1s/3d per sq. ft. Igranic were said to make a most reliable series of microphones. Some firm at Norwich was reputed to be excellent and cheap for rewinds'.[81] Even modern institutions must change with the times, and Lewis's CUWS clearly adapted to the changing technological and commercial environment with a new set of informal social practices.

Lewis brings us directly back too to the issue of the impact of state-of-the-art wireless technique in academic research. In his CUWS Presidential Address for 1933, he distinguished between the mere amateur and the serious enthusiast, sneering that 'if you derive your knowledge of radio developments from the Wireless World, World Radio and the other weeklies or more weaklys [sic] you must not suppose that your knowledge is up to date'.[82] Lewis's derisive joke speaks to the wireless enthusiast's ultra-modernity, and emphasises the extremely high levels of technical competence that he and his colleagues sustained and to which they urged others to aspire. This advanced expertise put him in an excellent position to contribute to the changes then taking place in the Cavendish nuclear science programme. As we have seen, part of the Cavendish response to inter-laboratory controversy in the mid-1920s was to develop new, more readily replicable forms of instrumentation. Commercially available wireless valves and complex circuits were combined to make relatively robust amplifiers for use in particle counters, for example.[83] In January 1932, these new electrical counters were crucial to the experiments which produced James Chadwick's discovery of the neutron. Later the same year, similar apparatus was used by CUWS member John Cockcroft and Ernest Walton to check the results of their experiments on the artificial disintegration of nuclei (Figure 11.1). With his skills and a reputation for being able 'to smell which valve or resistance in a set is giving trouble',[84] Lewis was an obvious recruit to the Cavendish electronics team. He had hoped to do research 'in anything other than radioactivity' (presumably radio, with Ratcliffe). But shortly before his graduation in 1930 he had been interviewed by Rutherford, who said: 'I am told you understand about these wireless valves. We are just beginning to use them in our alpha-ray work. If you get through your exams alright I would like you to join our group.'[85] With a DSIR studentship, Lewis duly entered the ranks of the research students at the Cavendish, joining Eryl

Fig. 11.1   John Cockcroft using electrical particle-counting equipment at the Cavendish Laboratory, 1932.

Illustrated London News, 11 June 1932, 970. Courtesy Mary Evans Picture Library.

Wynn-Williams and Francis Ward in the successful development of valve amplifiers for particle-counting.[86]

Over the next few years, Lewis's increasingly indispensable electronics skills brought him career advancement: a DSIR Senior Studentship (1932–4), a Gonville and Caius College Research Fellowship (1934), and university posts as Demonstrator (1935) and Lecturer (1937). Alongside his academic publications, he published several technical articles in *Wireless World* and the more specialist *Experimental Wireless and Wireless Engineer,* and followed debates in the *Proceedings of the Institute of Radio Engineers* and other professional journals.[87] As President of the CUWS he continued the programme of speakers from industry, the BBC, the NPL and the Post Office Research Department, and maintained an active itinerary of visits, including trips to the Marconi works at Chelmsford and the Research Laboratories of the General Electric Company at Wembley. But Lewis also brings us back to the question of the military connections of the Society. As we have already seen, Lewis had been a member of the Haileybury School OTC, and in 1931 was commissioned as a second lieutenant in the Territorial Army. As he pursued his research alongside Wynn-Williams and others at the Cavendish, he was simultaneously involved in the signals research section of the Cambridge University OTC. In his CUWS Presidential address for 1933, Lewis reviewed the history of the Society. He drew attention to its close connections with the OTC, and pointed out that Ratcliffe had at one time been simultaneously Commanding Officer of the OTC signals section and President of the Society, 'and now I have the honour to occupy the same position.' Lewis echoed the words of W.H. Eccles a decade earlier when he noted that:

> In these days when the word wireless suggests broadcasting or hams it is possible to forget that the pioneer personnel was mainly provided from naval & army signals; but one does not have to go far out of the broadcasting [realm] to realise that these still provide a solid background. The majority of the Presidents of this Society have been officers of the Royal Corps of Signals or of the OTC.[88]

Against a background in which pacifist sentiments were taking hold among leftist Cambridge undergraduates (the Cambridge Scientists' Anti-War group was established in 1932),[89] Lewis sought to strengthen links between the CUWS and the CUOTC. He led from the front, directing his electronics skills to the purposes of military modernity. Under contract from the War Office, he worked with fellow physicists John Kendrew and later Christopher Milner on the development of ultra-short-wave duplex telephony – walkie-talkies – in a makeshift laboratory in the OTC's hut on a rifle range on the outskirts of Cambridge (Figure 11.2).[90] With a characteristic flair for gadgetry and publicity, he and his cadre used the walkie-talkie set to provide intra-stadium communications during a 1938 White City athletics event with (according to Kendrew) 'quite entertaining' success.[91] But the technology was intended for military use (Figure 11.3), and by 1937, the Cambridge section advertised itself as having the only signals company in the whole of the OTC, and prided itself on its technical accomplishment and its 'up-to-date field sets'.[92]

Lewis and his peers could bring their skills with wireless valves and circuits to bear both in the academic laboratory and in the context of military innovation, and under his Presidency the CUWS sat between these two sites of manipulation of a thoroughly modern technology for thoroughly modern purposes.

Fig. 11.2 A member of the CUOTC Signals Section – probably John Kendrew – using the duplex radio-telephony set developed by Ben Lewis, c.1937.

Cambridge University Archives, CUOTC 7/8/iii. Courtesy Cambridge University Officer Training Corps.

Fig. 11.3 The Duke of Gloucester inspecting CUOTC wireless operation, c.1938.

Cambridge University Archives, CUOTC 7/8/iii. Courtesy Cambridge University Officer Training Corps.

Wireless and nuclear physics

'During the last ten years,' argued the journalist and cultural commentator Gerald Heard in a BBC broadcast in 1930, 'we have entered the Age of Radiation.' 'In twenty years,' he suggested, 'the understanding of radiation has completely changed our idea of what life is and what the solid world is':[93]

> It is the fundamental fact of our lives. We look through solids by radiation. We attack disease by radiation. The entire world can communicate without any physical contact by radiation … Perhaps you saw in the papers a few days ago a vivid illustration of this last change. The Rumanian government has had to deluge the ether with a barrage of sound to prevent some station in Russia shouting propaganda into every Rumanian listener's home. Such a crude fact shows startlingly how radiation, applied as broadcasting, has changed the world even since the war.

Radiation was 'therefore, of all the high roads of science the one which to-day is carrying the most traffic'. Heard's insightful characterisation of radiation's transformative effects across both the sciences and communication media beautifully captures contemporary recognition of its significance in interwar modernity. At one end of the spectrum, in the form of wireless technology and an emerging national broadcasting system, it re-shaped everyday life and habits and, more broadly, conceptions of national culture and identity. At the other, through the technologies and skilled techniques developed in a generation of wireless enthusiasts in the context created by broadcasting, it re-shaped the practice and content of the most recondite – and allegedly 'pure' – forms of science. The Cavendish Laboratory in the 1920s and 1930s was saturated with wireless culture, and not just in the ionospheric research group. In a BBC discussion in 1934, Patrick Blackett agreed with Julian Huxley that 'the needs of improving efficiency of radio transmission have led to very fundamental discoveries about the upper atmosphere':

> Yes. The work of Appleton and others on the highly conducting upper layers of the atmosphere has given us absolutely new knowledge about our own planet, and opened up a fascinating field of pure research. And the work arose directly out of the discovery that it is possible to send wireless messages round the earth.

Prompted by Huxley, Blackett extended the argument to nuclear science, the exemplar of 'purity'. Dismissing the 'sealing wax and string' mythology of the Cavendish as old hat, he observed that

> Lord Rutherford's own experiments require an apparatus of extreme complexity: innumerable valves and rows of thyratrons flashing, relays clicking, and so on. It looks rather like a cross between the advertisement lights in Piccadilly and the transmitting station of a modern battleship.[94]

The image is illuminating. Blackett captures the essence of practicalities of early 1930s' nuclear physics, and where its modernity actually lay: in cutting-edge wireless technique. One way and another, Rutherford's own career continued to cross and re-cross the worlds of wireless. He broadcast on the BBC several more times after his 1923 debut, became a member of its Panel of Advisers for Broadcast National Lectures in 1928, and joined its first General Advisory Board in 1935, where he helped shape policy.[95] In 1930 he made a transatlantic broadcast to the Royal Society of Canada via an 'all British route', and in June 1933 he again demonstrated his imperial credentials 'and the possibilities of long distance wire and radio telephony' when he broadcast from Cambridge to the Fifth Pacific Science Congress in Vancouver.[96] By this time, as a scientific statesman and recent President of the Royal Society (1925–30), he was becoming more concerned with the social ramifications of science in the face of the Depression, and in a number of public speeches began to reflect on the modern condition. Addressing the Institute of Mechanical Engineers in 1932, he situated the emerging discipline of nuclear physics alongside other noteworthy modern phenomena:

> This age will be for ever memorable for the development of new and swift methods of transport over land, sea, and in the air. The spectacular advances in the speed of the motor-car, and motor-boat, and still more of the flying machine, are familiar to you all and are triumphs of which the engineer may be proud.

Extending this characteristically modern trope of speed, Rutherford explicitly placed 'Atomic Projectiles and their Applications' – particularly the work of Cockcroft and Walton on the electrical acceleration of sub-atomic particles – in the context of recent, terrible memory of the power of the German 'Big Bertha' gun which had bombarded Paris during the Great War.[97] Outlining 'Trends in Modern Physics' to chemical engineers at London's Waldorf Hotel in 1934, he referred to the 'startling rapidity' of social and economic change:[98]

> In the last 30 years we had seen the advent of the motor car, of the flying machine, of wireless, and of broadcasting, not to mention the widespread use of automatic machinery. These great advances, whether connected with rapidity of transport or rapidity of transmission of speech, had had very great repercussions on the intellectual as well as the social structure of the world. These developments had all come, in a broad sense, as a consequence of scientific investigation and from the application of science to industry in manifold ways.

Here, he took the conventional line in which wireless and broadcasting were the products of science and its application. Two years later, surveying the 'Newer Alchemy' of transmutable elements in the new field of nuclear physics, he reversed the connection, and explicitly pointed out that the 'rapid advance in recent years of our knowledge of the transmutations of the ordinary elements has been largely due to the discovery of powerful methods of detecting and counting individual atoms', including 'amplifiers which are specially adapted for the purpose'.

Automatic registration and counting systems were important, and '[i]ngenious methods for this purpose have been devised by Dr Wynn-Williams and are in general use in many laboratories'.[99]

This description of the role of modern wireless technology and techniques in the shaping and stabilising of nuclear science was paralleled in other scientific fields well beyond the Cavendish and physics. As we have seen, the members of the CUWS were students enrolled on a very wide range of degree courses spanning the sciences, engineering and humanities, united more by their interest in wireless and wireless technique than by any disciplinary affiliation. Those techniques could be, and were, applied in other research fields – notably physiology. Following early efforts by his mentor Keith Lucas, Edgar Adrian built on the work of the visiting American Alexander Forbes – physiologist and wireless enthusiast – to introduce valve amplifiers into work at the Cambridge Physiological Laboratory in the early 1920s.[100] In the mid-1920s, drawing on work by Erlanger, Gasser and Newcomer, Adrian and a series of co-workers and visitors including Sybil Cooper, Yngve Zotterman and Detlev Bronk used a three-stage valve amplifier with a capillary electrometer to measure nerve impulses.[101] Joining Adrian's project, young researcher Bryan Matthews – interested in radio from his days at Clifton College and a member of the CUWS – developed more sophisticated and stable valve-amplification techniques and related recording instrumentation which allowed impulses from individual nerve fibres to be registered and recorded.[102] Matthews went on to work with D.H. Barron to develop still more sophisticated valve-amplification techniques for electro-physiological work. By 1935, he could write that

> [t]he revolution which the thermionic valve has brought about in instrumental methods and possibilities is influencing biological technique no less than that of physics … In electro-physiology the valve has opened up vast new territories which as yet have only begun to be explored.[103]

The London physiologist A.V. Hill – another former member of the Signals Section of the Cambridge OTC – noted in 1929 that with the aid of

> the tools which wireless engineers have provided, single nerve waves in single nerve-fibres can be amplified and recorded, and a whole new world of hurrying, scurrying activity is opened up to analysis … Not only can the separate waves in an afferent nerve-fibre be registered photographically, but they can even be transformed into sound and heard in a loudspeaker.

At a meeting of the Physiological Society, reported Hill, Adrian's American collaborator Detlev Bronk 'had a fine electrode thrust into an arm muscle, and the resulting action currents were led off to an amplifying system and thence to a loud speaker. As he varied the force exerted by his muscle, the pitch of the sound emitted by the loud speaker waxed and waned; at first a rattle like a machine gun, and finally a musical note; one *heard* a single human muscle-fibre varying the strength of its contraction.'[104] Like Rutherford's demonstrations of glowing valve amplifiers and clicking Geiger counters in his public lectures at the Royal Institution and elsewhere, this modern scientific theatre gave a vivid and persuasive display of the applications and transformative possibilities of the 'modern amplifying valve' in the 'pure' sciences.[105] Adrian himself confirmed the thoroughly constitutive role of wireless technique in electro-physiology in no uncertain terms:

> When the academic scientist is forced to justify his existence to the man in the street he is inclined to do so by pointing out the essential part played by academic research in the development of our modern comfort. It is only fair, therefore, to point out that in this case the boot is on the other leg and the academic research has depended on the very modern comfort of broadcasting.[106]

Modern wireless helped constitute new, modern sciences. And now we have a clearer understanding of why this was so. The widespread nature of wireless culture and its social and technical practices may help to explain the rapid development of experimental nuclear physics as an international discipline in the early 1930s, and further studies of wireless culture in its other leading institutions would doubtless help explain the dissemination of electronic techniques and the constitution of a relatively stable and reproducible experimental regime.[107]

At the cutting edge of wireless technique, and moving between the formality of speaker meetings and the informality of junk sales and discussions on scrounging, the CUWS brought together skilled researchers from diverse backgrounds and exposed them to the latest research from the BBC, the Post Office, the National Physical Laboratory, the DSIR's Radio Research Station at Slough, industrial labs like GEC and Marconi, and military signals establishments. Often through former Cambridge students and CUWS members now employed in these laboratories (like Falloon at Marconi), the CUWS inducted its members into networks which help explain patterns of scientific mobilisation for war in the later 1930s for these young modern technophiles, who intermixed the worlds of broadcast wireless and nuclear physics through their extraordinary facility with electronics and circuitry. Lewis and many other CUWS members would go on to play key roles at the wartime Telecommunications Research Establishment, birthplace of British radar; Lewis later headed Canada's postwar nuclear project.[108] Wynn-Williams would go on to play a leading role in the development of electronic code-breaking machines at Bletchley Park – and in the origins of modern digital computing. Several Cavendish postgraduates with electronics skills – among them James McGee, Eric White, Joseph Pawsey, Harold Miller, Frederick Nicoll, Bernard Crowther and John Cairns, almost all members of the CUWS – were recruited by the EMI Company in early 1930s for their research and development programme on television.[109] In more ways than one, the modern media world came out of the interwar Cavendish and its sub-culture of wireless enthusiasts and scroungers. And in that the 'modern' physics of the nucleus was fundamentally shaped by the techniques, technologies and spaces of modernity, we have a scientific modernism every bit as contingently 'modern' as its literary and artistic cultural counterparts.

Acknowledgements

I am grateful to Robert Bud, Frank James, Richard Staley and an anonymous referee for helpful comments and suggestions. I also thank Cambridge University Library for providing access to the papers of CUWS and CUOTC and Queen's University, Kingston, for the Lewis papers.

Notes

1 E. Rutherford, 'The Electrical Structure of Matter', *British Association for the Advancement of Science. Report of the Ninety-First Meeting. Liverpool – 1923*, (London: John Murray, 1924), 1–24, on 1.

2 A. Holt, 'The Liverpool Meeting of the British Association', *Nature*, 112 (6 October 1923): 523.

3 'Scientists in Conference. British Association at Liverpool. A Broadcast Address', *The Times* (13 September 1923): 10.

4 P.C. Mitchell, *My Fill of Days* (London: Faber & Faber, 1937); J. Hughes, 'Insects or Neutrons? Science News Values in Inter-War Britain', in *Journalism, Science and Society: Science Communication between News and Public Relations*, eds.

M.W. Bauer and M. Bucchi (New York and Abingdon: Routledge, 2007), 11–20.

5 'New Achievements of Science. Work of the British Association. Dr. Chalmers Mitchell's View', *Sunday Times* (23 September 1923): 18.

6 'Echoes and Gossip of the Day. The Disappointing Atom', *Liverpool Echo* (13 September 1923): 6.

7 'Heroic Age of Physics', *The Times* (13 September 1923): 11; 'Truth and Simplicity', *The Times* (14 September 1923): 11; 'Sir E. Rutherford's Address. Quickening Tide of Scientific Advance', *The Manchester Guardian* (13 September 1923): 8; 'The New Physical Synthesis', *Liverpool Daily Post and Mercury* (13 September 1923): 6.

8   C.P. Snow, *Variety of Men*, (London: Macmillan, 1967), 1; J. Hughes, 'Unity through Experiment? Reductionism, Rhetoric and the Politics of Nuclear Science, 1918–40', in *Pursuing the Unity of Science: Ideology and Scientific Practice from the Great War to the Cold War*, eds. H. Kamminga and G. Somsen, (London: Routledge, 2016), 50–81.

9   'A Miracle of Broadcasting. The B.B.C.'s Biggest Experiment', *Radio Times* (28 September 1923): 2.

10 'A Miracle of Broadcasting', 2.

11 For general introductions to these issues, see for example T. Misa, P. Brey and A. Feenberg, *Technology and Modernity*, (Cambridge, Mass.: MIT Press, 2003); B. Rieger, *Technology and the Culture of Modernity in Britain and Germany, 1890–1945*, (Cambridge: Cambridge University Press, 2005).

12 L. Beers and G. Thomas, *Brave New World: Imperial and Democratic Nation-Building in Britain between the Wars* (London: Institute of Historical Studies, 2012), 17–18. For a contemporaneous overview of the range of applications and impacts of wireless, see *The Marconi Book of Wireless*, (London: Marconiphone Co. Ltd., 1936).

13 A. Briggs, *The History of Broadcasting in the United Kingdom, Volume 1. The Birth of Broadcasting, 1986–1927* (Oxford: Oxford University Press, 1995 [1961]); A. Briggs, *The History of Broadcasting in the United Kingdom, Volume 2. The Golden Age of Wireless, 1927–1939* (Oxford: Oxford University Press, 1995 [1965]); D.L. LeMahieu, *A Culture for Democracy: Mass Communication and the Cultivated Mind in Britain Between the Wars* (Oxford: Clarendon Press, 1988).

14 M. Cuddy-Keane, 'Virginia Woolf, Sound Technologies, and the New Aurality', in *Virginia Woolf in the Age of Mechanical Reproduction*, ed. P. L. Caughie (New York and London: Garland, 2000), 69–96; D. Morat, ed., *Sounds of Modern History: Auditory Cultures in 19th- and 20th-Century Europe* (New York and Oxford: Berghahn, 2014); R.P. Scales, *Radio and the Politics of Sound in Interwar France, 1921–1939* (Cambridge: Cambridge University Press, 2016).

15 M. Pegg, *Broadcasting and Society, 1918–1939* (London: Croom Helm, 1983); P. Scannell and D. Cardiff, *A Social History of British Broadcasting, Volume 1, 1922–1939. Serving the Nation*, (Oxford: Basil Blackwell, 1991); K. Haring, *Ham Radio's Technical Culture* (Cambridge, Mass.: MIT Press, 2007); M. Bailey, 'The Angel in the Ether: Early Radio and the Constitution of the Household', in *Narrating Media History*, ed. M. Bailey (London: Routledge, 2009), 52–65; M. Andrews, 'Homes Both Sides of the Microphone: The Wireless and Domestic Space in Inter-War Britain', *Women's History Review*, 21 (2012): 605–621.

16 K. Bijsterveld, *Mechanical Sound: Technology, Culture, and Public Problems of Noise in the Twentieth Century* (Cambridge, Mass., MIT Press, 2008).

17 E. Pawley, *BBC Engineering* (London: BBC Publications, 1972); B. Hennessy, *The Emergence of Broadcasting in Britain* (Lympstone: Southerleigh, 2005); G. Bussey, *Wireless: The Crucial Decade. History of the British Wireless Industry 1924–34* (London: Peter Peregrinus, 1990).

18 N. Arceneaux, 'The Wireless in the Window: Department Stores and Radio Retailing in the 1920s', *Journalism & Mass Communication Quarterly* 83 (2006): 581–595.

19 K. Geddes and G. Bussey, *The Setmakers: A History of the Radio and Television Industry* (London: BREMA, 1991); B. Vyse and G. Jessop, *The Saga of Marconi-Osram Valve: A History of Valve-Making* (Pinner: Vyse Ltd., 2000); M. Glucksmann, *Women Assemble: Women Workers and the New Light Industries in Inter-War Britain* (London: Routledge, 1990).

20 J. Bunker, *From Rattle to Radio* (Studley: K.A.F. Brewin Books, 1988), 158ff.; K. Laybourn and D. Taylor, *Policing in England and Wales, 1918–39: The Fed, Flying Squads and Forensics* (Basingstoke: Palgrave Macmillan, 2011), 194ff.

21 R.E. Priestley, *The Signal Service in the European War of 1914 to 1918 (France)* (Chatham: W. & J. Mackay & Co., 1921); B.N. Hall, 'The British Army and Wireless Communication, 1896–1918', *War in History* 19 (2012): 290–321; A. Beyerchen, 'From Radio to Radar: Interwar Military Adaptation to Technological Change in Germany, the United Kingdom, and the United States', in *Military Innovation in the Interwar Period*, eds. W. Murray and A.R. Millett, (Cambridge: Cambridge University Press, 1996), 265–299.

22 N. West, *GCHQ: The Secret Wireless War 1900–86* (London: Weidenfeld and Nicolson, 1986) (but see also E. O'Halpin, 'Intelligence Fact and Fiction', *Intelligence and National Security*, 2 (1987): 168–171); R. Aldrich, *GCHQ* (London: Harper Press, 2010), 13ff.

23 H.J.S. Tworek, 'The Savior of the Nation? Regulating Radio in the Interwar Period', *Journal of Policy History*, 27 (2015): 465–491; S. Street, *Crossing the Ether: British Public Service Radio and Commercial Competition, 1922–1945* (Eastleigh: John Libbey Publishing, 2006) usefully reminds us that broadcasting was not wholly synonymous with the BBC in the interwar period.

24 On the BBC's close relationship to the state during the General Strike of 1926, for example, see Briggs, *The History of Broadcasting in the United Kingdom*, vol. 1, 329–351.

25 T. Hajkowski, *The BBC and National Identity in Britain, 1922–1953* (Manchester: Manchester University Press, 2010), esp. 109–134; D.R. Cohen, 'Annexing the Oracular Voice: Form, Ideology, and the BBC', in *Broadcasting Modernism*, eds. D.R. Cohen, M. Coyle and J. Lewty (Gainesville: University of Florida Press, 2009), 142–157.

26 J.M. MacKenzie, '"In Touch with the Infinite": The BBC and the Empire, 1923–53', in J.M. MacKenzie, ed., *Imperialism and Popular Culture*, (Manchester: Manchester University Press, 1986), 165–191; D. Headrick, *The Tentacles of Progress: Technology Transfer in the Age of Imperialism* (New York & Oxford: Oxford University Press, 1988), 126–134.

27 T.C. Campbell, *Wireless Writing in the Age of Marconi* (Minneapolis: University of Minnesota Press, 2006); M. Goble, *Beautiful Circuits: Modernism and the Mediated Life* (New York: Columbia University Press, 2010); D. Rando, *Modernist Fiction and News: Representing Experience in the Early Twentieth Century* (New York: Palgrave Macmillan, 2011); D. Trotter, *Literature in the First Media Age: Britain Between the Wars* (Cambridge, Mass; Harvard UP, 2013); K. Price, 'Gender and the Domestication of Wireless Technology in 1920s Pulp Fiction', in *Domesticity in the Making of Modern Science*, eds. D. L. Opitz et al. (Basingstoke: Palgrave Macmillan, 2016), 129–150.

28 LeMahieu, 'A Culture for Democracy'; J. Doctor, *The BBC and Ultra-Modern Music, 1922–36: Shaping a Nation's Tastes* (Cambridge: Cambridge University Press, 2000); T. Avery, *Radio Modernism: Literature, Ethics and the BBC* (Aldershot: Ashgate, 2006); M. Feldman, E. Tonning and H. Mead, eds., *Broadcasting in the Modernist Era*, (London: Bloomsbury, 2014).

29 A.C. Jones, 'Speaking of Science: BBC Science Broadcasting and its Critics, 1923–64' (unpublished PhD dissertation, University College London, 2010); A. Jones, 'Mary Adams and the Producer's Role in Early BBC Science Broadcasts', *Public Understanding of Science*, 21 (2011), 968–963; P. Bowler, *Science for All: The Popularization of Science in Early Twentieth-century Britain* (Chicago: University of Chicago Press, 2009), 209–214.

30 G. Beer, '"Wireless": Popular Physics, Radio and Modernism', in *Cultural Babbage: Technology, Time and Invention*, eds. F. Spufford and J. Uglow (London: Faber and Faber, 1996), 149–166.

31 J. Hughes, 'Craftsmanship and Social Service: W. H. Bragg and the Modern Royal Institution', in *'The Common Purposes of Life': Essays on the History of the Royal Institution of Great Britain, 1799–1999*, ed. F.A.J.L. James, (Aldershot: Ashgate, 2002), 225–247. For an analysis of Snow's 'two cultures' invoking the interwar Cambridge background to be discussed later in this paper, see G. Ortolano, *The Two Cultures Controversy: Science, Literature and Cultural Politics in Postwar Britain* (Cambridge: Cambridge University Press, 2009).

32 Doctor, *The BBC and Ultra-Modern Music*.

33 On 'bomb historiography', see J. Hughes, 'Radioactivity and Nuclear Physics', in *The Cambridge History of Science, Volume 5. The Modern Mathematical and Physical Sciences*, ed. M.J. Nye, (Cambridge: Cambridge University Press, 2003), 350–374; R. Staley, 'Trajectories in the History and Historiography of Physics in the Twentieth Century', *History of Science*, 51 (2013), 151–177.

34 M. Whitworth, *Einstein's Wake: Relativity, Metaphor, and Modernist Literature* (Oxford: Oxford University Press, 2001); G. Parkinson, *Surrealism, Art and Modern Science* (New Haven and London: Yale UP, 2008); Holly Henry, *Virginia Woolf and*

the Discourse of Science: The Aesthetics of Astronomy (Cambridge: Cambridge University Press, 2003); D. Bradshaw, 'The Best of Companions: J.W. N. Sullivan, Aldous Huxley, and the New Physics', Review of English Studies, 47 (1996), 188–206 and 352–368.

35 J. Hughes, 'Redefining the Context: Oxford and the Wider World of British Physics, 1900–1940', in Physics in Oxford 1939–1939: Laboratories, Learning, and College Life, eds. R. Fox and G. Gooday, (Oxford: Oxford University Press, 2005), 267–300.

36 R. Staley, 'On the Co-creation of Classical and Modern Physics', Isis 96 (2005): 530–558; R. Staley, Einstein's Generation: Origins of the Relativity Revolution (Chicago: University of Chicago Press, 2009); I. Clarke, 'Negotiating Progress: Promoting "Modern" Physics in Britain, 1900–1940' (unpublished PhD thesis, University of Manchester, 2012); I. Clarke, 'How to Manage a Revolution: Isaac Newton in the Early Twentieth Century', Notes and Records of the Royal Society, 68 (2014): 323–337; I. Clarke, 'The Gatekeepers of Modern Physics: Periodicals and Peer Review in 1920s Britain', Isis 106 (2015): 70–93; G. Gooday and D. J. Mitchell, 'Rethinking Classical Physics', in Oxford Handbook of the History of Physics, eds. J. Buchwald and R. Fox (Oxford: Oxford University Press, 2013), 721–764.

37 J. Hughes, 'Making Isotopes Matter: Francis Aston and the Mass-Spectrograph', Dynamis, 29 (2009): 131–165.

38 R. Clark, Sir Edward Appleton (Oxford: Pergamon, 1971), 20ff.; J.A. Ratcliffe, 'Edward Victor Appleton', Biographical Memoirs of Fellows of the Royal Society, 12 (1966): 1–21; J.A. Ratcliffe, 'The Early Days of Ionospheric Research: Early Ionosphere Investigations of Appleton and his Colleagues', Philosophical Transactions of the Royal Society of London, 280A (1975): 3–9.

39 K.G. Budden, 'John Ashworth Ratcliffe', Biographical Memoirs of Fellows of the Royal Society, 34 (1988): 670–711; M.A.F. Barnett, 'The Early Days of Ionosphere Research', Journal of Atmospheric and Terrestrial Physics, 36 (1974): 2071–2078; J.A. Ratcliffe, 'Experimental Methods of Ionospheric Investigation 1925–1955', Journal of Atmospheric and Terrestrial Physics, 36 (1974): 2095–2103.

40 A. Anduaga, Wireless & Empire: Geopolitics, Radio Industry & Ionosphere in the British Empire, 1918–1939 (Oxford: Oxford University Press, 2009); C.-P. Yeang, Probing the Sky with Radio Waves: From Wireless Technology to the Development of Atmospheric Science (Chicago: University of Chicago Press, 2013), 201ff.

41 R.H. Stuewer, 'Artificial Disintegration and the Cambridge-Vienna Controversy', in Observation, Experiment and Hypothesis in Modern Physical Science, eds. P. Achinstein and O. Hannaway, (Cambridge, Mass. and London: MIT Press, 1985), 239–307; Hughes, 'Unity through Experiment?'; M. Rentetzi, Trafficking Materials and Gendered Experimental Practices: Radium Research in Early 20th Century Vienna (New York: Columbia University Press, 2008).

42 Hughes, 'Unity through Experiment?'; Rentetzi, Trafficking Materials and Gendered Experimental Practices, 137ff.

43 J. Hughes. '"Modernists with a Vengeance": Changing Cultures of Theory in Nuclear Science, 1920–1930', Studies in History and Philosophy of Modern Physics, 29 (1998): 339–367.

44 J.L. Heilbron and R.W. Seidel, Lawrence and his Laboratory. A History of the Lawrence Berkeley Laboratory, Volume 1, (Berkeley: University of California Press, 1989); J. Hughes, The Manhattan Project: Big Science and the Bomb (Cambridge: Icon Books, 2003).

45 J. Hughes, 'Plasticine and Valves: Industry, Instrumentation and the Emergence of Nuclear Physics', in The Invisible Industrialist: Manufactures and the Production of Scientific Knowledge, eds. J.-P. Gaudillère and I. Löwy (London: Macmillan, 1998), 58–101.

46 G. Hartcup and T.E. Allibone, Cockcroft and the Atom (Bristol: Adam Hilger, 1984); Hughes, 'Plasticine and Valves'.

47 P. Galison, Image and Logic: A Material Culture of Microphysics (Chicago: University of Chicago Press, 1997), 433ff.

48 C.E. Ramsbottom, 'The Era of the Home Wireless Constructor', International Conference on 100 Years of Radio (London: Institution of Electrical Engineers, 1995), 114–119; Geddes and Bussey, Setmakers.

49 Briggs, The History of Broadcasting in the United Kingdom, vol. 1, 49.

50 Bussey, Wireless, 3.

51 Hennessy, *Emergence of Broadcasting in Britain*, 76. Founded in 1913, the Wireless Society of London became the Radio Society of Great Britain in 1922. See J. Clarricoats, *World at their Fingertips: The Story of Amateur Radio in the United Kingdom and a History of the Radio Society of Great Britain* (London: Radio Society of Great Britain, 1967).

52 Quoted in Bussey, *Wireless*, 11.

53 G. Hartcup, *The War of Invention. Scientific Developments, 1914–18* (London: Brassey's Defence Publishers, 1988), 152–156; N. Feather, 'Richard Whiddington, 1885–1970', *Biographical Memoirs of Fellows of the Royal Society*, 17 (1971): 741–756; W. Wilson, 'Owen Willans Richardson, 1879–1959', *Biographical Memoirs of Fellows of the Royal Society*, 5 (1959): 207–215; R. Whiddington, 'Frank Horton, 1878–1957', *Biographical Memoirs of Fellows of the Royal Society*, 4 (1958): 117–127.

54 For an excellent overview of scientists in the Great War, see R. Macleod, 'Scientists', in *The Cambridge History of the First World War, Volume 2. The State*, ed. J. Winter, (Cambridge: Cambridge University Press, 2014), 434–459.

55 Quoted in Bussey, *Wireless*, 1.

56 Briggs, *The History of Broadcasting in the United Kingdom*, vol. 1, 36–58.

57 Cambridge University Wireless Society minutes (17 October 1920), CUWS Minute Book 1920–1945, Cambridge University Archives UA Soc. 96. 1. 1, Cambridge University Library (hereafter 'CUWS minutes'); 'Cambridge University Wireless Society', *Wireless World*, 8 (1921): 715.

58 H. Strachan, *History of the Cambridge University Officers Training Corps* (Tunbridge Wells: Midas Books, 1976); J. Chadwick, 'Frederick John Marrian Stratton, 1881–1960', *Biographical Memoirs of Fellows of the Royal Society*, 7 (1961): 280–293. Stratton was mentioned five times in dispatches and subsequently awarded the DSO.

59 D. Wilson, *Rutherford: Simple Genius* (London: Hodder & Stoughton, 1983), 87ff, 339ff.

60 Clark, *Appleton*; Ratcliffe, 'Appleton'.

61 CUWS minutes, 29 November 1920.

62 CUWS minutes, 7 February 1921; A.R. Constable, 'Henry Jackson: Pioneer of Wireless Communication', *Bulletin of the British Vintage Wireless Society*, 25 (2000): 20–25.

63 CUWS minutes, 8 February 1922.

64 CUWS minutes, 22 January 1923.

65 CUWS minutes, 24 April 1921; T.J.N. Hilken, *Engineering at Cambridge University, 1783–1965* (Cambridge: Cambridge University Press, 1967), 155–158.

66 CUWS minutes, 16 October 1921; Chadwick, 'Stratton'; D. Portway, *Militant Don*, (London: Robert Hale, 1964), esp. 105–110; Strachan, *History of the Cambridge University Officers Training Corps* (n. 58). M.V. Wilkes, *Memoirs of a Computer Pioneer* (Cambridge, Mass.: MIT Press, 1985), 16.

67 CUWS minutes, 3 December 1923. On Osram, see Vyse and Jessop, *The Saga of Marconi-Osram Valve*.

68 Briggs, *The History of Broadcasting in the United Kingdom*, vol. 1; Bussey, *Wireless*; Hennessy, *Emergence of Broadcasting*.

69 A. Russell, *Tony D'Alton's Wireless* (London: Boy's Own Paper, 1920), 14; Haring, *Ham Radio's Technical Culture*; K.D. Brown, *Factory of Dreams: A History of Meccano Ltd, 1901–1979* (Lancaster: Crucible, 2007).

70 O. Lodge, *Ether & Reality* (London: Hodder and Stoughton, 1925); O. Lodge, *Talks about Wireless* (London: Cassell and Company, 1925).

71 CUWS minutes, 19 February 1923.

72 Briggs, *The History of Broadcasting in the United Kingdom*, vol. 1; Bussey, *Wireless*, esp. 91ff.

73 CUWS minutes, 10 November 1930.

74 'Shirley Falloon' [obituary], *The Times* (24 August 1992): 13.

75 G. Hartcup and T.E Allibone, *Cockcroft and the Atom* (Bristol: Hilger, 1984).

76 B. Lovell and D.G. Hurst, 'Wilfrid Bennett Lewis', *Biographical Memoirs of Fellows of the Royal Society*, 34 (1988): 453–509, 455.

77 CUWS minutes, 10 March 1930.

78 CUWS minutes, 13 November 1933.

79 CUWS minutes, 28 November 1932, 27 November 1933.

80 CUWS minutes, 22 October 1934.

81 CUWS minutes, 3 June 1935.

82 Lewis, 'Wireless Soc Presidential Address 1933', Box 35 folder 6, W.B. Lewis papers, Queen's University, Kingston, Ontario.

83 J. Hughes, 'Plasticine and Valves'.

84 Lovell and Hurst, 'Wilfrid Bennett Lewis', 460.

85 Lovell and Hurst, 'Wilfrid Bennett Lewis', 457.

86 W.B. Lewis, 'The Development of Electrical Counting Methods in the

Cavendish', in *Cambridge Physics in the Thirties*, ed. J. Hendry (Bristol: Adam Hilger, 1984), 133–136; C.E. Wynn-Williams, 'The Scale-of-Two Counter', in *Cambridge Physics in the Thirties*, 141–149; F.A.B. Ward, 'Physics in Cambridge in the Late 1920s', in *The Making of Physicists*, ed. R. Williamson, (Bristol: Adam Hilger, 1987), 77–85.

87 Lovell and Hurst, 'Wilfrid Bennett Lewis'.

88 W.B. Lewis, 'Wireless Soc Presidential Address 1933'; CUWS minutes, 23 October 1933.

89 G. Werskey, *The Visible College* (London: Allen Lane, 1978), 219, 223ff; M. Wilkins, *The Third Man of the Double Helix: An Autobiography* (Oxford: Oxford University Press, 2005), 35ff.

90 W.B. Lewis and C.J. Milner, 'A Portable Duplex Radio-Telephone', *The Wireless Engineer* (September 1936): 475–482; K.C. Holmes, 'Sir John Cowdery Kendrew', *Biographical Memoirs of Fellows of the Royal Society*, 47 (2001): 312–332.

91 'Wonderful Wooderson. Wireless Scores', *Daily Sketch* (2 August 1938): 15; R. Fawcett, *Nuclear Pursuits: The Scientific Biography of Wilfrid Bennett Lewis* (Montreal: McGill-Queen's University Press, 1994), 12.

92 Cambridge University OTC recruiting brochures, 1937 and 1938, cited in Strachan, *History of the Cambridge University Officers Training Corps*, 164–165.

93 G. Heard, 'The Age of Radiation', *The Listener* (30 July 1930): 174–175, on 174. On Heard, see A. Falby, *Between the Pigeonholes: Gerald Heard, 1889–1971* (Newcastle: Cambridge Scholars Publishing, 2009).

94 'Pure Science', in *Scientific Research and Social Needs*, J. Huxley ed. (London: Watts & Co., 1934), 203–224, on 209; see also P.M.S. Blackett, 'The Craft of Experimental Physics', in *Cambridge University Studies*, ed. H. Wright, (London: Ivor Nicholson & Watson, 1933), 67–96.

95 Wilson, *Rutherford*, 466–471.

96 Wilson, *Rutherford*, 467; Anonymous, 'Scientific Problems of the Pacific', *Nature*, 131 (1933): 831.

97 E. Rutherford, 'Atomic Projectiles and their Applications', *Proceedings of the Institution of Mechanical Engineers*, 123 (1932): 183–205, on 183–184.

98 E. Rutherford, 'Trends in Modern Physics', *Proceedings of the Society of Chemical Industry*, 16 (1934): 69–71, on 69.

99 E. Rutherford, *The Newer Alchemy: Based on The Henry Sidgwick Memorial Lecture Delivered at Newnham College, Cambridge, November 1936* (Cambridge: Cambridge University Press, 1937), 19, 25, 28.

100 J.C. Eccles, 'Alexander Forbes and his Achievement in Electrophysiology', *Perspectives in Biology and Medicine*, 13 (1970), 388–404; J.K. Bradley and E.M. Tansey, 'The Coming of the Electronic Age to the Cambridge Physiological Laboratory: E. D. Adrian's Valve Amplifier in 1921', *Notes and Records of the Royal Society*, 50 (1996): 217–228; J. Garson, 'The Birth of Information in the Brain: Edgar Adrian and the Vacuum Tube', *Science in Context*, 28 (2015): 31–52.

101 R.G. Frank, Jr., 'Instruments, Nerve Action and the All-or-None Principle', *Osiris*, 9 (1994), 208–235; A. Hodgkin, 'Edgar Douglas Adrian, Baron Adrian of Cambridge', *Biographical Memoirs of Fellows of the Royal Society*, 25 (1979): 1–73, 26ff.

102 J. Gray, 'Bryan Harold Cabot Matthews', *Biographical Memoirs of Fellows of the Royal Society* 35 (1990): 264–279.

103 B.H.C. Matthews, 'Recent Developments in Electrical Measuring Instruments for Biological and Medical Purposes', *Journal of Scientific Instruments*, 12 (1935): 209–214.

104 A.V. Hill, 'Experiments on Frogs and Men', *The Lancet*, 214, 10 August 1929, 261–267, on 265.

105 Hill, 'Experiments on Frogs and Men', 261–267, on 265.

106 E.D. Adrian, *The Basis of Sensation: The Action of the Sense Organs* (London: Christophers, 1928), 39.

107 Hughes, 'Plasticine and Valves'; J. Heilbron and R.W. Seidel, *Lawrence and his Laboratory: A History of the Lawrence Berkeley Laboratory* (Berkeley: University of California Press, 1989); L. Brown, *Centennial History of the Carnegie Institution of Washington, Volume 2. The Department of Terrestrial Magnetism* (Cambridge: Cambridge University Press, 2005).

108 Lovell and Hurst, 'Wilfrid Bennett Lewis',; Fawcett, *Nuclear Pursuits* (n. 90); C. Latham and A. Stobbs, eds., *Pioneers of Radar* (Stroud: Sutton Publishing, 1999).

109 R.W. Burns, *British Television: The Formative Years* (London: IEE, 1986).

12

Whose modernism, whose speed? Designing mobility for the future, 1880s–1945

Ruth Oldenziel

What modernism?

In 1897, Barcelona's Modernists met at the tavern *Els Quatre Gats*. The tavern's Catalan name, 'The Four Cats,' playfully suggested the four founders, while the feline reference was inspired by the famous cabaret-café *Le Chat Noir* in Paris. Most of all, the sign on the tavern's facade played on the Catalan slang for: 'just a couple of guys'. That expression of modesty was tongue in cheek. The Catalan men did not at all plan to be just a couple of guys. The painters and poets – ranging from Pablo Picasso to Francis Picabia who frequented the café and shuttled back and forth between the Catalan and French capitals – considered themselves the only 'people-in-the-know'. They were self-confident artists set on changing the world – in the name of modernism.[1]

On the occasion of the tavern's opening in 1897, co-owner and poster designer Ramon Casas painted a portrait of himself: *Ramon Casas and Pere Romeu on a Tandem* (Figure 12.1). He hung the self-portrait in the café as an artistic statement of the modernist movement. His inscription, 'to ride a bicycle, you can't go with your back straight', meant that to make progress one had to break with tradition. The year of the tavern's opening was also the height of the bicycle mania, when the progressive members of the bourgeoisie mounted their bicycles *en masse*. Casas' choice of the bicycle was a sign of the times.

Only four years later, Casas replaced the self-portrait with a new one: *Ramon Casas and Pere Romeu in an Automobile* (1901) (Figure 12.2).

Fig. 12.1 Catalan modernist and poster designer Ramon Casas' self-portrait with his patron, *Ramon Casas and Pere Romeu on a Tandem* (1897), painted at the occasion of the opening of the modernist café in Barcelona with the inscription, 'to ride a bicycle, you can't go with your back straight: to make progress you have to break with tradition'.

© Museu Nacional d'Art de Catalunya, Barcelona 2017. Photo: Calveras/Mérida/Sagristà.

Now, the painter presented himself and his patron as motorists instead of cyclists as the ultimate pose of being modern. In subsequent pairings of the two paintings, the tandem no longer represented modernism: one came to mean the old nineteenth century (bicycle); the other the new century (automobile). The representational shift seems picture perfect for a narration of modernism – too perfect, as it turns out.

The shift in pictorial power from bicycles to cars as the symbol of being modern was much more complex than Casas' act of replacement in Barcelona suggests. Staying modern from one moment to the next was not a matter of endlessly substituting one novelty with the next – a succession of improved modernisms or fads – as the swapping of the self-portraiture as a cyclist for a motorist implied. For one, several modernisms coexisted, overlapped and reinforced each other, as Mitchell

Fig. 12.2 In 1901, Casas replaced the original painting with a new self-portrait using the car as his vehicle of modernism: *Ramon Casas and Pere Romeu in an Automobile*.

© Museu Nacional d'Art de Catalunya, Barcelona 2017. Photo: Calveras/Mérida/Sagristà.

Ash argues (Chapter 1 in this volume): from a radical rejection of Victorian direct representation of nature to a technological (or techno-cratic) modernism that exploited nature's resources without concern for the next generation. Technological modernism included the self-referential metropolis that did away with distinctions between country and city (trams); day and night (electric street lighting); and distance. Modernism, as he reminds us, came in many forms. Technological modernism was one of them.[2]

On the whole, though, not all modernisms were created equal. What being modern meant in the interwar period was contested and unstable; and who controlled the narrative framework of modernism mattered a great deal for those who were not in-the-know like Picasso and Picabia. The interpretation of modernism as the easy cross-border movements by modernists – artists equally at home in *Els Quatre Gats* as in *Le Chat Noir* as well as the globetrotting engineers showing off their innovations at world fairs – ignores that most people and their ideas travelled far less seamlessly. Such a portrayal overlooks how some modernisms were decentred, delegitimised, and dismissed as old-fashioned, but others were promoted as forward-looking. The delegitimisation of not-being-modern (enough) or even old-fashioned could have long-term repercussions in terms of future investments, from political decision-making to allocating funds. In other words, whoever or whatever was designated as truly modern went further than just driving the discourse and could have huge material consequences.

Being modern: a balancing act

In the summer of 1897 when *Els Quatre Gats* opened, kings, queens, princes and princesses got on the bicycle. Traditionally, individual mobility was a class-bound activity. Ordinary people walked; the aristocracy rode horses.[3] Cycling represented a middle-class alternative to horse riding. Riding a bicycle at a leisurely pace in the countryside was the updated version of being an aristocrat.[4] By the 1880s, the iconic literary expression formerly for the steam locomotive now denoted bicycles: in travelogues, middle-class cyclists lovingly called their bicycles their 'iron horses' – from the French *cheval de fer* and Italian *cavallo di ferro* to the Dutch *ijzeren ros*, and in many other languages in between.[5]

By 1897, Europe's royal families had to admit they could no longer ignore the middle-class modernity of 'self-propelled voyagers'. Newspapers reported how everyone who mattered in high society seemed to ride modernity: the young middle-class men climbing the social ladder, freedom-yearning women, and progressive members of the aristocracy and royal families – in that order. For freedom-loving women, cycling symbolised political modernity. The American feminist Susan B. Anthony wrote in 1896 that cycling 'has done more to emancipate women than anything else in the world. I stand and rejoice every time I see a woman ride by on a wheel'. A year earlier, Frances Willard, the suffragist and founder of the Woman's Christian Temperance Union, published a bestseller on how she had learned to ride a bicycle from 'pure love of adventure ... from a love of acquiring this new implement of power'.[6] Although Europe's royal houses decided it was high time to ride the wave of modern mobility, cycling was still controversial for some – and women in particular. For example, on the eve of her coronation as queen of the Netherlands in 1898, young Wilhelmina wanted to cycle, but the government vetoed her wish 'in view of the future'. Their queens, the Dutch parliament felt, should not cycle.[7]

The cycling industry in particular promoted modernism's gendered version of women's individual mobility and freedom. In powerful poster campaigns, the bicycle business – the late nineteenth-century high-tech industry – masterfully managed to redefine cycling as suitable for middle-class (young) women. Advertising-dependent magazines, which had women as their main readership, sponsored literary fiction that helped to allay fears about their emancipation. Through attractive images in advertisements and formulaic fiction, the industry helped redefine women's cycling as an acceptable form of modern mobility throughout the 1890s.[8]

Nevertheless, cycling remained a balancing act. High society and middle-class upstarts considered working-class men's neck-breaking speed in bicycle races an inappropriate commercialisation of their privileged leisure activity. Not surprisingly, the Futurists, intent on finding ways to shock society, explored this working-class modernity to its fullest visual potential. When F.T. Marinetti rejected the classical culture of the nineteenth century in his 1909 Futurist Manifesto, he offered the 'dynamics of existence', a machine aesthetic represented by the speed of technological progress and brevity of modern life: a futurist 'total art' in painting, sculpture, architecture, film, textiles and music. The cyclist represented that future best: the most intimate marriage between man and machine, an aesthetic of energy and speed in paintings like Umberto Boccioni's *Dynamism of a Cyclist* (1913), Nathala Goncharova's *The Cyclist* (1913) and Mario Sironi's *The Cyclist* (1916). The rhythm, the speed and the mechanism of moving through the modern landscape with self-propelled velocity fascinated the futurists the most. By the eve of the First World War, the thriving literary production (cycling travelogues), the explorations of experimental artists (Futurists), and the sponsorship of commercial art (advertising posters) had helped to establish cycling as truly modern. The bicycle industry had been instrumental in culturally sponsoring cycling as a form of middle-class modernism through advertising and genteel fiction. This commercial sponsorship was lacking in the interwar period.

The same market players, civil-society actors and experts once involved in the bicycle business shifted to the car industry in the 1910s – and with them the sponsorship for a discourse of modern mobility.[9] Every carmaker from Opel to Peugeot had started out as bicycle manufacturer. Within a few years, cycling was declassed as unmodern and hopelessly old-fashioned. Modernist boosters argued that it was merely a matter of time before cycling would make way for the truly forward-looking mobility of the automobile. That did not happen. Modern mobility was not a matter of simply substituting one mode with the next, as Casas had suggested by switching his paintings. In fact, in subsequent decades, cycling would go on to boom as never before until well in the 1960s, unlike automobility, which took many decades to fulfil its promise as the way of the future. The interwar period represents an age of contestation, providing pointed discussions about the – unstable – meanings of modernity.

People of all stripes used bicycles for novel purposes, particularly when the bicycle industry reduced the prices still further in the 1920s, by tapping into the market of male skilled workers and civil servants.[10] On Sundays, workers gathered to explore the surrounding countryside. French hairdressers organised cycle excursions on their day of rest with their *Club Sportif de la Coiffure*; so did French butchers, policemen, railroad men, assistants in Parisian department stores like *Bon Marché* and clerks at major banks like *Société Générale*. Young folk in rural Finland, followed by farmers and their wives, began cycling to the fields, the next town, the church, the dance hall and the cinema. Lower- and middle-class people everywhere in the urbanising world explored cycling to tour for a day to the seaside, the park or the countryside.

People's most novel use of the bicycle, however, was to commute: workers cycled to work early in the morning and back home late at night, creating rush hours that observers proclaimed were a sight to be seen. In 1927, a comparative study showed how widespread bicycles had become. That year, every third resident in the Netherlands had a bicycle, closely followed by Sweden (every fourth), Denmark, Switzerland and Belgium (every fifth), Germany and France (every sixth), Britain (every seventh) and Italy (every thirteenth). The USA represented the exception in the industrial world.[11] Bicycles remained the most popular means of transport well into the early 1960s, outpacing cars and public transit. For most people, cycling offered the modern alternative to walking and public transit.[12]

Thus in a matter of twenty years, bicycles had shifted from a middle-class means of leisure and a working-class professional sporting device, to an efficient way for workers and civil servants to commute when cities were rapidly expanding. By all measures, the bicycle had been a hugely successful innovation. Culturally, however, its modernity as a novelty was contested during the interbellum and after.

Whose modernity?

In the fast-paced interwar world, modernism was an unstable sign almost by definition. From a Barcelona tavern to Berlin's streets, how-to-be-modern could be declassed, reclassed and recycled – quickly and simultaneously. In 1923, a journalist at the Berlin *Tagesblatt* wrote that the bicycle, 'condemned as unmodern, has suddenly become truly timely again. Bicycles have been pulled out of the junk room and the attic, and set free in service of transportation'.[13] Twenty years after the artists in the Barcelona *Els Quatre Gats* declared bicycles to be things of the past and cars the true vehicles of the future, the streets of Berlin were whirring with bicycles. The *Tagesblatt* journalist gushed, 'Berlin is now the city of bicycles. It has replaced Copenhagen, whose pedalling army was world famous'. In 1923, the year of hyperinflation, ordinary people in the capital of the Weimar Republic found cycling both a welcome alternative to walking and a cheaper solution than public transit when fares increased. They commuted by bicycle *en masse*.

New social groups began using bicycles in entirely novel ways, the *Tagesblatt* observed: 'If you stand in the morning and afternoon rush hours in the main suburban arteries, endless rows of cycling Berliners glide by. The bicycle is now the cheapest means of transportation for the civil servant, for business people and trippers … It is obedient to our own will, and independent of price or strike chicaneries', the journalist enthused.[14] In the midst of the economic depression – and after Budapest's public transit company raised ticket prices – in 1936 the shops in the Hungarian capital sold 12,000 bicycles within months. Every sixth person in Budapest had a bicycle in 1935. As a journalist observed, 'We know that until now the bicycle was the mode of transport less favoured by delivery people, but in recent months, the number of well-dressed cycling men and surprisingly also women has increased'.[15] Similar reports came from other cities. From Berlin to Budapest, the 1930s economic crisis further challenged government investments in both automobility and public transit.

The sight of ordinary people's modern speed and individual mobility amazed genteel observers, not only in urban centres of Germany, but also in those of colonial Africa. In 1930, compared with the end of the First World War, Kinsley Fairbridge found the change he observed in traffic truly 'startling' in the busy British colonial hub of Bulawayo in Rhodesia. 'The native servants, including delivery boys and labourers, ride good bicycles through the main thoroughfares, astonishing one who could think back to the times when the push bike was the aristocratic preserve of the white men'.[16] Journalists reported on the traffic census that counted fewer cyclists on Sunday than during the week. To the genteel readers, accustomed to seeing the well-to-do touring around the countryside on a bicycle for their pleasure, the workers, women and colonial subjects, who cycled daily to commute, was a truly amazing phenomenon. These cycling commuters were quite newsworthy.

By the 1930s, urban cycling had also come to symbolise the aspirations of both civil servants of the expanding commercial and government bureaucracies as well as working-class men. Cycling helped urban and agricultural workers expand their job opportunities and reach out-of-the way construction sites, farm fields and markets as never before all over the industrialising world.[17]

The artist who captured this new working-class modernism best was Bertolt Brecht. In his 1932 film *Kuhle Wampe, or: To Whom Belongs the World?* (*Kuhle Wampe oder: Wem gehört die Welt?*), the German Marxist poet, playwright and theatre director Brecht and the German-Bulgarian cineaste Slatan Dudow brilliantly explored the visual repertoire of the city bicycle to represent the proletariat during the depth of the Depression. In interwar Germany, embracing the industrial city and its residents stood in opposition to the mountain films genre that Leni von Riefenstahl and others exploited to the full.[18] *Kuhle Wampe* was the first – and as it turned out the last – overly Marxist-inspired movie in the Weimar Republic, produced by a communist-funded production company outside the commercial studio system before the Nazi takeover in 1933. To Brecht, the bicycle represented working-class modernism.

No filmmaker captured the visual modernism of bicycles and what that meant to workers as Brecht did. In *Kuhle Wampe* – named after a tent city on Berlin's outskirts where jobless, homeless families took refuge, but the words could also mean 'empty bellies' in local dialect – unemployed men frantically cycle around the metropolis in the futile pursuit of a (day) job. The bicycle is their most precious possession, enabling them to land a job. Another socialist director, the Italian Vittorio de Sica, in his 1948 *Bicycle Thieves* (*Ladri di biciclette*), would fully explore that theme two decades later.

The cinematographic effects of the bicycles that Brecht and Dudow employed had visual pedigrees. For over three decades, commercial artists to Futurist painters had explored bicycles' velocity visually. The speed of these modern mobility machines, when cars were still a rarity, fascinated them. Brecht's film, however, firmly planted the visual language in the realm of the working class. In his scenario, the kinetic dynamism of bicycles represented the workers' ability to move swiftly around the urban landscape. It was an enchanting image of modernity, but the whirring of the bicycle wheels also represents the futile frenzy of unemployed young men searching for work. Through fast-paced, sharp music, a polyphone prelude in marcato style, the film's composer Kurt

Eisler sought to contrast the images of whirring wheels and workers pushing themselves hunched over their handlebars. His musical score of disorienting and unrelenting dissonance also signalled the workers' futile pursuit for jobs during the depth of the Depression. The bicycle scenes conveyed that workers left no stone unturned to find a job. By experimenting in form and content, the socialist film makers emphasised that not the workers, but the capitalist system was to blame for people not having jobs.[19] Brecht recaptured the modernism of the Futurists two decades earlier; he also firmly portrays the bicycle as a modern working-class vehicle. At this point, journalists and filmmakers depicted cycling as a symbol of modern urban traffic. An emerging car-based coalition, however, came to delegitimise cycling as modern in the mid-1920s.

Speed and the invention of 'slow traffic'

In 1913, German traffic engineers categorised cycling as 'fast' traffic for their international colleagues who gathered in London. By 1926, cycling had lost that status of modernity in the minds of policy makers and engineers in the international traffic standard.[20] Since then, engineers have designated cycling as 'slow' traffic, lumped in the same category as animals, but separate from pedestrians. The Permanent International Association of Road Congresses (PIARC) hammered out international standards that are still with us today. Through PIARC, road and traffic engineers, automobile clubs, urban planners and political elites linked with the emerging automotive industry to create a counter-narrative for motorised speed as the true sign of modernism. They cast their modernist discourse in technocratic, neutral-sounding terms of speed to blueprint modernity for automobility. Every other traffic participant who deviated from this norm was deemed 'slow' and unmodern.[21]

That car-governed modernity was still highly debated in the interbellum – and would only become a reality after the Second World War. On both sides of the Atlantic during the 1920s, the presence of car drivers in the city was not a given, but rather controversial. Police departments, journalists and pedestrian organisations considered cars intruders of public space; speeding cars were killing machines.[22] From the mid-1920s, the car lobby sought to change motorists' bad reputation. The discourse on modernity – of pitching motorised traffic as the true way of the future – played a key role in that campaign. The debates on the modern always implied the opposite: 'fast' versus 'slow' traffic; motorised versus non-motorised circulation; flow versus disruption; modern versus old fashioned – in short, a distinction between future and past.

To help that future along, urban authorities and traffic engineering experts designed new traffic rules that favoured cars. Pedestrians, cyclists and horse-drawn vehicles ('slow' traffic) had to make way for cars ('fast' traffic). The rules began to impose fines on pedestrians for something they had always done, like crossing the street diagonally as the shortest distance between 'a' to 'b', to be dismissed as 'jaywalking'. Cyclists, first forbidden to ride too close to cars or on the left, now were forced to ride in single file or on the sidewalks. When fatal accidents skyrocketed, the blame and liability shifted from motorists to non-motorists. Car and public transit interests at times joined in the propaganda campaigns that cast pedestrians and cyclists as anti-modern – even illegitimate – users of the streets.[23] Urban planners – among them even socialist architects Congrès Internationaux d'Architecture Moderne (CIAM) dedicated to improving the lives of workers – also prepared for an urban future governed by cars; they ignored cycling's importance as a working-class modernity.[24] The marketing campaign to prepare for a car-governed future turned particularly intense when automobility was economically less viable during the Depression.

The US carmaker General Motors (GM) presented the most famous example of the car-governed promise for the future in the *Futurama* exhibit at New York's 1939 World Fair. The exhibit showed motorists of the future moving along an automated highway, where traffic rules facilitated fast-moving traffic and banned slow traffic as a principle of highway design. More than anything else, GM's *Futurama* exhibit sought to mobilize citizens' and state support for a highway system financed by public funds. The discourse of modernity was key in this campaign: car numbers were still exceedingly small as well as politically controversial, while pedestrians and cyclists dominated in many cities throughout the industrialising world, even in large US cities.

In short, during the 1930s the bicycle was – prematurely – cast as unmodern and as a thing of the past. Over the next decades, when urban planners found that cycling was not a dying art but even thriving in some cases, they were surprised. In rare cases, they would adjust their urban vision for the future – temporarily – with separate cycling infrastructures. In most cases, planners simply ignored their findings. Both world wars played an important role in establishing motorised vehicles as the key innovation in the road towards the future.

Shock of the old, or modern killing machines[25]

The two world wars are usually presented as the first truly modern wars, showing the tactical advantages of fossil-fuelled vehicles like cars, trucks and tanks over the hopelessly old-fashioned cavalry. Horses hark back to aristocratic European warfare that was about to be relegated to the basement of history with the collapse of many empires and their aristocratic classes. In the narration of modern warfare, the First World War was merely a dress rehearsal for the Second, when the killing machine achieved unprecedented efficiency. On this stage, bicycle infantries occupy a rather tenuous role.[26]

Early on, the military recognised bicycles were more fuel- and maintenance-efficient than horses. All the European armies as well as the American and Japanese maintained some bicycle units and even divisions. Armies investigated how to use pneumatic tyres coupled with shorter, sturdier frames. By the outbreak of the First World War, most armies had incorporated cycling soldiers, who, as messengers and scouts, took over the functions of dragoons, substituting bicycles for horses. France, for example, experimented with folding bicycles that soldiers slung across their backs. Each French line infantry and chasseur battalion had a cyclist unit for skirmishing, scouting and dispatch carrying.

The advantages of bicycles were many. Cyclists were a cost-effective alternative to horse-mounted troops, demanding relatively short training, being silent when on the move, and requiring little logistical support. Cycling units also could carry more equipment and travel longer distances than soldiers on foot. Bicycles offered a surprise factor over motorcycles, cars and trucks: they were quiet and flexible, and did not require precious petroleum or shipping. During both world wars, most armies gratefully included bicycles as tactical weapons in their arsenal.

This was also the case during the Second World War. Finnish counter-insurgents appreciated cycling units as flexible infantry troops. The Nazis mobilised cycling soldiers as a rapid deployment force for the Holocaust – a stealth weapon and efficient killing machine that barely left a trace.[27] Silent, rapid, off the grid and self-sufficient without gasoline, one such bicycle unit (no. 322) was responsible for liquidating 10,000 Jewish citizens in Poland by May 1942 alone.[28] Although the military embraced bicycles as tactical weapons, in most narratives tanks and trucks have come to represent modern warfare and the fossil-fuelled economy that fired the century.

In people's collective memory, bicycles have occupied a complicated position for civilians as well. In both warring and neutral countries, cycling was booming during times of crisis like the Depression and the world wars. In wartime, when governments established gasoline rationing for civilians, cycling offered ordinary citizens an efficient, independent form of mobility.[29] Yet in the collective memory of the war, cycling became associated with material shortages and destruction, retro-actively reinforced by the Cold War's ideological warfare, in which US promoters of the capitalist system used consumer goods like cars and kitchens to sway public opinion away from models of collective provision-ing.[30] Associated with war, material shortages and trauma, bicycles were classified as unmodern after the Second World War. To the surprise of many post-war planners, however, cyclists were not a dying breed as they had expected; cars did not simply replace bicycles, as Casas had suggested when he substituted one self-portrait of modernism with another. Indeed, since the 1970s, cycling has been embraced anew as modern and forward looking in a new cultural framework of sustainability – and is thriving again.

Conclusion

Mobilising modernity is as much about one's identity in the present as an aspiration for the future – about the road ahead, beyond the horizon. Building that road was not just a metaphor, but also a reality – including the question of who had the right to ride on it. In the interwar period, a car-based coalition blueprinted that future by typecasting cyclists, pedestrians and horse-drawn carriages as unmodern despite their dominance in the street. Projections guided budget allocations for road building and rules regulating traffic streams.

The story of the bicycle, in particular, is that of a technology once celebrated as the era's most modern machine around 1897, then contested during the interwar period, and dismissed as hopelessly old-fashioned in the 1950s, before becoming modern again as the green machine since the 1970s. Focussing on the bicycle lets us view what being modern has meant, how it was contested and who controlled its narrative. The story of cycling represents particularly well how tech-nologies that were once modern could just as easily go out of fashion, before being reinvented as innovative again. In the cultural definition of technological innovations, modernism as a discourse played a key role.[31]

The 1900 Paris exhibition is often seen as symbolising the moment automobiles entered the twentieth century as the trailblazers of individual mobility and technological modernity. Cars may have taken up an extraordinary amount of street space, commanded unusual cultural power, made a lot of noise and caused havoc in their own way. Nevertheless, their share in the total traffic remained exceedingly low, around 3 to 4 per cent well into the late 1940s – accessible only to the well-to-do. Cycling rather than motoring dominated the streets from Barcelona, Berlin and Budapest to Bulawayo for at least half a century. Nevertheless, it has been conventional to point to cars rather than bicycles as the symbol of modernity in the twentieth century.

The distorted view of history comes from a misreading of data from the USA, where bicycle-production figures plummeted from sky-high levels and car sales increased dramatically. This was not true for the rest of the world, where bicycles were long cherished as symbols of modernity. The bias is also a product of a country able to control the narrative as the dominant nation in the world since the Second World War. During the Cold War, policy projections in America's car-governed present seemed to signal Europe's inevitable future at a time when the streets were teeming with pedestrians and cyclists.[32]

In this scenario, the key aspect of modernity was – like Casas' paintings – one of replacement: cars simply took over the bicycle as the vehicle of modernity. The story of being modern, however, is one of contestation and framing. In today's post-industrial cities, the bicycle is back on the political and cultural agenda as the vehicle of lifestyle choice, suggesting again a forward-looking – and sustainable – future. Ultimately, the ups and downs of cycling serve as a reminder that we need to redefine the twentieth-century representation of modernism not within a single frame but as contested concept, by hanging the two paintings of modernity side by side.

Notes

1   I thank Lewis Pyenson who alerted me to the two paintings. Susan Roe, *In Montmartre*; William H. Robinson; Jordi Falgàs, Carmen Belen Lord and Josefina Alix Trueba; Cleveland Museum of Art; Metropolitan Museum of Art, *Barcelona and Modernity: Picasso, Gaudi, Miró, Dalí* (New Haven: Yale University, 2006); Esther Anaya and Santiago Gorostizo, 'The Historiography of Cycling Mobility', *Mobility in History* 5 (2014): 37–42.

2   See Mitchell G. Ash, 'Multiple Modernisms in Concert: the Sciences, Technology and Culture in Vienna' Chapter 1, in this volume.

3   Duncan R. Jamieson, *The Self-Propelled Traveller: How the Cycle Revolutionized Travel* (London: Rowman and Littlefield, 2015).

4   Writing about cycling adventures was also a form of self-discovery for the adventurous middle class, enacting the tradition of the bourgeois *Bildungsreise* as the crowning achievement of becoming a cultured European. Ruth Oldenziel and Mikael Hård, *Consumers, Users, Rebels: The People Who Shaped Europe* (London: Palgrave, 2013). Jamieson, *The Self-Propelled Traveller*.

5   Frédéric Héran, *Le retour de la bicyclette: Une histoire des déplacements urbains en Europe, de 1817 à 2050* (Paris: Éditions La Découverte, 2014); Jamieson, *The Self-Propelled Traveller*.

6   Frances E. Willard, *A Wheel within a Wheel: How I Learned to Ride the Bicycle, With Some Reflections Along the Way* (New York, Chicago: F.H. Revell company, 1895).

7   Since then, Wilhelmina and her daughter and granddaughter have all made a point of having their photographs taken while cycling, to show they were modern monarchs in touch with ordinary people. Giselinde Kuipers, 'De fiets van Hare Majesteit. Over nationale habitus en sociologische vergelijking', *Sociologie* 6, no. 3 (2011): 3–25.

8   Ellen Garvey, 'Reframing the Bicycle: Advertising-Supported Magazines and Scorching Women', *American Quarterly* 47 (1994): 66–101.

9   David V. Herlihy, *Bicycle. The History* (New Haven and London: Yale University, 2004). Cycling also became highly contested. Socialists and communists often pointed out in local and national parliaments, which they had just entered for the first time after the First World War, that the luxury taxes levied on cyclists were unfair, punishing disproportionately those who depended on bicycles for work. Adri A. Albert de la Bruhèze and Ruth Oldenziel, 'Who Pays, Who Benefits? Bicycle Taxes as Tools of the Public Good, 1890–2012', in *Cycling and Recycling: Histories of Sustainable Practices*, eds. Ruth Oldenziel and Helmuth Trischler (London: Berghahn, 2015), 73–100.

10 The most comprehensive study on urban cycling, see: Adri A. de la Bruhèze and Frank C.A. Veraart, *Fietsverkeer in praktijk en beleid in de twintigste eeuw. Overeenkomsten en verschillen in fietsgebruik in Amsterdam, Eindhoven, Enschede, Zuidoost Limburg, Antwerpen, Manchester, Kopenhagen, Hannover en Basel* (Eindhoven: Stichting Historie der Techniek, 1999). For the extended and revised volume, see: Ruth Oldenziel et al., eds., *Cycling Cities: The European Experience* (Eindhoven: Foundation of the History of Technology and LMU Rachel Carson Center, 2016). See also: *Stadt- und Landesplanung Bremen, 1926–1930* (H.M. Hauschild, 1931); Nan van Zutphen, 'Sociale geschiedenis van het fietsen te Leuven, 1880–1900', in *Jaarboek 1979 Vrienden Stedelijke Musea. Fiets en film rond 1900: moderne uitvindingen in Leuven*, eds. Nan van Zutphen et al. (Leuven: Crab, 1981), 11–256; Carlos Héctor Caracciolo, 'Bicicleta, circulación vial y espacio público en la Italia Fascista', *Historia Critica* 39 (2009): 20-42. Increasing numbers of studies present the bicycle as the first vehicle of individual mass transport in the 'ride to modernity'. Manuel Stoffers, Harry Oosterhuis and Peter Cox, 'Bicycle History as Transport History: The Cultural Turn', in *Mobility in History: Themes in Transport*, eds. Gijs Mom, et al. (Neuchatel: Editions Alphil-Presses Universitaires Suisses, 2010), 265–74. How the automobile became the dominant mode of transportation entrenched in a political-industrial-technical-cultural complex, see: Matthew Patterson, *Automobile Politics: Ecology and Cultural Political Economy* (Cambridge: Cambridge University Press, 2007). On utility bicycles for commuting, see: Frank C.A. Veraart, 'Geschiedenis van de fiets in Nederland, 1870–1940' (MA thesis, TU Eindhoven, 1995); Michael Taylor, 'The Bicycle Boom and the Bicycle Bloc: Cycling and Politics in the 1890s', *Indiana Magazine of History* 104, no. 3 (2008): 213–40. On working-class and rural leisure cycling see: Richard Holt, 'The Bicycle, the Bourgeoisie and the Discovery of Rural France', *British Journal of Sports History* 2 (1985): 127–39; Tiina Männistö-Funk, 'The Crossroads of Technology and Tradition: Vernacular Bicycles in Rural Finland, 1880–1910', *Technology and Culture* 52, no. 4 (2011): 733–54. After 1922, cheap and stripped German bicycles flooded the Dutch market See: Veraart, 'Geschiedenis van de fiets', 83–84.

11 Ruth Oldenziel and Adri A. de la Bruhèze, 'Contested Spaces: Bicycle Lanes in Urban Europe, 1900–1995', *Transfers* 1, no. 2 (2011): 31–49.

12 Oldenziel et al., *Cycling Cities*. See also Bruhèze and Veraart, *Fietsverkeer*. See also Karl Hodges, 'Did the Emergence of the Automobile End the Bicycle Boom?', in *Proceedings Fourth International Cycle History Conference* (San Francisco: Bicycle Books, 1993), 39–42; Hans Joachim Schacht, *Der Radwegebau in Deutschland* (Halle: Schriften des Seminars für Verkehrswesen and der Martin-Luther-Unversität Halle-Wittenberg, 1937); Herlihy, *Bicycle*, 328; Anne-Katrin Ebert, 'Cycling towards the Nation: The Use of the Bicycle in Germany and the Netherlands, 1880–1940', *European Review of History—Revue européenne d'Histoire* 11, no. 3 (2004): 347–64; Tiina Männistö-Funk, 'Gendered Practices in Finnish Cycling, 1890–1939', *Icon* 16 (2010): 53–73; Herlihy, *Bicycle*, 328. In the 1930s British parliamentary debates, government officials claimed at least 9,000,000 cyclists.

13 [Cycling in Berlin], *Berliner Tagesblatt* (2 September 1923) supplement in Iain Boyd Whyte and David Frisby, *Metropolis Berlin: 1880–1940* (Berkeley: University of California Press, 2012), 554.

14 Whyte and Frisby, *Metropolis Berlin: 1880–1940*.

15 N.N., 'Tébolyda és a zsebtolvajok paradricsoma lett a "reformok" következményeként a villamosból [because of the "reforms" the tram became a madhouse and a paradise for pickpockets]', *Népszava*, July 4, 1936, as quoted by Kátalin Tóth, 'Budapest: Subversive Cycling', in *Cycling Cities: Hundred Years of European Policy and Practice, 1920–2015*, eds. Ruth Oldenziel, et al. (Eindhoven: Foundation of the History of Technology and LMU Rachel Carson Center, 2016).

16 Quoted in Terence O. Ranger, 'Bicycles and the Social History of Bulawayo', in *Short Writings from Bulawayo*, ed. Jane Morris (Hillside, Bulawayo: amaBooks, 2003), 76–81.

17 Oldenziel and Hård, *Consumers, Users, Rebels*, Chapter 4.

18 I am indebted to Seth Peabody for this insight.

19 Bruce Arthur Murray, *Film and the German Left in the Weimar Republic: From Caligari to Kuhle Wampe*, first edition (Austin: University of Texas Press, 1990)., 216–24; Franz A. Birgel, 'Kuhle Wampe, Leftist Cinema, and the Politics of Film Censorship and Weimar Germany',

Historical Reflections 35, no. 2: 40–62. Murray, *Film and the German Left in the Weimar Republic*, 223.

20 [August] Bredtschneider and [Paul] Kunitz, 'Traffic and Administration: Regulations for Fast and Slow Traffic on Roads no. 46', in *III Congress. London 1913; Second Section. Traffic and Administration: 7th Question* (Paris: PIARC, Permanent International Association of Road Congresses, 1913).

21 Ruth Oldenziel, 'Accounting Tricks: How Cyclists and Pedestrians were Thrown under the Bus', paper presented at Philadelphia T2M, September 2014.

22 Michael John Law, 'Speed and Blood on the Nypass: The New Automobilities of Inter-war London', *Urban History* 39, no. 3 (2012): 490; Peter D. Norton, 'Streets Rivals: Jaywalking and the Invention of the Motor Age Street', *Technology and Culture* 48, no. 2 (2007): 331–59; Peter Norton, *Fighting Traffic: The Dawn of the Motor Age in the American City* (Cambridge, MA: MIT, 2008); Uwe Fraunholz, *Motorphobia: Anti-Automobiler Protest in Kaiserreich und Weimarer Republik, Kritische Studien zur Geschichtswissenschaft* (Göttingen: Vandenhoeck & Ruprecht, 2002); Kurt Moser, 'The Dark Side of "Automobilism", 1900–30: Violence, War and the Motor Car', *The Journal of Transport History* 24, no. 2 (2003): 238–58; Patrick Fridenson, 'La société francaise et les accidents de la route, 1890–1914', *Ethnologie francaise* 21, no. 3 (1991): 307–13. Jeffrey H. Jackson, 'Solidarism in the City Streets. La Société protectrice contre les excès de l'automobilisme and the Problem of Traffic in Early Twentieth-Century Paris', *French Cultural Studies* 20, no. 3 (2009): 237–56.

23 Norton, *Fighting Traffic: The Dawn of the Motor Age in the American City*; Peter Norton, 'Four Paradigms: Traffic Safety in the Twentieth-Century United States', *Technology and Culture* 56, no. 2 (2015): 319–34; Barbara Schmucki, 'Against "the Eviction of the Pedestrian". The Pedestrians' Association and Walking Practices in Urban Britain after World War II', *Radical History Review*, no. 114 (2012): 113; 37; Muhammed M. Ishaque and Robert B. Noland, 'Making Roads Safe for Pedestrians or Keeping Them out of the Way: An Historical Perspective on Pedestrian Policies in Britain', *Journal of Transport History* 27, no. 1 (2006):

115–37; Joe Moran, 'Crossing the Road in Britain, 1931–1976', *The Historical Journal* 49, no. 2 (2006): 477–96; P.W.J. Bartrip, 'Pedestrians, Motorists, and No-Fault Compensation for Road Accidents in 1930s Britain', *The Journal of Legal History* 31, no. 1 (2010): 45–60. Fridenson, 'La société francaise et les accidents de la route, 1890–1914'.

24 Ruth Oldenziel, 'CIAM and Cycling: Designing for Working-Class Mobility, 1930–1935', *Journal of Urban History* (manuscript).

25 The title is inspired by David Edgerton, *The Shock of the Old. Technology and Global History since 1900* (London: Profile Books, 2006). He emphasises the need to consider innovations not from the point of view of inventions but from their continued use. That is also the argument here.

26 Dismissed as old-fashioned by the 1960s in the West, bicycles became highly successful as the 'weapon of the people' in Vietcong and North Vietnamese Army guerrilla. In response, the US Pentagon investigated the history of cycling. R.S. Kohn, *Bicycle Troops* (Booklife, 2011). Alexander V. Koelle, 'Pedaling on the Periphery: The African American Twenty-fifth Infantry Bicycle Corps and the Roads of American Expansion', *The Western Historical Quarterly* 41, no. 3 (2010): 305–26; Kay Moore, *The Great Bicycle Experiment: The Army's Historic Black Bicycle Corps, 1896–97* (Missoula, Mont.: Mountain Press Pub. Co., 2012). D.R. Maree, 'Bicycles in the Anglo-Boer War of 1899–1902', *Military History Journal* 4, no. 1 (1977), 15–21.

27 Each of the German Army's light infantry battalions (*Jäger*) had one (*Radfahr-Kompanie*) at the outbreak of the war – eighty in total divided into eight *Radfahr-Bataillonen* (bicycle battalions).

28 Konrad Kwiet, 'From the Dairy of a Killing Unit', in *Why Germany? National Socialist Anti-Semitism and the European Context*, ed. John Milfull (Providence: Berg, 1993), 75–90; 'Auftakt zum Holocaust. Ein Polizeibatallion im Osteinsatz', in *Der Nationalsozialismus: Studien zur Ideologie und Herrschaft*, eds. Hellmuth Auerbach, et al. (Frankfurt am Main: Fischer Taschenbuch Verlag, 1993), 191–210. Soon the Nazis realised that to kill Europe's Jewish citizens faster, they had to develop a killing machine on a larger scale by creating an assembly line of murder.

29 Oldenziel et al., *Cycling Cities*.

30 Catherine Bertho Lavenir, 'Scarcity, Poverty, Exclusion: Negative Associations of the Bicycle's Uses and Cultural History in France', in *Cycling and Recycling: Histories of Sustainable Practices*, Ruth Oldenziel and Helmut Trischler, eds. (London: Berghahn, 2015), 91–113; Ruth Oldenziel and Karin Zachmann, eds., *Cold War Kitchen: Americanization, Technology, and European Users* (Cambridge, MA: MIT Press, 2009).

31 Elizabeth Shove, 'The Shadowy Side of Innovation: Unmaking and Sustainability', *Technology Analysis & Strategic Management: Innovation, Consumption, and Environmental Sustainability* 24, no. 4 (2012): 363, 367, 372–373; Ruth Oldenziel and Helmuth Trischler, 'How Old Technologies Became Sustainable. An Introduction', in *Cycling and Recycling: Histories of Sustainable Practices*, eds. Ruth Oldenziel and Helmuth Trischler, *Environment and History: International Perspectives* (London: Berghahn, 2015), 1–19.

32 Jean-Pierre Bardou, *The Automobile Revolution: The Impact of an Industry* (Chapell Hill, N.C.: University of North Carolina Press, 1982); Theo Barker, *The Economic and Social Effects of the Spread of Motor Vehicles: An International Centenary Tribute* (Basingstoke: Macmillan, 1987); Nick Georgano, *The American Automobile: A Centenary 1893–1993* (London: Prion, 1992).

Section 4
Life, biology and the organicist metaphor

13
Ludwig Koch's birdsong on wartime BBC radio: knowledge, citizenship and solace
Michael Guida

> One needs to cling to everything that is beautiful and uplifting, and, above all peaceful in a mechanical age.
>
> Emma Turner[1]

This chapter is about the recording of British birdsong in the late 1930s and its broadcasting throughout the Second World War. It will examine the scientific and cultural contexts that surrounded the capture of birdsong with the emerging techniques of field recording, and its dissemination through the established medium of wireless radio. The focus is the work of the German nature sound recordist Ludwig Koch, who presented his unique British birdsong recordings on air with the BBC throughout wartime and by the late 1940s was a household name.[2] John Burton, who was in charge of the BBC wildlife sound library at the Natural History Unit from 1962 to 1988, has written about Koch's work as a foundation of the library's collection.[3] Others have looked briefly at how Koch's recording efforts gave scientists the opportunity to study birdsong in a new way, breaking away from problematic transcription traditions of old.[4] The aim here is to look at Koch's work as a cultural endeavour with particular meanings during wartime, while placing it within the framework of ornithology and nature study of the day.

Koch's work was promoted and supported by several leading naturalists and ornithologists, Julian Huxley and Max Nicholson in particular. Their ideas about how best to develop knowledge and understanding of birdsong are explored alongside Koch's own enthusiasms for birdsong as a pleasure that all Britons could access in the crisis of war.

This exploration concentrates on the use of the broadcasting medium of radio to send birdsong into the homes of the nation. It therefore draws upon primary sources that include, not only the publications of Koch, Nicholson and Huxley, but also the materials of the BBC Written Archive and historical writing about BBC broadcasting. Listening to the radio became part of the construction and dissemination of national identity in the 1920s and 1930s,[5] and during wartime the practice became central to creating a sense of unity and securing morale.[6] It is in these contexts that birdsong took its place next to the chief output during wartime, that of popular music.[7]

I argue that while Koch and his scientific colleagues believed that the best way to appreciate and build knowledge about birdsong was to listen carefully, the pleasures to be had were available to everyone, without training. During wartime, there was a change in emphasis in which birdsong was associated with patriotism, and knowing one's country through its natural sound signatures was invested with ideas of citizenship. Birdsong too became a symbol of all British natural heritage that was being fought for. Koch's recordings themselves were seen in the years after the war as a national treasure to be preserved. Though never a scientist, Koch established a reputation for exactitude and passion in nature study that led in part to the creation of *The Naturalist* programme in 1946. Koch's work was grounded in a fundamental principle of broadcasting that John Reith had articulated at the inception of the BBC in the 1920s. This was the idea that the modernising and mechanising world, that wireless radio was becoming part of, needed moments of nature's calm to counteract the crash and bustle of progress. In this sense, Koch's birdsong programmes conveyed the riches of culture and knowledge, but also offered solace and peace that only the sounds of nature could.

With this in mind, first there will be a survey of Koch's recording work, which formed the basis of BBC broadcasts, and how he and his collaborators thought about the best ways to appreciate birdsong. Second, there will follow an analysis of Koch's broadcasts, their public reception and significance. This will lead to a discussion of some of the meanings and uses of birdsong on the radio.

Listening carefully to recorded birdsong

Ludwig Koch trained as a musician and *Lieder* singer in Frankfurt. His love of music and performance strongly influenced his interests in natural history. In the early 1930s, he worked as a director at Electrical and

Musical Industries (EMI), where he developed gramophone recordings for 'cultural' and 'educational' purposes.[8] It was here that Koch formulated the idea of a 'sound-book', which comprised text, photographic plates and gramophone recordings in one product. With the ornithologist Oscar Heinroth he published his first sound-book dedicated to birdsong in 1935 called *Feathered Mastersingers (Gefiederte Meistersinger)*.[9] Using acoustic and the new Neumann electrical recording gear, Koch wanted to break away from the 'musical notations and curves which meant nothing either to a scientist or to a bird-lover'.[10] He felt, too, that the translation of bird sounds into words 'such as *tu, tu, tu* or *tse tse tse* will never bring to the ears of the average listener the sweetness of the song of the wood-lark or the characteristic note of the marsh-tit'.[11] In Britain, in 1922, the zoologist Walter Garstang had published his affectionate *Songs of the Birds* using these modes of representation, alongside his poems about the songs of some twenty birds.[12]

Koch is credited with making the first recording of a bird, in 1889, on the Edison phonograph his father had bought back from the Leipzig fair.[13] It was his caged Indian Shama that he recorded as an eight-year-old boy. Yet, years later Koch became convinced that caged birds did not sing in the same way as those in the wild. By conducting experiments with captive song thrushes, blackbirds and blackcaps, he showed that mimicry would often make song very different from that encountered in the wild.[14] Until Koch began recording birds in their natural surroundings, only Carl Reich's recordings of his captive German blackbird, sprosser, thrush and nightingale were available in the HMV catalogue.[15]

As a German Jew, Koch moved to London to escape the Nazi threat and made a point of naming his first British sound-book in 1936 *Songs of Wild Birds*; these birds were free, not captive.[16] With Max Nicholson's expert ornithological text, and an introduction from the high-profile science populariser and secretary of the Zoological Society of Great Britain, Julian Huxley Koch produced a sound-book collection of 14 birds. Many of the recordings were of the familiar fluttering life that most people knew: the blackbird, song thrush, chaffinch, great tit, robin, wren, hedge-sparrow, turtle dove and wood pigeon. But there were more unusual treats too in the green woodpecker, willow-warbler, white throat, and the iconic sounds of the cuckoo and nightingale.[17] However well-known such birds were to the British public, listening experiences would have been brief and fleeting for most. Rarely was a bird heard in full song, with all its variations. With the sound-book, songs in all their complexity were captured for delight and study at home. Koch's recordings were the first occasion that bird-lovers and ornithologists could get to know their bird heritage in a new way.

The purpose of this collection was to allow listeners to concentrate on and study a single songster, one at a time, Koch encouraging listeners to pay careful attention in order to derive pleasure from the experience of a bird in song and to recognise its character and nuances. To this end, Koch created sound specimens of each bird, striving to isolate individuals as far as he could, which required painstaking persistence in the field. The collection did not seek to provide comforting atmospherics or bucolic montages. The second sound-book, *More Songs of Wild Birds*, almost provided this with a 'medley of birds' voices'; however, these recordings, which included a little owl, rook, jay and 'dawn choir', are there to afford 'a pleasurable exercise for the bird-lover who will have the opportunity of distinguishing one song from another.'[18] The intention was to educate the listener and Max Nicholson provided detailed listening notes as a guide to understand characteristics within and between species:

> (2 min 30 sec) The *cuckoo* is now heard calling. During the first fifteen seconds he utters a dozen cries without a break, then he flies away
> (2 min 45 sec) and is heard faintly for a few seconds in the background. Now at
> (2 min 55 sec) he is back again, and two other voices – a *chiffchaff's* and a *woodpigeon's* – can be heard faintly in the background.
> (3 min 00 sec) Now he is calling more deliberately, and in the first ten seconds of the third minute he only utters six notes. At this stage the *chiffchaff* becomes rather more distinct, singing a double note as monotonous in its own way as the cuckoo's, and just before the record ends a faint
> (3 min 11 sec) *blackbird* song is heard.[19]

Nicholson warned against sentimentalising the beauties of birdsong by encouraging listeners to build up their knowledge to avoid 'false emotions or beliefs which might hinder a true appreciation'.[20] Developing a 'trained ear for bird music' was what he suggested to avoid such pitfalls.[21] In fact, such close attention was required if listeners were to discover the 'magic power and vitality' of the voices of nature.[22] Distractions ruined the 'spell of bird music' and full appreciation could be best had by listening in silence, seated comfortably with the lights dimmed, Nicholson tutored.[23] Careful listening was required for knowledge-making but it was also the gateway to pleasure-making.

The sound-books of birdsong sold well, surprising the publisher Witherby,[24] the first remaining in print for at least 17 years and eight impressions, two during the Second World War. Ornithologists loved these unique collections,[25] but the sound-books were well-reviewed by the general press too. *The Listener,* the high-brow magazine of the BBC, said that Koch's first set of discs offered 'a new vista of delight and knowledge to everyman'. Moreover, the reviewer felt there was something special about the sounds Koch had put on disc that distinguished them from the common currency of popular music: 'They are worth a dozen of the music everyone knows. They are worth twelve hundred cage-birds'.[26] The comment points to the standing of wild birdsong in culture in relation to what some saw as the triviality of popular music flooding from the wireless and gramophone players.

These recordings were quite unique at this time and formed the basis of Koch's BBC radio broadcasts throughout the war. Birds and their song had been the subject of programmes on the wireless before 1939 though. Huxley had broadcast a successful six-part series in 1930 called *Bird Watching and Bird Behaviour*, with a spin-off book,[27] and the naturalist and anthropologist Tom Harrisson had broadcast his three-part *Watching Birds* in 1936. Both had used their programmes to experiment with the idea of getting listeners involved, making observations and sending in data or specimens.[28] An interest in birds was increasingly widespread by the beginning of the Second World War, the naturalist James Fisher claiming that he knew of all kinds of people who were at it: 'a late Prime Minister, A Secretary of State, a charwoman, two policemen, two kings, one ex-king, five Communists, four Labour, one Liberal, and three Conservative Members of Parliament, the chairman of a County Council, several farm labourers … at least forty-six school masters, and an engine-driver'.[29] The observation of bird behaviour was becoming part of a British mode of thinking about the nation, its values and how one was part of it.

Wartime broadcasts for 'all classes of society'

In March 1936, Koch was approached by Mary Adams, one of the BBC's first science programme producers,[30] just a month after he had arrived in London. Adams wanted to hear his collection of continental birdsong recordings, on a tip-off from Max Nicholson.[31] However, Koch was busy pursuing his new collection of British birdsong recordings, which became his first sound-book later that year. Koch's first appearance on air was in 1936 on *Radio Gazette*, a news programme that experimented with new recording techniques including the Blattnerphone recording machine.[32]

Here he described the challenges of capturing the sound of birds in song on a bulky recording system with lengths of microphone cable, and went on to play what he said was the first recording of a green woodpecker that had taken him more than two weeks to make during the spring of that year.[33]

When war began, Koch busied himself with the business of birdlife. Two weeks before the evacuation of allied forces from the beaches of Dunkirk in 1940, a letter from Koch appeared in *The Times*. Koch encouraged readers to find solace in the beauty of birdsong: 'War or no war, bird life is going on and even the armed power of the three dictators cannot prevent it. I would like to advise everybody in a position to do so, to relax his nerves, in listening to the songs, now so beautiful, of the British birds. If one watches carefully, one can be sure of surprises'.[34] The purpose of the letter was in fact to alert readers to the phenomenon of mimicry between species. He had recently heard a 'blackbird mimicking the yaffle call of the green woodpecker' and he urged readers to send him 'any observations of unusual bird calls or song'. War or no war (the letter sits among others debating munitions supplies and urgent pleas for funds from the Red Cross), Koch was determined to recruit the public to help better understand the curiosities of spring-time bird behaviour, which might also serve to take their minds away from wartime worries.

After a period of internment in 1941 on the Isle of Man as an 'enemy alien' Koch joined the European Service of the BBC on the suggestion of Julian Huxley.[35] Koch never stopped recording, though the sound of aircraft overhead often made the work impossible:[36]

> I was allowed [by the BBC] to make all kinds of recordings. I visited a number of factories to explore unusual noises, but amid the din of machinery I longed for the sounds of nature, and persuaded my superiors that this was the right moment to show the enemy, by recording all kinds of farm animals, that even bombing could not entirely shatter the natural peace of this island.[37]

The farm animal recordings do not seem to have been aired, but one five-minute piece from 1942 called *Early Morning on a Hampshire Farm* suggests that Koch wanted to bring the pastoral into the front room at a time when cities in England, Wales and Scotland were still very much under threat from German air attacks. The 'sound picture', as he called it, was an absorbing and detailed study of the clucking of hens, the voices of sparrows, ducks, geese, a donkey, a horse, piglets and pigs, cows and finally an extended piece featuring individual sheep, lambs and then the flock bleating a bucolic chorus for the last minute.[38]

Koch was given regular slots on the radio, from 1941 and to the end of the war, on *Children's Hour, Country Magazine* and then his own series of solo shows. According to the *Radio Times*, during the war Koch and his recordings appeared on air on 32 occasions. *Country Magazine* was conceived as a wartime programme, which by 1946 had an audience of almost seven million listeners.[39] The programme closed with one of Koch's sound-pictures of the countryside.[40] *Children's Hour* had been running since radio began and went out during wartime just before tea-time. By appearing on *Children's Hour* Koch had the ears of millions when many under 16s had been evacuated from cities and vulnerable coastal towns in the south and east of England. Very popular with adults as well as children, the programme acted as a daily point of contact between displaced family and friends.[41]

Almost all of Koch's programmes on *Children's Hour* featured birds and their song. For his young listeners, Koch managed to convey the way he worked in the field – the early rising, the pursuit of a bird and the exquisite pay-off of witnessing a good singer. This excerpt from a 1943 script gives a sense of how Koch would take his audience with him using vivid and involving descriptions to bring the outdoors to life, while communicating details of the life of the skylark:

> Let us creep closer, but carefully, not to do any harm to the birds' grass nests in a hollow in the ground. It is still pitch dark, before dawn; to hear the first lark you must get up with the lark. No noise, but many miles away in a hamlet I can hear faint barking and lowing. We are close to the skylark, it starts rising again. Here it is:-
>
> (Sky Lark)
>
> That was a very good performance and an extraordinarily good singer.[42]

These *Children's Hour* programmes can be seen as complementary to 'nature study' science material within the Broadcasts to Schools classroom initiative.[43] Beyond the everyday, the status of Koch's birdsong broadcasts was indicated by their inclusion in Christmas Eve programming in 1941, when *Children's Hour* comprised a piece from Koch called 'Listen to Our Song-Birds in Winter', followed by a 'Visit to the Church of the Nativity in Bethlehem' and, finally, prayers.[44]

Koch went on to host five-, ten- and then fifteen-minute solo shows from 1943 to 1945. These programmes punctuated more conventional BBC programming that included talks, plays, light music and the BBC orchestra. He combined a typically playful narrative with the didactic in *The Nuthatch Sings in February* broadcast in 1943 on Sunday 7 February from 6.55 to 7pm: 'My particular nuthatch was living in a woodpecker's hole in a large chestnut tree. The entrance to the hole was too large for him so he narrowed it by filling it up with mud.' He then went on to introduce the listener to six different calls and songs, starting with the warning note as Koch approached the tree ('it sounds, misleadingly, rather peaceful'), the angry call as Koch approached the tree, the mating note ('rather like boys whistling to each other'), and the trill which is sung 'on different levels, from a soft piano to a wide, carrying forte', and so on.[45] There was depth and detail in Koch's broadcasts as well as good-humoured entertainment.

The public response to Koch's broadcasts must be pieced together from various sources. Koch, reflecting in his 1955 memoir on his public audience, believed that his birdsong programmes had piqued interest across all lines of age, class and gender:

> But among my listeners there are obviously a great number of adepts, men and women of all ages, and of all classes of society, whose thousands of letters and whose eager reactions when listening to my lectures have proved to me that all this hard and infinitely patient and subtle work was not in vain ... they have fallen to the lure of nature's most lovely and lively gift, the huge variety of birds and their songs and calls.[46]

Throughout the war, Koch's reputation was developing. A letter in the *Radio Times* from Dora Read in West London in 1943 found that hearing birdsong on the radio could rejuvenate war-workers: 'Many thanks for letting us hear the wonderful birdsong, full of hope and peace to come. Millions of us, used to rambling before the war, are now in factories doing war work. Let us hear more of Ludwig Koch's birds!'[47] There were enthusiastic reports in regional newspapers about his lectures in schools and colleges (Royal College of Music, Eton College, Bradford Girls Grammar School)[48] and the Taunton Field Club[49], and the Bristol Naturalist Society hosted lectures from 'the great ornithologist'.[50] In 1944, the *Western Times* told of Miss Wyness who gave a talk about British birds, using Koch's gramophone discs, at her Women's Institute meeting in Dolton village hall, Devon.[51] This activity was instructional, but it was motivated by a love of birds and the way their song made people feel.[52]

Critics were almost unanimous in their appreciation of Koch's broadcasts, although his extraordinary way of speaking English attracted scorn from one critic,[53] and some concern within the BBC for its lack of clarity for children.[54] His style was to almost sing his lines with his marked German accent. It is hard to imagine though that birdsong could have meant as much to listeners without Koch's distinctive, and quite foreign, on-air persona, which in all likelihood drew listeners closer to the radio, preparing them for the recordings Koch was about to play.

The BBC archives give few clues about what programme producers saw in Koch's work and broadcasts. He was evidently a man of dedication and tenacity, whose perfectionist tendencies made little allowance for wartime conditions. His steady stream of requests to pursue new recording projects of all kinds throughout the war could be met with exasperation at their ambition and cost.[55] BBC executives and producers would have been reassured by Koch's postbag responses, however. If his work during the war was trialled first with young listeners and parents tuned in to *Children's Hour*, his appearances on *Country Magazine* and then being given solo programmes indicate an increasing confidence in his appeal and also his utility during the war.

Certainly, no one else had assembled a collection of field recordings of British birds ever before. Koch was of interest because his work got programme-makers at the BBC thinking in new ways about what broadcasting could be. The BBC producer Mary Adams was interested because she felt Koch's recordings would make some of her science broadcasts more appealing and digestible to a broad audience.[56] Koch's work was highly relevant to Marie Slocombe who was beginning to establish, with Tim Eckersley, the BBC sound archives and what were called 'effects' in the late 1930s.[57] Tom Harrisson was a fan of Koch's sound-world. As the radio critic for *The Observer* he had taken the opportunity to make mischief with the efforts of the BBC's sound effects library.[58] He complained that the BBC was using an impoverished repertoire of recorded sounds, with 'a snatch of B.B.C. Seagull, which I have heard represent "the sea" over and over again – once four times in a day.' He singled out Koch's 'aural documentary' collection, underused and undervalued by 'governors, directors and producers', as the benchmark for what could be achieved with the right kind of skill and aptitude. Koch's reputation after the war was ascending in such a way that in 1946 Desmond Hawkins, who went on to found the BBC Natural History Unit, created *The Naturalist*, partly as a vehicle for Koch's recordings and thoughts about the natural world. Hawkins called *The Naturalist* 'a programme of science and observation'.[59]

The meanings and uses of Koch's birdsong in wartime Britain

Koch's broadcast work during wartime fulfilled many needs and functions, while the meanings of birdsong were complex and loaded with cultural and political symbolism. To conclude, there will be a focus on the opinions of the senior establishment scientists and ornithologists, Huxley and Nicholson, and of the writer and naturalist James Fisher, as they all voice a professional interest as well as personal passion for birds and their song. First, I argue that engagement with birdsong while listening to Koch's broadcasts can be seen as an act of citizenship, in knowing one's country. Second, I explore the ideas that hold up birdsong as a marker of British nature and a source of patriotic pride that is worth fighting for. Lastly, I show how Koch's broadcasts stemmed from John Reith's founding vision for radio, in which the sound of birdsong could counter the noise and pressure of modern life.

Citizenship through knowing

The emphasis that Koch's sound-books placed on careful listening was still a condition of the proper appreciation of birdsong in his radio programmes that followed during the war. However, Koch did not intend his broadcasts to be solely an education in Britain's birdlife, especially if that excluded some listeners. The virtuoso performances, the intriguing variations and the effect of birdsong on human mood were the real attractions for Koch and he wanted everyone to enjoy the boost. Birds for him were 'the most attractive, variegated, amusing, uplifting and artful of all living things'.[60] It was important to Koch that this delight could be had without any particular training or prior knowledge – only interest and attention was required, not study. Julian Huxley, too, saw birdsong as an egalitarian joy that anyone could appreciate. 'I suppose that birds give more pleasure and interest to humanity … than all the other groups of the animal kingdom taken together', Huxley said in his introduction to Koch's first sound-book. Huxley believed that through 'their beauty, vivacity, by their songs and freedom of flight, by their migrations and their domestic arrangements, they make an obvious appeal to the layman, however uninstructed'.[61] The sense of hearing, Huxley argued, accessed and seduced the emotions in ways that intellectual engagement could not: 'The associations called up by sound seem to share with those aroused by smell the properties of fullness, immediacy, and emotional completeness to an extent not aroused by those dependent on sight or intellectual comprehension'.[62] In this case then, all Britons from all backgrounds could be touched deeply by birdsong. The effect of birdsong was all the more potent at a time when city populations were exposed to a new threatening soundscape of sirens and bombing and the challenging sensory conditions imposed by the blackout.

Both Huxley and Nicholson facilitated Koch's work in part because of their interest in popularising bird-work for scientific purposes through networks of amateurs. From the 1910s Huxley had been encouraging amateur ornithologists to 'direct their emerging observational networks at problems of scientific moment'.[63] Nicholson, with others, was instrumental in the creation of the British Trust for Ornithology in 1933, which aimed to 'set up a chain of organised bird-watchers' and to render obsolete the 'old fashioned splendid isolation of the birdwatcher'.[64] By the late 1930s, networks of amateur observers recruited though announcements in the press and on national radio were in place across the country.[65] This seemingly democratic citizen-science was a means by which ordinary people could get to know the land, the countryside and, by extension, the nation. In his six-part radio series during 1930, Huxley explained this idea to Britons in a programme called *The Pleasures of Bird-Watching*: 'to go on a country walk and see and hear different kinds of wild birds is thus to the bird-watcher rather like running across a number of familiar neighbours, local characters, or old acquaintances'.[66] One became part of the community of nature by proceeding through it and taking an interest. Birdsong itself was an 'expression' of the nation, Huxley said:

> The yellow-hammer's song seems the best possible expression of hot country roads in July, the turtle-dove's crooning of midsummer afternoons, the redshank's call of sea-breeze over saltings and tidal mudflats, the robin's song of peaceful autumnal melancholy as the leaves fall in a sun which has lost its warming power.[67]

It is possible to consider listening to birdsong on the radio in the same way as bird-watching at large because both involved a community of amateur listeners around Britain who were participating in and reaffirming their social and national identities. In listening to and getting to know Koch's birdsong, a small but significant ceremony of wireless citizenship was enacted. In wartime such participation could feel like patriotism, or even legitimate war-work, for those on the homefront.[68] As a foreign national in Britain, Koch himself had enhanced his credentials as a guest citizen by solidifying on gramophone the national heritage of birdsong, ensuring it was preserved, shared and studied.

Fighting for British nature

Patriotic and scientific interests were together at play when British birds were declared the best singers. Koch had demonstrated on air in 1944 during *The Song Thrush is Silent in August* 'the great superiority of the British over the German song-thrush whom I also know well'. He played first his German recording, then his British one, asking listeners to make- up their own minds.[69] During his first spring in Britain in 1936, while strolling in the college gardens of Cambridge University, Koch straight away 'had the impression that both the blackbird and the song-thrush sang more beautifully than I had heard them do in Germany'.[70] His refugee status in the safety of Britain may well have influenced how he heard British birds, but he was not the only one who held such views. Seasoned ornithologists like Max Nicholson had made similar claims in the 1930s, asserting that in no other country in the Northern or Southern hemispheres was birdsong as powerful, varied and pleasing as in England. That so many resident species were 'good songsters' common to gardens and familiar to ordinary people, combined with the long song season and temperate climate, made 'England a paradise for bird-song'.[71]

If British birds were so highly prized then they would need to be cherished and defended, and one ornithologist writing on behalf of all Britons expressed that particularly clearly. James Fisher published his highly successful Pelican paperback called simply *Watching Birds* in the midst of war. Fisher was optimistic that such a book could improve the lives of ordinary people during wartime. Writing just after the Battle of Britain, in November 1940, he placed birds at the centre of the conflict:

> Some people might consider an apology necessary for the appearance of a book about birds at a time when Britain is fighting for its own and many other lives. I make no such apology … Birds are part of the heritage we are fighting for. After this war ordinary people are going to have a better time than they have had; they are going to get about more … many will get the opportunity hitherto sought in vain, of watching wild creatures and making discoveries about them. It is for these men and women, and not the privileged

few to whom ornithology has been an indulgence, that I have written this little book.[72]

Fisher offered the book to the public because he felt the study of birds concerned many ordinary people 'who meet each other in the street' but it was, as he said, a book of 'science' not 'aesthetics'.[73] It was not an easy read. The text covered anatomy, migration, habitats, territory and courtship, with technical illustrations and no photography, yet Fisher's book went on to sell over three million copies and is credited for enthusing a whole generation of the public into an appreciation of birds.[74] Perhaps to own and browse through this little book was itself a patriotic impulse. Like Fisher had done with his book, Koch with his recordings had captured, protected and given authority to Britain's bird culture. Koch's recordings acted to preserve part of Britain's precious sonic character, resulting in a collection of more than 500 recordings that was purchased by the BBC in early 1948.[75] Beyond a move to archive a special collection of natural history sounds for scientific research, the purchase recognised a national treasure of British identity, if not an international one geared to peace and cooperation, as Julian Huxley, then the director-general of UNESCO, indicated by his resolution to have it preserved.

Birdsong employed as a 'silence'

Koch's broadcasts undoubtedly drew power from their evocation of peaceful rural myths of the imagination that were widely propagated during the war. The gentle age-old rhythms of the southern English countryside, where relatively few now lived, were promoted in recruitment posters, depicting pastures and leafy villages, intended to galvanise men to defend the nation.[76] Celebrations of rural heritage were featured on BBC radio in programmes such as *Country Magazine* (that Koch presented on), *The Countryman in Wartime* and *Your Garden in Wartime*. Another, *The Land We Defend*, pictured Britain as one vast and pretty village populated by lovers of nature and countryside.[77] This kind of programming was part of the BBC's effort to maintain morale, radio being conceived of as an 'instrument of war'[78] as well as a medium of 'solace'.[79] Perhaps the most important aspect of Koch's birdsong broadcasts in wartime was the connection to what Sonya Rose has referred to as the 'authentic nation' – the English countryside that stood for peace, tranquillity, stability and harmony.[80] Birdsong was all of these things and could itself be representative of the 'authentic nation', the keynote of which was quietude. That birdsong could be seen as both a distinct sonic expression of the character of the nation and, at the same time, be an emblem of peace and tranquillity was a duality that had been incorporated within the philosophy of BBC radio when it was a new public communication medium in the 1920s.

It was John Reith himself, director general of the BBC from its earliest years until 1938, who argued that birdsong was the kind of sound that was 'not incompatible with the conditions of silence'.[81] He had promised in his 1924 manifesto, *Broadcast Over Britain*, that broadcasting would delight urban and city dwellers with the 'many voices of Nature'. Reith was responding to fears and criticism about what the medium of radio – with its invisible, speed-of-light electromagnetic pulses travelling through bodies and buildings – might be doing to humans.[82] He was also seeking to counter concerns that the talk and music flooding into homes through the wireless might standardise taste and thought through its mass address.[83] Moreover, 'the broadcasting craze' was seen by some commentators as a new and inescapable noise of the modern technological realm.[84] Reith wanted his public service broadcasting to be transcendent, not simply a cultural utility.

Quite by chance, Reith found an opportunity to conduct a technical experiment with his new medium that had the potential to answer such criticism. At the suggestion of the cellist Beatrice Harrison, a live broadcast of a nightingale singing while she played outside in her Surrey garden was attempted over several evenings in May 1924.[85] The experiment worked and the iconic sound of the nightingale was conveyed to homes around the country. The public were enraptured. There were said to be a million listeners that May.[86] Reith was convinced that broadcasting birdsong and other sounds of nature was a way to soften anxieties and fears about the new public service. Birdsong would be a tiny part of the broadcast output but he believed it would have a resounding symbolic value. For Reith, it was evidence that broadcasting could convey magic beyond its human emanations to put listeners 'in touch with the infinite'. When he described the nightingale's song as 'something of the silence which all of us in this busy world unconsciously crave and urgently need', he acknowledged that broadcasting needed to demonstrate qualities that distinguished it from the cacophonies of a mechanising civilisation.[87]

Julian Huxley had written in 1930 that in the modern world birds 'have a place in civilisation as well as in wild nature'.[88] Max Nicholson too argued in his text accompanying Koch's sound-book that the civilised and eternal presence of birdsong could counter human barbarism in wartime:

We may be uncertain whether London and Paris and Berlin will be reduced to heaps of ruins by the misuse of scientific weapons in the interests of mutual destruction, but we can be sure that in any case nightingales will sing in Surrey every May, and golden orioles will still flute with civilised perfection in German and French spinneys, regardless of human barbarism or of human achievements.[89]

For Ludwig Koch, the enchantment and solace of hearing a bird in song was the antithesis of the crash of industrialised warfare. Birdsong was natural order and beauty set against the ugliness of man-made chaos, a civilisation gone wrong. Birdsong could best encapsulate the peaceful national character of the British because this sound came from the shires and hills that stood in permanent and quiet authority. Koch's birdsong was a modern sound though, recorded on shellac disc, broadcast to the entire nation through the airwaves of the BBC. Through the loudspeaker, alongside music and speech, birds were heard in homes of the town and city where they could be considered afresh. Listeners to Koch's programmes were familiar with common birds from the park, garden and kitchen window, yet his sound-world revealed a greater repertoire of song, explained its purposes, while heightening the status of birdsong as a national signal of bounty and peace.

Notes

1   Emma Turner, *Every Garden a Bird Sanctuary* (London: Witherby, 1935), quoted in Tom Harrisson, 'Birds and Man', *The Listener* (29 April 1936): 808. Turner was an ornithologist and bird photographer, celebrated for her 1911 image of a nesting bittern in Norfolk.

2   The radio historian Seán Street has argued that Koch was as well-known to the public after the war as David Attenborough is today. See his radio feature about Koch on BBC Radio 4, Archive Pioneers, *Ludwig Koch and the Music of Nature* (15 April 2009): www.bbc.co.uk/archive/archive_pioneers/6505.shtml, accessed June 2018.

3   John Burton, 'The BBC Natural History Unit Wildlife Sound Library 1948–1988', *Wildlife Sound* 12 (2012): 19.

4   David Rothenberg, *Why Birds Sing* (London: Penguin, 2005), 58–60; Joeri Bruyninckx, 'Sound Science:

Recording and Listening in the Biology of Bird Song, 1880–1980' (PhD thesis, University of Maastricht, 2013).

5   Paddy Scannell and David Cardiff, *A Social History of British Broadcasting. Volume One 1922–1939, Serving the Nation* (Oxford: Basil Blackwell, 1991), 10, 277–303; Thomas Hajkowski, *The BBC and National Identity in Britain, 1922–1953* (Manchester: Manchester University Press, 2010).

6   Sian Nicholas, *The Echo of War: Homefront Propaganda and the Wartime BBC, 1939–1945* (London: Palgrave Macmillan, 1996), 2.

7   Christina Baade, *Victory Through Harmony: The BBC and Popular Music in World War II* (New York: Oxford University Press, 2012).

8   Ludwig Koch, *Memoirs of a Birdman* (London: Phoenix House, 1955), 25.

9   Koch, *Memoirs of a Birdman*, 25–29.

10 Koch, *Memoirs of a Birdman*, 25.

11   Koch, *Memoirs of a Birdman*, 25.
12   Walter Garstang, *Songs of the Birds* (London: The Bodley Head, 1922), 15. Garstang's work was an interesting attempt to understand birdsong (without the use of recording) by deploying the scientific analysis of written and musical transcriptions combined with his own poetic experiments to study fully the expression of bird emotions in song.
13   Burton, 'The BBC Natural History Unit', 19.
14   Koch, *Memoirs of a Birdman*, 26.
15   As early as 1911, the HMV catalogue listed in the Whistling section a 'unique bird record', which was Carl Reich's caged nightingale: *The Gramophone Company Record Catalogue* (February to July 1911): 67. The 1914 catalogue listed all four birds: *His Master's Voice Catalogue of Records* (1914): 139.
16   E.M. Nicholson and Ludwig Koch, *Songs of Wild Birds* (London: Witherby, 1936/1951).
17   The British Library's online wildlife sound collection includes all of Koch's recordings published in 1936 and 1937: http://sounds.bl.uk/Environment/Early-wildlife-recordings, accessed June 2018.
18   E.M. Nicholson and Ludwig Koch, *More Songs of Wild Birds* (London: Witherby, 1937), inside front cover.
19   Nicholson and Koch, *Songs of Wild Birds*, 197.
20   Nicholson and Koch, *Songs of Wild Birds*, 185.
21   Nicholson and Koch, *Songs of Wild Birds*, 65.
22   Nicholson and Koch, *Songs of Wild Birds*, 185.
23   Nicholson and Koch, *Songs of Wild Birds*, 186.
24   BBC Written Archive, S26/1/6 Koch, correspondence file 6, Witherby statement, December 1936.
25   Book reviews, *British Birds* (1 December 1937): 239.
26   Book Chronicle, *The Listener* (4 November 1936): 877.
27   Julian Huxley, *Bird Watching and Bird Behaviour* (London: Chatto and Windus, 1930).
28   In 1916, Julian Huxley wrote a two-part article in *Auk*, suggesting that the 'vast army of bird lovers and bird-watchers today in existence' could be directed to channel their enthusiasms into solving, with biologists, fundamental problems of science: Julian Huxley, 'Bird-Watching and Biological Science (Part 1)' *Auk* 33 (1916): 142–161. In Tom Harrisson's programme about the little owl, he asked listeners to document behaviour and send in owl pellets: BBC Written Archive, RConti1, Talks, Tom Harrisson file 1A 1932–36, little owl script.
29   James Fisher, *Watching Birds* (London: Penguin, 1941/1946), 13.
30   Allan Jones, 'Science in the 1930s and the BBC: Competition and Collaboration'. Paper presented at On Archives, University of Wisconsin-Madison, USA, (6–9 July 2010): 5.
31   BBC Written Archive, S26/1/1, correspondence file, letter (17 March 1936).
32   Asa Briggs, *The History of Broadcasting in the United Kingdom,* Volume II, The Golden Age of Wireless (London: Oxford University Press, 1965), 99.
33   Desmond Hawkins, 'A Salute to Ludwig Koch', 12-inch LP (London: BBC Radio Enterprises, 1970). This LP, narrated by Desmond Hawkins who established the BBC's Natural History Unit, gives a good impression of Koch's personality and broadcasting style. It can be heard online: https://archive.org/details/ASaluteToLudwigKoch, accessed June 2018.
34   Ludwig Koch, 'A Blackbird Mimic', *The Times* (13 May 1940): 4.
35   Koch, *Memoirs of a Birdman*, 70.
36   Koch, *Memoirs of a Birdman*, 73.
37   Koch, *Memoirs of a Birdman*, 71.
38   British Library, Ludwig Koch sound pictures, 'Early Morning on a Hampshire Farm', gramophone 1LL0003863, April, 1942.
39   BBC Written Archive, N2/25 North Region, Country Magazine, memorandum from John Polworth (8 March 1946).
40   Koch, *Memoirs of a Birdman*, 71.
41   Nicholas, *The Echo of War*, 45.
42   BBC Written Archive, Ludwig Koch scripts, 'Listen to Our Songsters' (13 June 1943).
43   For a broad review of school broadcasting see David Crook, 'School Broadcasting in the United Kingdom: An Exploratory History', *Journal of Educational Administration And History* 39 (2007): 217–226.
44   *Radio Times*, BBC Home Service, Children's Hour (19 December 1941): 14.
45   BBC Written Archive, Ludwig Koch scripts, 'The Nuthatch Sings in February' (7 February 1943).
46   Koch, *Memoirs of a Birdman*, 179.

47 Tweet-Tweet, *Radio Times* letter (19 February 1943).

48 British Library, Ludwig Koch papers, LKC2–4, press cuttings.

49 'Recording Birdsongs', *Taunton Courier and Western Advertiser* (14 April 1945): 2.

50 'Items of Local News', *Western Daily Press* (11 March 1944): 5.

51 'From Towns and Villages in West Country', *Western Times* (6 October 1944): 5.

52 Street has argued that Koch's recordings and the way he presented them in his broadcasts were special for the potent emotions they could evoke: Seán Street, *The Poetry of Radio: The Colour of Sound* (Oxford: Routledge, 2012), 100–103.

53 'Critic on the Hearth', W.E. Williams, *The Listener* (11 February 1943): 188. Williams, the editor-in-chief of Penguin books, felt that Koch's accent was 'the kind to which the B.B.C. should not issue a certificate of air-worthiness.'

54 British Library, Ludwig Koch papers, broadcasting scripts LKA9, letter from Derek McCulloch (11 June 1943). The *Children's Hour* director told Koch that he had received 'instructions from above' to tell him that some children found it difficult to follow his voice.

55 BBC Written Archive, LI/239/2 1948–51, Left Staff, Marie Slocombe memo (1 January 1943).

56 Jones, 'Science in the 1930s and the BBC', 7–9.

57 See Seán Street's account on BBC Radio 4 *Archive Hour*, 'Marie Slocombe and the BBC Sound Archive' (1 September 2007), where he explains how Slocombe established the BBC Sound Archive when few were interested, beginning with the voices of Winston Churchill, George Bernhard Shaw and G.K. Chesterton: www.bbc.co.uk/archive/archive_pioneers/6502.shtml, accessed June 2018.

58 Tom Harrisson, 'Radio', *Observer* (3 January 1943): 2.

59 Burton, 'The BBC Natural History Unit', 21.

60 Koch, *Memoirs of a Birdman*, 11.

61 Nicholson and Koch, *Songs of Wild Birds*, xiii.

62 Nicholson and Koch, *Songs of Wild Birds*, xiv.

63 David E. Allen, 'Amateurs and professionals', in *The Cambridge History of Science: The Modern Biological and Earth Sciences*, eds. Peter J. Bowler and John V. Pickstone (Cambridge: Cambridge University Press, 2009), 32.

64 Nicholson quoted in Helen Macdonald, '"What Makes You a Scientist is the Way You Look at Things": Ornithology and the Observer 1930–1955', *Studies in History and Philosophy of Biological and Biomedical Sciences* 33 (2002): 55–56.

65 Macdonald, '"What Makes You a Scientist"', 56.

66 Huxley's series was published the same year in book form, see Julian Huxley, *Bird-Watching and Bird Behaviour* (London: Chatto and Windus, 1930), 5.

67 Huxley, *Bird-Watching and Bird Behaviour*, 7.

68 Macdonald in 'What Makes You a Scientist' has argued that critical observation and knowledge building performed by organised citizen-scientists was considered to be a way to participate in legitimate war-work because the constructive thought involved in such work was able to repel wartime mass-fear, which otherwise might destabilise the nation, 63–64.

69 BBC Written Archive, Koch script, 'The Song Thrush is Silent in August' (18 September 1944).

70 Koch, *Memoirs of a Birdman*, 36.

71 Nicholson and Koch, *Songs of Wild Birds*, 26.

72 Preface to first 1941 printing in Fisher, *Watching Birds*.

73 Fisher, *Watching Birds*, 14.

74 Stephen Moss, *A Bird in the Bush: A Social History of Birdwatching* (London: Aurum Press, 2004), 168. See also the *Oxford Dictionary of National Biography* for this bold claim.

75 Burton, 'The BBC Natural History Unit', 21.

76 See Frank Newbould's poster series *Your Britain, Fight for it Now* (Baynard Press, 1942), Imperial War Museum catalogue number Art.IWM PST3641. The myths of the English rural are well examined in Alun Howkins, 'The Discovery of Rural England', in *Englishness: Politics and Culture 1880–1920*, eds. Robert Colls and Philip Dodd (London: Bloomsbury, 2014), 97–111.

77 Nicholas, *The Echo of War*, 233.

78 Nicholas, *The Echo of War*, 2.

79 Briggs, *The Golden Age of Wireless*, 13.

80 Sonia Rose, *Which People's War? National Identity and Citizenship in Wartime Britain*, 1939–1945 (New York: Oxford University Press, 2003), 212–213.

81   John Reith, *Broadcast Over Britain* (London: Hodder and Stoughton), 221.

82   Arthur Burrows, *The Story of Broadcasting* (London: Cassell, 1924), preface and 75.

83   Kate Lacey, *Listening Publics: The Politics and Experience of Listening in the Media Age* (Cambridge: Polity Press, 2013), 83.

84   See, for example, letter, *John O' London's Weekly* 9 (1923): 625.

85   Patricia Cleveland-Peck, *The Cello and the Nightingales: The Autobiography of Beatrice Harrison* (London: John Murray, 1985), 131–133.

86   Such was the public enthusiasm for this experiment that broadcasts continued every May for the next 11 years, and from 1936 to 1942 the bird alone took the microphone for ten minutes between eleven and midnight. See Richard Mabey, *Whistling in the Dark: In Pursuit of the Nightingale* (London: Sinclair Stevenson, 1993), 102–103.

87   Reith, *Broadcast Over Britain*, 221–223.

88   Huxley, *Bird-Watching and Bird Behaviour*, 116.

89   Nicholson and Koch, *Songs of Wild Birds*, 183.

14

'More Modern than the Moderns': performing cultural evolution in the Kibbo Kift Kindred

Annebella Pollen

Introduction

The legacy of evolutionary theory had a profound effect upon art and culture in the early twentieth century, not least via the various manifestations of evolutionary ideologies that were widely embraced across the political spectrum and sharpened in the popular mind by the events of the First World War. As Gillian Beer has argued, evolutionary theory was 'a form of imaginative history' that impacted upon notions of the past and visions of the future, and acted as a guiding metaphor for a wide range of cultural applications that reached far beyond biology.[1] Drawing on works that assess the intersection of scientific ideas in application, from Stephen Jay Gould to Oliver Botar, this chapter explores the ways that popular scientific ideas about the life force, degeneration, cultural evolution and the biogenetic law were disseminated and incorporated into the symbols, philosophies and practices of experimental woodcraft campaign groups in interwar England.

Woodcraft

During and after the First World War, when a stream of pacifist troop leaders split from the British Boy Scouts, disillusioned with what they perceived to be its increasing militarism, they frequently sought to return to the founding ideals of artist and naturalist Ernest Thompson Seton.[2]

English by birth but American resident for most of his life, Seton had devised a system of 'Woodcraft Indians' training for boys as a broadly socialist educational scheme that combined knowledge of nature with so-called picturesque practices, loosely inspired by idealised Native Americans drawn from myth and literature.[3] These ideas had, in part, inspired British scouting endeavours but some felt that they had been pushed too far aside in the dominating disciplinary structure of drills and 'preparedness' in Baden-Powell's imperialist project. One seceding group in particular incorporated aspects of Seton's methods into an ambitious utopian outdoor movement that embraced both sexes and all ages. The Kindred of the Kibbo Kift, founded in 1920 and led by charismatic artist and author John Hargrave, comprised several hundred seekers and reformers including utopian socialists, former suffragettes and Theosophists, and aimed for nothing less than world peace, to be achieved through an eclectic blend of camping, hiking and handicraft.[4] Despite their relatively small numbers and the eccentricity of their methods, the Kindred's uncompromising vision for the new world that they expected to lead was total, encompassing bodies and language, design and dress, music and performance, education and economics, spirituality and science.

This chapter explores the range of ways that science – broadly understood – was used as creative inspiration and intellectual justification for woodcraft philosophies, and how the cultural programme that Hargrave and his largely self-educated followers devised was underwritten by popular understandings of evolutionary biology. As such, this chapter explores the ways that scholarly thinking was received and applied outside of metropolitan elites. In the context of woodcraft groups, it also examines how such disciplinary concepts could also be reinterpreted into a distinctive set of cultural activities aimed at putting scientific theory into practice, as the groups styled themselves to be producers as well as consumers of new ideas, and ultimately as leaders and instructors-in-training for the new world to come. As this book as a whole argues, engagement with science was shared by a range of artists and writers experimenting with new cultural forms in the early years of the twentieth century; the confluence of these aspects was a key component in the experience of modernity. A central aspect of Kibbo Kift's project, among all woodcraft groups, is that such themes were not only discussed in group communication; scientific concepts and metaphors were also demonstrated. Internalised ideas were exteriorised on clothes, paraded on banners and performed in rituals. Moreover, especially in terms of the concept of cultural evolution, Kibbo Kift sought not only to disseminate the concept but also to embody and become its progressive outcome, through the attainment of the perfected physiques and higher consciousness required for the founding of their new 'race of intellectual barbarians'.[5]

Eugenics

While the words 'Kindred' and 'Kibbo Kift' – the latter drawn from an archaic Cheshire term meaning 'Proof of Strength'[6] – together signalled an organisation grounded in solid brotherhood, the curious name also subtly revealed the group's core eugenic ambitions. Formed with the aim of establishing a new 'confraternity' of elites, comprised of fit, trained, virile and beautiful men and women who would marry, reproduce and thereby establish a 'heritage of health' for future generations, Kibbo Kift was borne of a broader anxiety that so-called civilisation had become physically and culturally degenerate.[7] Hargrave had outlined the present state of destitution, as he saw it, in a nearly 400-page-long sprawling text of 1919 entitled *The Great War Brings it Home: The Natural Reconstruction of an Unnatural Existence*. This ambitious, angry volume, which incorporated practical woodcraft and camping techniques with spiritual guidance and political polemic, had been written in the fertile space between Hargrave's return from war service as a stretcher-bearer in the disastrous Dardanelles campaign and the establishment of his independent alternative outdoor movement. Echoing a concern that was shared by many in the wake of the scale of war fatalities, Hargrave stated, 'Our best blood soaks into the sand of Suvla Bay, and into the mud and grass of Flanders. We have weeded out all our weaklings by medical examination, and *they* are left at home – *to breed*!'[8] He argued:

> Every effete civilisation must crumble away. The only hope is that a new and virile offshoot may arise to strike out a line of its own … nowadays, owing to the fact that modern civilisation has penetrated throughout the world, there are no 'Barbarians' to sweep us away. Therefore the cure must be applied internally – and *we must produce the 'Barbarian' stock ourselves*.[9]

Hargrave initially planned a training scheme that could be implemented through the channels of socialist political parties and existing progressive organisations, rather than founding a new group of his own to lead the challenge. In a chapter entitled 'What is Being Done', he catalogued a range of reform institutions that were 'all set upon the same trail', that is, 'to counteract the evils of over-civilisation'. These included existing youth movements, open-air schools and the Eugenics Education Society.[10]

The Eugenics Society, established in 1907, aimed to improve 'racial health' – a term which was used synonymously with national health or the health of the human race – through better breeding, and was based on a concern that the general physical fitness of the British population was diminishing the reproductive quality of the 'stock'.[11] In practice, across its various outposts, eugenics encompassed a range of so-called 'positive' and 'negative' ambitions, from the promotion of improved sex education to – at the other extreme – the objectification of bodies and the promotion of sterilisation among the deeply questionable classification of the physically and mentally 'unfit'.[12] Richard Soloway has argued that eugenics offered, as an ideology, 'a biological way of thinking about social, economic, political, and cultural change'; it was also one that lent scientific credibility to middle- and upper-class anxieties and fears.[13] Although most knowledge of eugenics is now coloured by the extreme consequences of its deployment in the Nazi regime, historians of the Eugenics Society have carefully noted that its early aims were sufficiently diverse for it 'to be harnessed to different ideological beliefs, ranging from the ultraconservative to the social reformist and socialist'.[14]

Hargrave's self-styled 'natural eugenics' certainly fell into the social amelioration category. While his statements of disgust for the frail and the ill are unpleasant to a modern reader, his proposed scheme was ultimately benign. It aimed to educate men and women in the positive values of mental health and physical 'hardihood' so that their preferences would become 'instinctive'. The consequent improvement of the 'blood-line', as he put it, could thus be inherited by future generations without the need for further intervention.[15] Although eugenics' varied ideologies were expressed rather differently by its varied supporters, as Lucy Bland and Lesley A. Hall have noted, for many 'eugenics was part of a general bundle of "modern" ideas about the reform of society'.[16] Many of those that we would now consider to be left-leaning radical thinkers, including novelist H.G. Wells, sexologist Havelock Ellis and biologist Julian Huxley, were, as Tim Armstrong has put it, 'modernist eugenicists'.[17] Each was also a member of Kibbo Kift's Advisory Council, and their ideas are directly traceable in Kibbo Kift philosophies.

Popular science

Hargrave's establishment of his Kibbo Kift movement after his forced ejection from the scouts was marked by the drafting of an ambitious covenant. This required its members to commit to major political plans, such as reorganisation of industry on a non-competitive basis, synchronised international disarmament, the establishment of a single international currency, and a world council to include 'every civilised and primitive nation'.[18] Authorisation that these aims were serious and that the miscellaneous, amateur and modestly educated signatories were fit to carry out such world-changing tasks was given by an impressive list of names at the top of Kibbo Kift letterheads and other promotional material. Leading thinkers including politicians and Nobel Prize winners and lent their names to the Kibbo Kift project, if not their active involvement.[19] In practice, most seem to have lent support by letter alone. There is no record that the council ever met in person and few attended Kibbo Kift events. Nonetheless eminent scientists such as Huxley and J. Arthur Thomson, as well as popularisers of biology in application (including Ellis, Wells, anthropologist Alfred Cort Haddon and polymath Patrick Geddes) all lent scholarly respectability to the project. The interests of other prominent figures who were approached – from biologist Ray Lankester, populariser of the term 'degenerate', to playwright and eugenicist George Bernard Shaw – further underscore the particular timbre of Kibbo Kift's intellectual basis as an ambitious reform project based on scientific ideals.[20]

Why was the discipline of science such a focus, and its practitioners and popularisers so eagerly courted by Kibbo Kift as legitimators? The role that science could play in an organisation largely aimed at cultural reform through outdoor living and handicraft production may not appear to be immediately obvious, but its applications were numerous. In the first instance, scientific method was regularly evoked in Kibbo Kift literature as a measure of seriousness of mind, precision and critical rigour. Although the internal contradictions of Kibbo Kift's own methods for achieving their ultimate objective – merely the unification of the human race – might not have stood up to microscopic scrutiny, the group was determined that its philosophies were perceived as robust, and the use of scientific research terminology conferred this status. As one Kinsman, Idrisyn Oliver Evans, put it, writing under the adopted Kin name of Blue Swift in a 1927 article entitled 'K.K. and Scientific Method' in the Kibbo Kift periodical *The Flail*, 'the world's problems can only be solved by a rigorous determination to see facts as they are, to discover and verify and expound the truth. This attitude is difficult to attain, but it may be reached by the cultivation of the scientific spirit of free enquiry'.[21]

Very few Kibbo Kift members held professional qualifications, not least in science, although all members were rooted in a broad culture of middle-class self-improvement. Most members worked full-time in white-collar occupations, with office workers and teachers within commuting distance of central London forming a significant proportion of the group's demographic. The Kibbo Kift project was also very much part of this self-educative drive, with a circulating library of philosophical texts available for loan, and recommended reading lists, book reviews and potted summaries of selected scholarship regularly provided in the group's internal publications. Evans was a key figure in Kibbo Kift and his background – as a civil servant with a keen amateur interest in psychoanalysis, archaeology and science fiction, among other subjects – was typical of Kibbo Kift aspirational membership.[22] Evans authored a monthly feature entitled 'An Epitome of Science' in Kibbo Kift periodical *Nomad* during the mid-1920s, whose coverage included such ambitious topics as Trigonometry, Non-Euclidean Geometry and Hyper-Space. Evans explained the purpose of his summaries to the Kindred as follows:

> As most of the Kindred have to spend the greater part of their lives 'earning a living' in some task of no special intellectual value, they are only able to give the study of science a very limited attention, and are therefore not usually able to make themselves familiar with the literature of their subject.[23]

For Kibbo Kift members, modelled on Wells's fictional New Samurai ideal of a scientifically minded voluntary elite, the relevance of these for Kibbo Kift practices should have been evident.[24] Evans spelled it out for those in any doubt: 'You cannot pitch a tent without bringing into play a number of mechanical forces, make a cup of tea or boil and egg without making use of chemical reactions, or lead a K.K. Tribe without using practical psychology'. Evans complained that, 'as a matter of general practice in everyday experience we are content with just as little science as will serve our immediate purposes – and often enough used in a very rule-of-thumb manner'.[25] The 'Epitome of Science' series was designed to go much further. Like Hargrave, Evans acted not just as a receiver of intellectual ideas but also as their interlocutor. His interpretations were themselves largely synthesised from popularisers of science, history and culture, for example from J.G. Fraser to H.G. Wells.[26] In Evans's view, the research scientist was 'the highest type civilisation has produced'.[27]

Science was thus understood as the apotheosis of intellectual culture – a culture from which many members could only view from a distant position – and thus professional scientists' benediction of the Kibbo Kift project, alongside scientific enthusiasts' discussion of their ideas, conferred some of this elite status by proxy.[28]

Evans insisted, 'Each kinsman is expected to have some practice of creative art; it is of equal importance that he should have some experience of scientific work'.[29] The level of expertise that Kibbo Kift members were expected to acquire was demanding, and included a 'general knowledge of the following theories: Evolutionary, Nebular, Atomic'.[30] A series of pictorial Badges of Knowledge – a form adapted from scouting structures – was awarded to adults those who had mastered fields as varied as Oceanography, Embryology, Radiography and Astronomy. How laboratory science was to be studied by untrained amateurs on limited incomes was dealt with practically. 'Infra-Atomic Physics needs too much apparatus for the ordinary student', members were warned, 'while Relativity demands too great a mastery of Mathematics'. Instead, Kinsfolk were encouraged to begin closer to home by making 'careful and systematic observations in their own locality in such subjects as Natural History, Ecology, Geology, and Meteorology'; they were also encouraged to engage in the less equipment-heavy studies of Anthropology, Sociology, Psychoanalysis and even the Occult.[31] This was science very broadly understood. The merging of practices across pure, applied and social categories, alongside studies that might be more comfortably placed in the humanities, served a practical purpose for those without necessary equipment or training but also fitted with Kibbo Kift's larger ambitions: to unite fractured areas into integrated wholes.

Holism and vitalism

As part of their far-reaching system for world unity, jointly inspired by Wellsian dreams of a world state and League of Nations practical plans for international reconstruction, Kibbo Kift sought the union of art, science and philosophy. Hargrave argued that knowledge had become splintered into separate, specialist domains. With characteristic ambition, he saw their unification as an essential Kin duty. In his dizzying exposition of his movement's aims, *The Confession of the Kibbo Kift*, published in 1927, Hargrave paraphrased scientist Claude Bernard at the end of the previous century. Bernard had yearningly predicted, 'There will come a day when physiologists, poets and philosophers will all speak the same language and understand one another'; for Hargrave this time had come with Kibbo Kift.[32] What we might describe today as an interdisciplinary aim was more of a mystical mission for the group; art, science and philosophy were considered to be the three core branches of the Tree of Knowledge, and their holy coalition was described as the manifestation of the 'Sancgraal' of the Knights of the Round Table, hidden in plain sight.[33]

In the Kinlog, the vast illuminated logbook of the movement, launched in 1924 to record the official history of Kibbo Kift for the benefit of future generations, Hargrave outlined their principles in the form of a twentieth-century Book of Kells (Figure 14.1). Amidst saga metre text and in a typical medieval-modernist artistic style, an illustration shows a classically robed artist, a gas-mask-clad scientist and a bearded philosopher collectively grasping the 'Three-Edged Sword of Truth' that will bring them together (Figure 14.2). Below this, a green-hooded figure representing a Kinsman emerges from a tangled flow of ideas, supported by a wreath of figures that reveal Kibbo Kift's intellectual debts and inspirations. Among the ancient Greek, Egyptian and Chinese masters and sages depicted, one figure stands out temporally and stylistically – that of Henri Bergson.[34]

Bergson's Neo-Vitalist writings, in *Creative Evolution* and other texts, had achieved mass popularity in the early years of the twentieth century, not least because his metaphysical concept of an unknowable and invisible force at the root of all living things seemed to offer the hope of a romantic reinsertion of spirit in an otherwise disenchanted, mechanistic world.[35] Kinswoman Kathleen M. Milnes, an art teacher and the Kinlog 'Scriptor' (calligrapher and illustrator), was a particular devotee of Bergson's theories and wrote about them effusively in her personal Kibbo Kift log. Bergson was also recommended reading in Kin educational guidance. Even when not named as such, a powerful philosophy of Neo-Vitalism was evident across Kibbo Kift thinking.[36] At its most banal this was manifest in the regular use of the term 'vital' in group literature, as linguistic shorthand for the dynamism, progress and energy that Kibbo Kift venerated and saw its membership embodying. At a more profound level, the influence of new vitalistic ideas infused Kibbo Kift's worldview and gave it philosophical justification.

Neo-Vitalism has been described by Oliver Botar and Isabel Wunsche as one of a number of 'biocentric' systems of knowledge and ideas popular in the late nineteenth and early twentieth centuries; they position it alongside organicism, holism, monism and Neo-Lamarckism.

Fig. 14.1   Kinlog cover by John Hargrave, 1924.

© Kibbo Kift Foundation, with kind permission of Museum of London.

Fig. 14.2   Kinlog interior, illustrated by Kathleen Milnes.

© Kibbo Kift Foundation, with kind permission of Museum of London.

While each had distinctive variations, all biocentric models held in common:

> the privileging of biology as the source for the paradigmatic metaphor of science, society, and aesthetics; a consequent, biologically based epistemology; an emphasis on the centrality of 'nature,' 'life,' ... the self-directedness and 'unity' of all life; a valorization of the quasi-mystical feeling of unity with all nature ... a stress on flux and mutability in nature rather than stasis; and a concern for 'wholeness' as opposed to reduction at all levels.[37]

Biocentrism, or 'biologistic Neo-Romanticism', is located by Botar in particular in the *Lebensphilosophie* of Nietzsche and Bergson, and in the work of what he describes as 'scientists with philosophical pretentions', in which he includes Ernst Haeckel, Élisée Reclus and Patrick Geddes.[38] All vitalistic philosophers were, as Richard Lofthouse has put it, also 'impatient of traditional epistemological boundaries'.[39] While Kibbo Kift's intellectual bases were more eclectic than those drawn solely from biocentric sources, the names and approaches highlighted here each had a direct intersection with the group. Most literally, the holist biogeographers and educational reformers Reclus and Geddes lectured at Kibbo Kift meetings and led tours of prehistoric sites for members. Nietzsche's *Übermensch* ideal is evident in Kibbo Kift desires for human perfectibility and the cultivation of a self-conscious elite. Several of Haeckel's ideals, including the concept of the Monad as a concept to communicate the unity of existence (as opposed to the duality of mind and spirit), were key to Kin thinking – the Monad symbol even formed the basis of the group's much-utilised insignia, the Mark.

Monism, Haeckel's nineteenth-century concept of nature as a singular whole, had started as a secular, materialistic creed in its first formation but by the end of Haeckel's life it had become a vitalistic one.[40] In later Monism and its related philosophical territories, all was one; all was spirit. In Neo-Vitalist thinking more broadly, even objects previously thought inanimate were understood to be charged with life force. In a 1925 article entitled 'A Short Exposition of the Philosophic Basis of the Kibbo Kift', Hargrave took up these ideas that bridged science and spiritual philosophy and put them into the service of Kibbo Kift. He celebrated the breakdown of the tripartite classification system of animal, vegetable and mineral, arguing that vitalistic forms of thinking confirmed instead a 'blood relationship and atomic-kinship with Birds, Beasts, Flowers, Rocks, Stars and the Energy of all of the Suns'. Drawing from Kibbo Kift advisory councillor J. Arthur Thomson's assertion that 'phrases such as "dead" matter and "inert" matter have gone by the board' in modern science, Hargrave argued that the latest thinking in the discipline now served to verify the 'ancient seers and philosophers'.[41]

For Kibbo Kift, the assertion that matter could be composed of the same energy that underpinned all living things reinforced their philosophy of world peace and cosmic unity. In a striking passage that demonstrated spiritual immanence and atomic kinship at play in an evocative list of modern miscellany, Hargrave asserted that this new way of thinking:

> means that teapots, chairs, mud, electric light bulbs, fingernails, hammers, steam engines, mountains, hats, shoes, needles, tram tickets, lilies, telephones, tents, dynamos, walking sticks, cow dung, churches, iron foundries, neckties, cats, human beings, steel plates, bricks and mortar, glass, sealing-wax, trees, thoughts, tables, music, flowers and flower-pots, clouds, gutter-gratings, books, food, buttons, machine guns, beads, rain, clocks, boots, ferro-concrete, eggs, sunlight, coal, stars, solar systems, slugs, pictures, maggots, wheel bolts, smells, darkness and light, collar-studs, speech, seeds, birds, bootlaces, insects, skeletons, pepper-corns, babies, Space, Time, Matter, all religions, all Spirits, all Matter(s) … all, all, are actually the ONE GREAT POWER.[42]

Within the sphere of biocentrism and its understanding of the complex interconnectedness of all things, new metaphysical understandings of the concepts of 'nature' and 'life' emerged. However indirectly these *Lebensphilosophie* and *Naturphilosophie* ideas travelled to Kibbo Kift, a pantheistic understanding of the natural world provided a core underpinning to the group's beliefs. Anna Bramwell, in her history of ecological thought in the twentieth century, has documented the increasing tendency, post-Darwin, for God to be replaced by Nature; a personified force that she describes as 'somewhat dominating'; this was 'a Nature expected to educate and guide humanity'.[43] Life, too – frequently capitalised and personified with divine qualities, as in many Kibbo Kift references – became less a descriptive term to summarise the passage from cradle to grave and more a stand-alone philosophical category.[44] Herbert Schnadelbach, describing its application in the German context, notes that the life-concept was an 'attack on a civilisation which had become intellectualistic and antilife, against a culture which was shackled by convention'. It stood for 'what was "authentic", for dynamism, creativity, immediacy'. 'Life', he notes, became the slogan of the youth-movement, and of educational, biological and dynamic reforms.[45]

An early organisational banner of a Lodge within Kibbo Kift makes this philosophy visible through an extraordinarily daring image for its time (Figure 14.3).

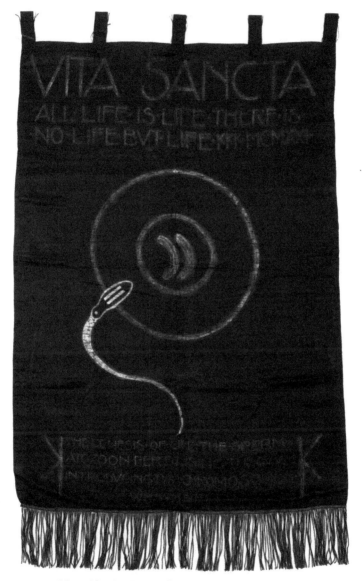

Fig. 14.3 Kibbo Kift Vita Sancta banner, 1921.

© Kibbo Kift Foundation. With kind permission of Museum of London.

Executed in gold paint on black satin, under the heading Vita Sancta, the mystical banner depicts, according to its inscription, 'The genesis of life: a spermatozoon fertilising the ovum introducing two chromosomes'. In this image, the moment of fertilisation is venerated, even fetishised; elsewhere in Kibbo Kift insignia, sperm penetrated eyes of Horus and mystical suns, not to indicate sexual licence but to celebrate philosophical *élan vital*. The banner's subheading, 'All Life is Life: There is no Life but Life' is almost comical in its circularity but its purpose was to communicate the group's vitalistic beliefs in a striking motto. The style of the refrain undoubtedly owed something to Hargrave's talents in publicity; he earned his bread-and-butter income throughout the interwar period as a copywriter and commercial artist for a major advertising agency and knew the power of symbol, stunt and slogan. Other statements and chants that the Kibbo Kift corralled around were similarly cryptic but equally inflected with the union of physics and metaphysics, from campfire songs of praise for 'Energy, Energy, Ceaseless Energy' to the succinct spiritual encapsulation of interconnectedness, 'One is One'.

Kibbo Kift's arcane language and occult rituals suggest a highly idiosyncratic formation that could be dismissed as of only marginal relevance to the wider world. It is worth remembering, however, that alongside Kibbo Kift's documented appeal to major public figures in the arts, humanities and sciences of the period, much of the material which seems outlandish to twenty-first-century eyes had significant status in the mainstream at its point of publication. Haeckel's ideas, as Bramwell has pointed out, had mass distribution and influence: 'For self-educated working men, his two-shilling works with titles such as The Riddles of the Universe, or The Wonders of Life … were a life-line to political awareness through scientific knowledge'.[46] Bergson's books were bestsellers; indeed Bergson himself became something of a celebrity. Perhaps the book that came most frequently recommended as a Kibbo Kift model for understanding the world was H.G. Wells's biologically informed teleology, *The Outline of History*, which had been immediately successful on its release in 1919; by 1922 it had sold over a million copies.[47] Thompson's *Outline of Science* of 1921 aimed to build on Wells's success and sold half a million copies in its first five years. These ideas were not marginal, even though their reinterpretations and adaptations in Kibbo Kift were often unusual. They fitted into a broader popularity of science in application and a willing embrace of a range of modernist ideas that were circulating in the utopian space of the immediate post-war years when culture needed to be remade from top to bottom.

Evolution

Perhaps the most popular of all scientific ideas in application was that of evolution. By the 1920s some aspects of mid-nineteenth-century evolutionary theory had become accepted as orthodoxy but the power of the evolutionary idea outside of science showed little sign of diminishing. Indeed, as Gillian Beer has argued, 'evolutionary ideas are even more influential when they become assumptions embedded in the culture than while they are the subject of controversy'.[48] Evolutionary metaphors prevailed across all aspects of early-twentieth-century cultural life and the mulitvalency of the theory could lend it to a range of wildly divergent readings. Evolutionary thinking was also repeatedly invoked in the Kindred's writings; as an article on the subject stated categorically, 'Evolutionary Theory has always been recognised as fundamental to the whole Kin philosophy'.[49]

Evolution was used for a variety of purposes, including to justify the small size of the group against its competitors. As Hargrave put it, had not the simplicity and complexity of the tiny adaptable amoeba outsmarted the lumbering labyrinthodons?[50] Natural selection was also used as a means for explaining the necessity for physical and mental development:

> In this struggle for existence, plants and animals have developed weapons and means of protection that will help them to survive, and the great weapon and protection of Man in this struggle is his mind. If we let our bodies get unfit, if we let our brains slack, we shall be destroyed and our place will be taken by others. Therefore we of the Kibbo Kift camp out and keep fit, and we keep our minds alert by study.[51]

Evolutionary themes, more or less explicitly, were also visible across the Kibbo Kift's striking designs for ceremonial and propaganda purposes. Most literally, Darwin appeared on a series of parade banners of the Great Seers and Thinkers, as one of an eclectic community including Tolstoy, Plato, St Francis of Assisi, William Penn and Walt Whitman. Elsewhere Darwin was listed among Kibbo Kift's 'Heroes of World Service' and was claimed, somewhat retrospectively, as 'as strenuous a Woodcrafter as anyone could wish'.[52] The group's passionate interest in the evolution of humanity was repeatedly expressed in Kibbo Kift imagery: encounters between hooded kinsmen and stooped prehistoric figures in leopard skin robes were regular motifs. Perhaps the most

Fig. 14.4 Kibbo Kift 'Touching of the Totems' rite, 1925.

Collection of A. Pollen.

passionate of these encounters was the one established between the Kibbo Kift and the Piltdown Man. The 1912 discovery in rural Sussex of fragments of bone purported to be the jaw of the earliest inhabitant of England was considered by many to represent the 'missing link' in the evolutionary chain, providing incontrovertible proof of a relationship between present-day humans and their ape ancestors. Shown to be an elaborate hoax in the 1950s, in the 1920s an unwitting Kibbo Kift made and paraded plaster cast reproductions as totemic objects in their camp rituals and staged pilgrimages to the field where *Eoanthropus Dawsoni* was found (Figure 14.4).

  Although Kibbo Kift did not explicitly distinguish it as such, much of the group's evolutionary thinking drew from Neo-Lamarckian models, which proposed an evolutionary development that was willed and shaped by human creative intervention. This was not an evolution thrust upon humanity by outward forces, as in Darwin's theory of natural selection, but a rather more appealing and flattering model that gave a central role to human will and intention. Monists had asserted that man was a voluntary co-operator in the service of evolution, and that evolution could be transformed and improved by 'a conscious upward striving towards a higher condition, a pressing forward towards an ideal'.[53] Lamarck's principle that learned behaviour could be passed on as an inherited characteristic gave humanity an appealingly determining role in the development of the species.

Evans explained that Kibbo Kift's use of evolutionary theory:

> justifies the view of the Kin that progress is possible; and that it will need effort; that it can only arise through unprecedented ideas; that such ideas must deal with social and political questions; and that to produce and apply them … will be necessary. Our methods are in line with those through which evolution, biological and social, [has] taken place, and we therefore may regard them as likely to be productive of result.[54]

Here, evolution is not a natural trajectory that has occurred passively but a system that could be harnessed to one's own advantage. Evolution could and should be bent to one's will as a duty; as Hargrave had also noted, elsewhere, 'leaving things to evolution' is 'not in the tradition of full-blooded men'.[55]

Biogenetic law

Scientific thinking was understood to be at the very forefront of intellectual practice and thus offered an engagement with the biggest and newest of ideas. In addition to being hungry to engage with the latest thinking, however, Kibbo Kift practices also looked 'back', as they saw it, either to historic cultures (variously medieval or prehistoric) or – more problematically – to temporally coeval cultures inside and outside of Western Europe (those labelled 'folk' and/or 'savage'). Kibbo Kift's ambition was not to revive these cultures – revival was, in fact, taboo – but to reinvigorate them through the prism of modern experience, in order to achieve what H.G. Wells had called 'the next stage of history'.[56] Scientific thinking and its natural partner – the latest technological products – provided the ideal symbolic structure for communicating this complex retro-futurist trajectory. Kibbo Kift members saw no contradiction between their plans to construct an air squadron and the practice of traditional handicrafts, or the frank discussion of the newest forms of birth control and the use of archaic forms of language. The intersection of past, present and future is of key significance to understanding Kibbo Kift's philosophy; to look back was not to reject the modern world but to revisit cultural history in order to develop the group into something they hoped would be distinctively avant-garde. As Hargrave put it, in a typically ostentatious battle cry, perhaps overcompensating for being outside the closed circle of elite intellectual culture: 'We are more modern than the moderns'.[57]

This backward–forward trajectory was especially important in relation to the education of children. Built on Hargrave's scheme of 'Tribal Training', Kibbo Kift's educational programme was underpinned by the recapitulation theories of American psychologist G. Stanley Hall.[58] Hall applied an adaptation of Ernst Haeckel's biogenetic law, which proposed that ontogeny (organism growth) recapitulates phylogeny (evolutionary history of the species). Recapitulation as a concept can be found in application across a broad range of non-scientific disciplines during and after the nineteenth century; indeed Beer has described this formula as 'one of the most powerful new metaphors of the past 150 years'.[59] Stephen Jay Gould has also noted its enduring pervasiveness, despite it being fundamentally incorrect; he suspects that 'its influence as an import from evolutionary theory into other fields was exceeded only by natural selection'.[60] In Hall, Gould has argued, 'recapitulation reached the acme of its influence outside biology'.[61] Hall's theory was outlined at length in his 1906 book *Youth: Its Education, Regimen, and Hygiene*; children, as apparently natural 'savages', he asserted, needed to re-enact a sequence of stages of cultural evolution in order to become fully rounded beings.[62]

In Kibbo Kift's adaptation, children were taught the scientific development of world culture from its very earliest stages, as in Wells's *Outline of History*. It was said to be 'vitally important for the youngest child to be taught that the world began as a blazing ball of gas, and he must go on from that'.[63] Child development was then mapped onto a seven-stage linear understanding of cultural history. So-called 'cultural epoch' curricula had been pioneered as an educational experiment in the mid-nineteenth century.[64] In Kibbo Kift's formulation, applied in a clutch of experimental open-air and woodcraft schools in the interwar years, children were not only to study the aspects of history deemed to be appropriate to their developmental stage; they were to physically inhabit characteristics of the development of culture, as it was then understood.[65] Kibbo Kift's idiosyncratic model positioned 'prehistoric' and 'primitive' life at one end of the line, with 'modern day', 'present day' and then 'Kibbo Kift' as the three final stages of completion. Each stage had to be enacted through practical craft projects and prescribed picturesque perfor-mances, from the making of fire in the early stages to jazz dancing and committee meetings in the modern years.

The application of biologically informed recapitulation theory to youth training was not Hargrave's invention; it had informed the training

Fig. 14.5 John Hargrave with children at Tribal Training Camp, 1928.
Photograph by Angus McBean. Collection of A. Pollen.

of boys and young men in both Baden-Powell's and Seton's groups. When blended, in Kibbo Kift, with metaphysical Neo-Vitalist thinking and a Futurist-Primitivist visual style, however, the collective result was certainly an innovative – if sometimes bewildering – melange of art, science and philosophy. The learning-by-doing aspect of Kibbo Kift's performances was part of the group's commitment to direct action over discursive deliberation. This was in part borne of impatience with aspects of the modern world that were perceived to be 'overcivilised'. While the group identified with the 'scientific mind' rather than 'unthinking' mass mind, formal education was seen as inferior to knowledge that had been developed through 'instinct' and practical skills.[66] Theory for its own sake was largely dismissed as 'intellectual botheration'.[67] The importance of science was never for its own sake; biology was understood in Kibbo Kift's life reform project as the direct means by which physical, social and cultural betterment could be brought into being.

Critiques

The twenty-first-century status of the Kibbo Kift Kindred as largely forgotten might appear to establish damning proof of their project's inadequacy. Despite their high-profile public support, given the actual size of their numbers and resources, the scale and eclecticism of their ambitions seemed doomed to fail.[68] The group's fortunes were also hostage to the vicissitudes of the singular, top-down leadership of Hargrave, who dramatically shifted ideological direction a number of times during the 1920s. By the early 1930s, Hargrave had transformed Kibbo Kift into a political-economic campaign group and many aspects of the group's original ethos, practice and membership fell away as the original campers, artists and idealists were expected to become the Green Shirts, an urban, uniformed street-marching campaign group pressing for Social Credit for all. Even without this fundamental change in purpose, disciplinary developments in science, psychology, education and anthropology had negatively affected some of the core principles on which Kibbo Kift had been based, suggesting that their biocentric cultural project could only have ever been short-lived.

Although Kibbo Kift's Vita Sancta banner showed their veneration of fertilisation and their knowledge of at least some aspects of chromosome theory, Mendelian genetics, which demonstrated that important determining genes were present at the point of conception, undermined the ontogenic ideas behind recapitulation that the Kindred simultaneously held dear. Thompson, one of Kibbo Kift's scientific figureheads, had already cast doubt on recapitulation in his *Outline of Science* but the group seem to have read this text rather selectively.[69] Recapitulation's precarious footing in biology necessarily led to some knock-on questioning of its premise in the educational and psychological domains during the 1920s.[70] This also happened to some extent in woodcraft circles, where it was recognised that some practising the method knew far more about the practical psychology of the child than they did about culture or history.[71] At the same time as science was questioning biogenetic law, the cultural evolution model that formed its partner in recapitulation theories of education was also subject to institutional critique by new challenges in anthropology. Franz Boas, for example, showed the fallacy of race as a biological category based on studies of the cranium; he also argued for the development of tribal cultures to be studied in relation to their specific cultural contexts.[72] Although Hargrave was exposed to Boas's publications (he modelled drawings on illustrations in Boas's *Primitive Art* of 1927, for example) he seems not to have grasped the implications of Boas's findings for his own instruction – that so-called primitive cultures are just as flexible, dynamic and developmental as European cultures. As Kevin Armitage has noted, it was with more than a touch of irony that cultural theories of evolution unravelled: 'recapitulation gained prominence as a scientific justification for pedagogical methods meant to unify unbound human nature with modern civilization but was undone by the epitome of controlled civilised objectivity, scientific inquiry'.[73]

Other scientifically informed ideologies that had influenced Kibbo Kift at the outset were also subject to some significant realignment over time. The Eugenic Society achieved some public acceptability until its application by Nazis at the end of the 1930s made it publicly unpalatable and philosophically untenable.[74] Vitalistic beliefs in invisible forces and life's fundamental mystery struggled to maintain validity in science in the light of the discoveries made by Mendel. Neo-Vitalism had already begun to shift sideways into mystical philosophical circles by the 1920s. This was the manoeuvre made, for example, by leading vitalist Hans Driesch, and the same move was completed by Hargrave by the Second World War.[75] As Lofthouse has noted, 'vitalism flourished within a context much broader than the supposedly limited designs of "neutral" scientific enquiry. It was merely one strand in a thick rope comprising creative and emergent evolutionism, cosmic teleology, psychology, psychical research, the paranormal, the occult, eastern religion and spiritualism'.[76] Kibbo Kift's theories were similarly entwined in these ideas. Although they considered themselves to be modernists in their embrace of evolution, their application of it was often more poetic than precise and more spiritual than systematic; its greatest utility was as linguistic metaphors and visual tropes. At worst, as Evans suggested, woodcraft practitioners could be 'modernists with a greater theoretical admiration for Science with a capital S than readiness to submit to its austere disciplines, and careless as to whether their views are really scientific as long as they *sound* "scientific" and "evolutionary"'.[77]

Perhaps the final word on the subject should be given to Leslie Paul, a former Kibbo Kift member and a keen proponent of evolution in the 1920s in his own organisation, The Woodcraft Folk. Informed by precisely the same intellectual currents as Hargrave, he largely modelled his organisation on Kibbo Kift, from which it had begun.[78] Paul had written in 1926:

> On the basis of biology and evolution is built the philosophy that underlies both our educational methods and the charter of the youth movement. We believe that man must use himself consciously as a tool of evolution. That is, he must regard evolution as a process that touches him and his kind intimately, and that we are masters of our fate only when we assist our own becoming, and the evolution of the race[79]

By 1951, however, Paul had fundamentally changed his position. With the benefit of hindsight, he reflected:

> What strikes me about all this to-day is its irrelevance. It is doubtful if man is physically evolving any longer, it is certain that it is a dubious intellectual trick to apply the doctrines of physical evolution to human societies and cultures. Even if man is still evolving, no one can say with any certainty what acts of man will aid his evolution or hold it back (assuming it is possible to do either) … Unless man gives up thinking and moralizing, and goes back to an animal state in which the pure law of survival can operate again (if it ever really operated as Darwin supposed) then he must make decisions upon quite other grounds than 'evolutionary' ones. Evolutionary theory is irrelevant to the human situation, and only spurious philosophizers pretend otherwise.[80]

In conclusion, then, although its moment was short-lived, popular science in general and evolutionary biology in particular was used by woodcraft groups in the 1920s to inspire and defend a diverse range of ideologies and practices, from eugenic body culture and experimental educational policy to pantheistic religion. At their fullest flowering in the ideas and activities of Kibbo Kift, the application of science to cultural domains could achieve a hallowed status. Unlike later New Age oppositions to science (or 'scientism', as it is sometimes denigrated), in the 1920s Kibbo Kift saw science as an essential aspect of their philosophical make-up, part of an indivisible triumvirate with art and spirituality, that demonstrated their forward-thinking modernism. The study of the application of science in woodcraft groups offers a productive – if highly idiosyncratic – means of exploring the ways that scientific ideas in the early twentieth century were received and reinterpreted outside an intellectual elite. Significantly, concepts of chromosomes, apes, amoeba and atoms were lived as well as read in Kibbo Kift. Through evolution's practical and performative embodiment, they believed that theory would be made flesh and culture would progress.

Notes

1. Gillian Beer, *Darwin's Plots: Evolutionary Narrative in Darwin, George Eliot and Nineteenth-Century Fiction* (Cambridge: Cambridge University Press, 2000), 6.

2. Several groups based on woodcraft principles split from the British Boy Scouts for pacifist reasons. The first, the Order of Woodcraft Chivalry, is not

discussed in detail here. For more information on this organisation, see Derek Edgell, *The Order of Woodcraft Chivalry 1916–1949 as a New Age Alternative to the Boy Scouts* (Lampeter: Edwin Mellen Press, 1992).

3   Seton's methods can be seen in his own writings, for example, Ernest Thompson Seton, *Woodcraft and Indian Lore*, (Doubleday, Doran and Company, 1930). A useful recent interpretation can be found in David L. Witt, *Ernest Thompson Seton: The Life and Legacy of an Artist and Conservationist* (Layton, Utah: Gibbs Smith, 2010).

4   The archival papers of the Kindred of the Kibbo Kift are split between the London School of Economics Library Special Collections and the Social History Collections of the Museum of London. The first substantial sociological study of Kibbo Kift can be found in Mark Drakeford, *Social Movements and Their Supporters: The Green Shirts in England* (Basingstoke: Macmillan, 1997). A full-length account of the group's cultural history is provided in Annebella Pollen, *The Kindred of the Kibbo Kift: Intellectual Barbarians* (London: Donlon Books, 2015).

5   This phrase comes from a 1923 silent newsreel on the subject of Kibbo Kift. The title card reads, 'Meeting of new Kindred who aim at a race of Intellectual Barbarians. Mr H.G. Wells is a member of this camping fraternity, who combine the ideals of Scientists and Red Indians'. Great Missenden – The 'Kibbo Kift'. 1923. Topical Budget [news reel]. British Film Institute 626:1.

6   Roger Wilbraham, *An Attempt at a Glossary of Some Words Used in Cheshire*, second ed. (London: T. Rodd, 1826).

7   The literature is too extensive to be summarised here. Texts that offer useful context for Kibbo Kift attitudes to degeneration in the social body include Richard Overy, *The Morbid Age: Britain and the Crisis of Civilization, 1919–1939* (London: Penguin, 2009) and Ina Zweiniger-Bargielowska, *Managing the Body: Beauty, Health and Fitness in Britain, 1880–1939* (Oxford: Oxford University Press, 2010).

8   John Hargrave, *The Great War Brings It Home: The Natural Reconstruction of an Unnatural Existence* (London: Constable and Company Ltd, 1919), 51.

9   Hargrave, *The Great War Brings It Home*, 21.

10   Hargrave, *The Great War Brings It Home*, 68–74.

11   Nancy, Stepan, *The Idea of Race in Science: Great Britain, 1800–1960* (London: Macmillan, 1982), xviii.

12   These 'positive' and 'negative' approaches are outlined in Overy, *The Morbid Age*, 93–135.

13   R.A. Soloway, *Demography and Degeneration: Eugenics and the Declining Birthrate in Twentieth Century Britain* (Chapel Hill, NC: University of North Carolina Press, 1990), xviii.

14   Lucy Bland, and Lesley A. Hall, 'Eugenics in Britain: The View from the Metropole', in *The Oxford Handbook of the History of Eugenics*, eds. Alison Bashford and Philippa Levine (Oxford: Oxford University Press, 2010), 216.

15   Hargrave, *The Great War Brings it Home*, 364–367.

16   Bland and Hall, 'Eugenics in Britain', 217.

17   Tim Armstrong, *Modernism: A Cultural History* (Cambridge: Polity, 2005), 74–78.

18   A copy of the Kibbo Kift covenant is reprinted in the first issue of the first Kibbo Kift magazine, *The Mark* 1, no. 1 (June 1922): 16.

19   Other significant names on the Kibbo Kift Advisory Council include writers Maurice Hewlett and Rabindranath Tagore, campaigners and suffragists Henry W. Nevinson, Mary Neal and Emmeline Pethick Lawrence, and Norman Angell, MP. For a fuller discussion of the Advisory Council, see Pollen, *The Kindred of the Kibbo Kift*.

20   Lankester, Shaw and Haddon were all approached after the 1923 Althing (Annual General Meeting). Record of the Fourth Althing, 1923. Private papers of Hazel Powell, with thanks.

21   I.O. Evans [Blue Swift], 'K.K. and Scientific Method', *The Flail: An Independent Kibbo Kift Magazine*, 4, no. 1 (Summer 1927): 139–141.

22   In addition to authoring an important overview of the woodcraft movement, Evans published an adaptation of H.G. Wells's *Outline of History* for children as well as a book of scientific predictions. He later became the principal translator of Jules Verne's science fiction into English. I.O. Evans, *Woodcraft and World Service* (London: Noel Douglas, 1930); I.O. Evans, *The Junior Outline of History* (London: Denis Archer, 1932); I.O. Evans, *The World of Tomorrow: A Junior Book of*

Forecasts (London: Denis Archer, 1933). For Evans and Verne, see Brian Taves, '"Verne's Best Friend and His Worst Enemy": I.O. Evans and the Fitzroy Edition of Jules Verne', *Verniana: Jules Verne Studies* 4 (2012): 25–54. For copies of Evans's science fiction writing in the 1930s in *Tomorrow: The Magazine of the Future*, thanks to Charlotte Sleigh.

23 I.O. Evans [Blue Swift], 'Book Here: Books for a Kibbo Kift Library', *Nomad* 11, no. 1 (April 1924), 133–134.

24 The concept of New Samurai appears in H.G. Wells, *A Modern Utopia* (London: Penguin, 2005 [1905]). The concept was revived in H.G. Wells, *Men like Gods* (London: Cassell and Co., 1923). For further discussion of the intersection between Wells's works and Kibbo Kift, see Annebella Pollen, 'Utopian Futures and Imagined Pasts in the Ambivalent Modernism of the Kibbo Kift Kindred', in *Utopia: The Avant-Garde, Modernism and (Im)Possible Life*, eds. by David Ayers, Benedikt Hjartarson, Tomi Huttunen and Harri Veivo (Berlin: De Gruyter, 2015).

25 I.O. Evans [Blue Swift], 'The Tree of Knowledge: An Epitome of Science', *Nomad* 3, no. 2 (August 1924): 177–178.

26 Jonathan Rose briefly considers Hargrave as reader and interpreter of intellectual culture in Jonathan Rose, *The Intellectual Life of the British Working Classes* (New Haven and London: Yale University Press, 2001), 454–455. Joel S. Kahn, considering the ethnographic thinking of Hargrave, has suggested that he can be fit into the Gramscian model of the organic intellectual, as Kibbo Kift texts are neither hegemonic nor subaltern. Joel S. Kahn, *Modernity and Exclusion* (London: Sage, 2001), 33–34.

27 Evans, *Woodcraft and World Service*, 132.

28 Hargrave and the Kibbo Kift were dismissed as class deficient by those in more privileged positions. Rolf Gardiner, friend of writers W.H. Auden and D.H. Lawrence, and one-time member of Kibbo Kift, was wealthy and Cambridge University-educated; he snobbishly dismissed Kibbo Kift as 'Suburbia Incarnate' and complained, 'White Fox [Hargrave] has not got any, not one single first class brain or personality in the Kindred'. He noted that Kibbo Kift ideas were 'the product of a mentality that has grown up in Balham and been fed on Methodism, the Cinema, The Wonders

of Science (popularly explained) and the novels of Mr. H.G. Wells. They vulgarise the whole realm of experience'. Rolf Gardiner to Jack Winter (6 December 1925) Rolf Gardiner papers, Cambridge University Library Special Collections C6/3/1. Poet W.H. Auden described Hargrave as 'terribly lower middle-class': David Bradshaw, 'New Perspectives on Auden: Rolf Gardiner, Germany and the Orators', *The W. H. Auden Society Newsletter* 20 (2000): 20–28.

29 Evans, 'K.K. and Scientific Method', 140.

30 These skills are outlined in the Kibbo Kift's 'Seven Degrees' system of recapitulation. Stanley Dixon collection, courtesy of Gill Dixon. Private collection of Tim Turner, with thanks.

31 Editorial, *Wandlelog*, 2, no. 2 (Spring 1926): n.p.

32 John Hargrave, *The Confession of the Kibbo Kift* (Glasgow: William Maclellan, 1979 [1927]), 48.

33 John Hargrave, 'Letters to the Kindred', *Nomad* 11, no. 1 (April 1924): 125–127.

34 John Hargrave and Kathleen M. Milnes, *Kinlog: Being the Annals of the Kindred Called the Kibbo Kift*, 1924–1982, Museum of London Kibbo Kift collection. L198/H.1.

35 Henri Bergson, *Creative Evolution*, trans. Arthur Mitchell (London: University Press of America, 1983 [1911]).

36 While Hargrave and Kibbo Kift did not use the term to describe their own beliefs, Hargrave would have been familiar with the concepts through his reading. In the early 1920s there was also an attempt to establish a new, formal society called The Neo-Vitalists. Led by Douglas Renshaw, this group was allied with a range of reformist causes. Renshaw attended Kibbo Kift events and Hargrave lectured to his group in 1923.

37 Oliver A.I. Botar and Isabel Wunsche, 'Introduction: Biocentrism as a Constituent Element of Modernism' in *Biocentrism and Modernism*, eds. Oliver A.I. Botar and Isabel Wunsche (Farnham: Ashgate, 2011), 5.

38 Oliver A.I. Botar, 'Defining Biocentrism' in Botar and Wunsche, *Biocentrism and Modernism*, 15.

39 Richard A. Lofthouse, *Vitalism in Modern Art, C. 1900–1950: Otto Dix, Stanley Spencer, Max Beckmann and Jacob Epstein* (Lewiston: Edwin Mellen Press, 2005), 20.

40 Anna Bramwell, *Ecology in the 20th Century: A History* (New Haven and London: Yale University Press, 1989), 54.

41 John Hargrave, 'A Short Exposition of the Philosophic Basis of the Kibbo Kift', *Nomad* 2, no. 12 (May 1925), 281–284.

42 The Lodge of Instruction, Script III, Rune III. Museum of London, 2012.72/9.

43 Bramwell, *Ecology in the 20th Century*, 47.

44 Lofthouse, *Vitalism in Modern Art*, 30.

45 Herbert Schnadelbach, *Philosophy in Germany 1831–1933*, trans. Eric Matthews (Cambridge: Cambridge University Press, 1984), 139, quoted in Botar, 'Defining Biocentrism', 16–17.

46 Bramwell, *Ecology*, 41.

47 William T.H.G. Ross, *Wells's World Reborn: The Outline of History and Its Companions* (Selinsgrove, PA: Susquehanna University Press, 2002.

48 Beer, *Darwin's Plots*, 2.

49 I.O. Evans [Blue Swift], 'Evolution and the Kindred', *The Flail*, 7, no. 1 (Spring 1928): 263–265.

50 John Hargrave, 'Chief Ritesmaster's Speech', c. 1931. Kibbo Kift Collection, London School of Economics Library YMA/KK/20.

51 I.O. Evans [Blue Swift]. 'The Piltdown Skull', *Wandlelog*, 1, no. 1 (Winter 1924–5): n.p.

52 I.O. Evans, *Heroes of World Service* (London: Murray & Co., 1930), 50.

53 Joseph le Comte, *The Monist*, 1890–1, vol. 1, 334–335, quoted in Bramwell, *Ecology in the Twentieth Century*, 49.

54 Evans, 'Evolution and the Kindred', 265.

55 Hargrave, *The Confession*, 157.

56 H.G. Wells, *The Outline of History*, vols. I and II (New York: Garden City Publishing, 1920).

57 John Hargrave, 'Now for 1932!', *Broadsheet* 70 (January 1932), 1.

58 Hargrave's elaborate recapitulation schemes were laid out in several of his books including John Hargrave, *Tribal Training* (London: C. A. Pearson, 1919) and Hargrave, *The Great War Brings it Home*.

59 Gillian Beer, *Open Fields: Science in Cultural Encounter* (Oxford: Oxford University Press, 1996), 123.

60 Stephen Jay Gould, *Ontogeny and Phylogeny* (Cambridge, Massachusetts: Harvard University Press, 1977), 115.

61 Gould, *Ontogeny and Phylogeny*, 143.

62 G. Stanley Hall, *Youth: Its Education, Regimen, and Hygiene* (New York: D. Appleton and Co, 1906).

63 Arthur B. Allen [Lone Wolf], 'Kibbo Kift and the Teaching of World History', *The Flail* 1, no. 1 (Autumn 1926): 33.

64 Gould, *Ontogeny and Phylogeny*, 150.

65 Kibbo Kift was affiliated with several independent progressive schools run on open-air or woodcraft principles in the 1920s. These included Matlock Modern School in Derbyshire; The Woodcraft School, Winscombe; Badminton School, Bristol; Friar Row School, Caldbeck; Priory Row, Kings Langley; and The Garden School, Great Missenden. A Kibbo Kift 'Teacher's Gild' was established in 1927 to consolidate and develop progressive educational schemes through an affiliation of sympathetic teachers, both inside and outside the Kindred.

66 John Hargrave, 'The Attitude of the Kibbo Kift towards the Labour Party: A Brief Statement', *Broadsheet* 4 (October 1925): n.p.

67 Hargrave was highly dismissive of 'townbred intellectuals … who discussed ideas interminably'. See, for one example, his unpublished autobiography written during the Second World War. LSE Hargrave Collection, Box 28: n.p.

68 There were never more than 500 active members of Kibbo Kift at any one time, and membership frequently hovered at around 200. The group had no property, land or substantial means of income.

69 J. Arthur Thompson, *The Outline of Science* (London: George Newnes, 1921).

70 Gould, *Ontogeny and Phylogeny*, 153.

71 See, for example, Evans's review of Dorothy Revel's educational text, *Cheiron's Cave: The School of the Future* in *The Flail* 2, no. 9 (Winter 1928), 33–34, and the detailed demolition of woodcraft recapitulatory methods in Evans, *Woodcraft and World Service*, 128–132.

72 An outline of Boas's approach to anthropology is given in Jerry D. Moore, *Visions of Culture* (Lanham: Altamira, 2004).

73 Kevin C. Armitage, '"The Child is Born a Naturalist": Nature Study, Woodcraft Indians and the Theory of Recapitulation', *The Journal of the Gilded Age and the Progressive Era* 6, no. 1 (2007): 69–70.

74 Bland and Hall, 'Eugenics in Britain', 223.

75 Hargrave maintained an interest in the occult his whole life. After the demise of his movements in the late 1930s he established himself as a psychic healer and traded in healing paintings called

'psychographs'. He also wrote several articles in the 1940s for *The Occult Observer* journal produced by London's Atlantis Bookshop.

76 Lofthouse, *Vitalism in Modern Art*, 23.

77 Evans, *Woodcraft and World Service*, 128.

78 The Woodcraft Folk was established in 1925 after a group of socialists and co-operators in Kibbo Kift had clashed with Hargrave and walked out. They continue in 2018 with 15,000 members. For more on the break, see Pollen, *The Kindred of the Kibbo Kift*.

79 Leslie Paul, *The Child and the Race* leaflet, quoted in Leslie Paul, *Angry Young Man* (London: Faber and Faber, 1951), 109.

80 Paul, *Angry Young Man*, 109–110.

15

Organicism and the modern world: from A.N. Whitehead to Wyndham Lewis and D.H. Lawrence

Craig Gordon

Written in 1925, Alfred North Whitehead's *Science and the Modern World* proposes the apparent paradox that 'Science is taking on a new aspect which is neither purely physical, nor purely biological. It is becoming the study of organisms'.[1] Confronting the well-entrenched opposition between physical and biological systems whose philosophical, scientific, social and cultural significance is practically incalculable, he isolates the organism, typically the purview of the latter half of this divide, as a model providing the conceptual tools necessary to radically transform, if not abolish, this very opposition. In announcing the problem of organicity that becomes increasingly central to his emerging philosophical project from at least 1919 onwards,[2] he contends that our understanding of organism lies at the heart of a profound scientific and philosophical shift – one that requires us to abandon rigid distinctions between the physical and the biological, the organic and the inorganic. Indeed, if one takes seriously the title of the text in which this claim appears, he goes so far as to suggest that rethinking the problem of organism is crucial not merely to the technical operations of philosophers and scientists, but also to an understanding of the culture of late modernity – the modern world – itself. In that context, I seek in this chapter to trace some of the implications of this contention as it pertains both to the culture of literary modernism and its critical reception, taking the latter as an index of the broad cultural significance of the interdisciplinary nexus of shared concerns that Whitehead identifies. What do different ways of understanding organism tell us about being modern? And to what extent does the pursuit of the particular mode of organicity that animates Whitehead and a substantial group of philosophers, biologists and literary writers working across the closing decades of the nineteenth century and the early decades of the twentieth century constitute an important attempt to articulate new modes of modern being? In pursuing these questions, I will explore the impact of biological and philosophical modes of organicism upon two divergent examples of literary modernism: the work of Wyndham Lewis and D.H. Lawrence. In the case of Lewis, I will be interested primarily in his assessment of the liabilities he associates with popularised versions of organicist philosophy, whereas Lawrence will provide an example of a more affirmative response to the sort of organicism articulated by Whitehead and others. In both cases, I shall pay particular attention to the ways in which interactions with organicism shape the conceptualisation of individuality – a problem that is shared by many of Lewis's and Lawrence's modernist compatriots.

My starting point in this respect will be the hypothesis that the centrality of organicity to late-modern culture posited by Whitehead amounts almost to a truism, but that it does so precisely to the extent that various modernist organicisms refuse, or are seen to refuse, the specific model of organism that Whitehead seeks to articulate, remaining instead explicitly or implicitly within the ambit of a notion of organism that descends from Immanuel Kant's *Third Critique* and enters Anglo-American literary culture largely by way of Samuel Taylor Coleridge. Organism, in short, is understood as a closed, self-generating, autotelic totality, whose parts are subsumed within the whole, their identities reciprocally constituted by their functional integration in the whole. This organismic model makes its impact felt upon the understanding of everything from aesthetic form to the form and function of society. Raymond Williams's cautionary introduction to the *Keywords* entry for *organic* provides a productive starting point in this regard. 'Organic,' he writes,

> has a specific meaning in modern English, to refer to the processes or products of life, in human beings, animals, or plants. It has also an important applied or metaphorical meaning, to indicate certain kinds of relationship and thence certain kinds of society. In this latter sense it is an especially difficult word, and its history is in any case exceptionally complicated.[3]

Whereas Williams proceeds from the narrowly biological provenance of the term to its metaphorical extension to forms of subjectivity and society, I will move initially in the opposite direction. For when it comes to modernist culture, part of the historical complication to which Williams refers must surely be the large extent to which the force of organicism becomes a function of its status as a negative determinant of modernity – a basis of response or reaction. Insofar as the typical markers of late modernity include increasingly intense industrialisation, urbanisation and mechanisation, and the experience of modernity is frequently associated with subjective and social fragmentation, and subordination to the determining rhythms of an instrumental order, organicism offers itself as a common form of response, drawing on notions of the organism as a self-sufficient, systemic whole governed by autotelic patterns of growth and creation to provide the resources necessary for a countervailing commitment to integration and spontaneity. Exploring the 'alienating and destructive tendencies of modern labour' as a crucial component of late modernity, for example, Morag Shiach suggests that 'anxiety about the subjective and social costs of mechanisation is [frequently] met, and to some extent answered, by a vigorous organicism'; 'Modernist texts', she writes, 'so often strive to give the fullest possible expression to the destructive tendencies of modernity while simultaneously working to transcend these in the organic or vital energies of aesthetic form'.[4] While the specific kinds of organicism Shiach explores are varied and complex, the more general form of the phenomenon to which she draws our attention – the organicist impulse to 'transcend' modernity's destructive power – is often taken to be a gesture of reaction or retreat, as in Williams's telling comment regarding the Leavisite commitment to organic community. 'If there is one thing certain about "the organic community"', he writes, 'it is that it has always gone'.[5] Seen from this angle, organicism might appear to be central to the culture of late modernity only as a persistent object of nostalgic aspiration.

In this context, Whitehead's claims for organicism as an affirmative condition for a still emergent form of modernity seem, perhaps, particularly surprising insofar as the lexicon of organicism has not fared especially well in historical accounts of early-twentieth-century culture, and of modernist culture more specifically. The *organic* and its cognates have tended to signify reactionary impulses within late-modern culture – at best a nostalgic reflex and at worst the embrace of a dangerous irrationalism. In literary and cultural criticism, the organic has frequently been read as an index of ideological naturalisation, and has been associated most closely with the aesthetics of New Critical formalism and the moral criticism of F.R. Leavis, both of which draw upon broadly Kantian notions of organic wholeness. As Tilottama Rajan suggests,

'Since the demise of the New Criticism, the word "organic" has fallen into disrepute on both aesthetic and political grounds'. Understood to 'signify a whole greater than the sum of its parts which is totalitarian with respect to these subaltern parts, as well as a self-developing entity whose unfolding through a kind of entelechy confers a certain inevitability on the manner of its growth', it is a 'concept whose conservative social consequences become entrenched in the mid-nineteenth century, but whose initial aesthetic elaboration can be traced to the infamous "Romantic ideology"'. It is in the New Criticism, she argues, 'that the theory of organic form as the reconciliation of opposites, and the notion of a whole or structure as "parts arranged in their proper order," receive their definitive modern restatement'.[6] With regard to biological science (in both the discourse of the early twentieth century and subsequent accounts of the period) organicism suffers from its frequent association with vitalist reactions to mechanist and reductionist orthodoxy – typically viewed as an essentially unscientific response to the mechanistic implications of scientific development.[7] Writing, in 1936, within a tradition of organicist biology that he seeks to distinguish from vitalism, for example, Joseph Needham encapsulates this perception neatly. 'The older vitalism', he claims:

> could hardly be acquitted of leanings toward romantic animism, it *hoped* that rigid causal analysis would fail, whereas mechanism hoped that it would win … The motivation of biological mechanism was thus progressive, vigorous, and youthful, a seeking for independence and mastery. Vitalism had more affinity with the religious attitude of creaturely dependence upon a higher power, and in its emphatic affirmation of the complexity of the phenomena of life manifests something of that numinous respect for the otherness in things, which properly belongs to religious experience.[8]

Nor is vitalism the only 'bad company' organicism keeps in the early twentieth century. As Scott Gilbert and Sahotra Sarkar have argued, the disrepute into which organicism has fallen is inevitably connected to the incorporation of certain forms of organicism or holism into the scientific programme of the Third Reich. The Nazis, they write, 'espoused holism as a major part of their "Aryan science" … [and] saw holism (either of the vitalist or organicist variety) as a counter to the notion of nature as a "machine"'.[9] Anne Harrington also explores this connection between Nazism and organicism in her extensive account of German holistic science from the late nineteenth to the early twentieth century.

Without minimising the importance of the linkage, however, she offers the important caveat that:

> The 'racialising' of German holism and its partial absorption into the politics and mythology of National Socialism is an important part of the larger story of German holism … Nevertheless, even if we know how part of the story I tell in this book is going to 'come out,' it is important that we resist 'discovering' the outline of a terrible future in holism's past or imagining that all holistic, vitalistic, or teleological views of nature are part of a larger 'destruction of reason' that can be tracked in some straight, degenerating line from the romantics to Hegel to Nietzsche to Hitler.[10]

Extending Harrington's qualification, I would suggest that the widespread suspicion – scientific, political and aesthetic – of late-modern organicism depends in no small part on the reduction of organicist thought to one of its variants. Alongside or beneath the persistently post-Kantian coordinates through which modernist notions of organicity and organicism tend to get understood, the period encompassing roughly the final two decades of the nineteenth century and the first three decades of the twentieth century witnesses the unfolding of an alternative conception of the organic much more closely aligned with the concerns animating Whitehead's later work. The reconceptualisation of organism I have in mind, here, is less a unified position than a set of shared matters of concern that orient the interanimation of certain aspects of the period's philosophy, biological science and literary discourse. We find it exemplified in the biological work that extends from the vitalism of Hans Driesch, the emergent evolutionism of C. Lloyd Morgan, and J.S. Haldane's neo-vitalism, to the organicist biology of Joseph Needham and J.H. Woodger, and on to the theoretical or systems biology of Ludwig von Bertalanffy. In philosophy, key examples are embodied not only by Whitehead's work, but also that of Henri Bergson and Samuel Alexander, not to mention the philosophically oriented writings of various of the biologists mentioned above. And in literature, important examples of this line of thought range from Oscar Wilde and Edward Carpenter to Wyndham Lewis, D.H. Lawrence and Virginia Woolf. This list of names, of course, is necessarily partial, but it serves to suggest the broad contours of the convergence of interest I seek to explore. More importantly, the form of organicism that emerges from these writers' work seeks neither simply to pit the organic against the mechanical, nor to rely upon the sort of closed, fusional totalisation with which competing models of organicity are frequently associated, exploring instead forms of assemblage that organise the relationships between inanimate matter and living creatures through which individuals are constituted. In so doing, it provides significant resources for reassessing modernist attempts to theorise the role of aesthetics in mediating different conceptions of individuality and forms of sociopolitical organisation.

This conjunction lies at the heart, for example, of Wyndham Lewis's *Time and Western Man* (1927), the massive volume he devoted to what he describes as the all-pervasive 'Time-Cult' – the rubric under which he gathers the philosophical work of Bergson, Whitehead and Alexander, not to mention the literary work of certain contemporaries such as Ezra Pound, Gertrude Stein and James Joyce. This engagement with organicism is significant, in no small part, due to Lewis's intensely critical approach to its popular appeal. Indeed, I turn to Lewis not because he positively articulates the sort of organicist position in which I am interested, but because his critique of the role played by organicist thought in the popular imagination further clarifies the cultural coordinates that have shaped its reception, and isolates the crucial aspects of late modernity to which it opens productive avenues of response. As his notion of the 'Time-Cult' suggests, Lewis's account of the philosophical positions he addresses pays comparatively little attention to their specifically biological provenance in favour of their accounts of temporality. Nonetheless, his argument frequently links his principally temporal focus to organicism, as in the typically strident declaration that

> The Time-doctrine, first promulgated in the philosophy of Bergson, is in its essence … anti-physical and pro-mental. A great deal of partisan feeling is engendered in the course of its exposition: and all that feeling is directed to belittling and discrediting the 'spatializing instinct' of man. In opposition to that is placed a belief in the *organic* character of everything.[11]

There are at least a couple of important points to remark in Lewis's unrelenting focus on what he sneeringly refers to as the fascination with the 'organic character of everything'. Despite his hostility to the organic, the organicist impulse he identifies is somewhat atypical insofar as he characterises it not as a reaction to but rather as a symptom of modernity – if only, for Lewis, the weak form of modernity that is fashion or fad. Or, perhaps more precisely, he suggests that the popular appeal of organicism is based in a reactionary impulse, but that the consequences of the resulting fascination with the organic are symptomatic of the modernity he seeks to critique. 'From a popular point of view', he contends, 'the main feature of the space-time doctrines [of Bergson] … is that they offer, with the gestures of a saviour, *something* (that they call "organism," and that they assure us tallies with the great theory of Evolution – just to cheer us up!) – something *alive*, in place of "mechanism": "organism" in place of "matter"'.[12] Here the popular enthusiasm for organicism is tied to a familiar anxiety regarding the aggressive encroachment of mechanism and its determining force.

Less familiar are the consequences Lewis will draw from this response, based as they are on the 'pro-mental' tendency he identifies with organicism. On its face, this tendency appears simply to align with the attack on mechanism – asserting the agency of mind contra the inert matter of the machine – yet for Lewis it signals, to the contrary, a sort of hypertrophic extension of mentality that ultimately functions to evacuate *mind* as a category. Under the auspices of organicism, 'Dead, physical, nature comes to life. Chairs and tables, mountains and stars, are animated into a magnetic restlessness and sensitiveness, and exist on the same vital terms as men. They are as it were the lowest grade, the most sluggish, of animals. All is alive: and, in that sense, all is mental'.[13] And, for Lewis, once this levelling extension of the mental has been achieved, *mind*, as the defining characteristic of human individuality, is emptied of its significance. 'It is in the interests of "equality," it is in conformity with the "democratic" principle', he writes:

> that 'mind' is to be suppressed or annihilated. On the same principle we are to be converted into machines or into 'events' in place of persons (the 'person,' the free-man of antiquity, is not for the likes of us), and we are to accustom ourselves to regard our personalities as the 'continuous transition of one physical event into another.' In this way we get rid of that embarrassing thing, the 'mind,' which gives us (compared to mere tables, chairs or even vegetables and dogs) a rather aristocratic colour.[14]

The vertiginous path of Lewis's argument thus seems to arrive at a reversal of the typical characterisation of vitalist organicism – whether as the basis for criticism or celebration – as a form of spiritualist response to the mechanistic reduction performed by materialist science and philosophy. On this account, organicism becomes instead a species of mechanism.

Without seeking neatly to resolve the tensions within an argument that is far from neat, I would simply remark that it is important to remember that *Time and Western Man* is in many respects less a rigorous engagement with Whitehead's and Bergson's philosophical projects, than a critique of the implications of the popular reception of their work. The 'object of this book', he suggests in the Preface, 'is ultimately to contradict, and if possible defeat, these particular conceptions upon the popular ... plane, where they present themselves, as it is, in a rather misleading form'.[15] To the extent that Lewis's Time-Cult is the product of precisely this sort of popular reception, he is particularly concerned with that characterisation of organicism which would reduce it to the unqualified celebration of temporal flux and the primacy of idiosyncratic, irrational (or intuitive) forms of subjective experience that are taken to be its correlate. For Lewis, the thorough-going subjectivism popularly associated with the Time-Cult – cast in service of subjective autonomy – is all too consonant with what he views as the homogenising effects of mass democratic culture, which, he argues, fatally erodes the autonomous critical capacity of individuals precisely by inducing them to misrecognise as indices of authentic individuality what are, in fact, commodified aspects of a 'group personality'.

Lewis's roughly contemporaneous *The Art of Being Ruled* (1926) formulates the argument as follows:

> When people are encouraged, as happens in a democratic society, to believe that they wish to 'express their personality,' the question at once arises as to what their personality is. For the most part, if investigated, it would be rapidly found that they had none. So what would it be that they would eventually 'express'? ... It would be a *group personality* that they were 'expressing' – a pattern imposed on them by means of education and the hypnotism of cinema, wireless, and press. Each [individual] would, however, be firmly persuaded that it was 'his own' personality that he was 'expressing' ... The truth is that such an individual is induced to 'express his personality' because it is desired absolutely to standardise him and get him to rub off (in the process of the 'expression') any rough edges that remain from his untaught, spontaneous days ... [D]rawn into one orbit or another, he must in the contemporary world submit himself to one of several mechanical socially organised rhythms.[16]

We find in this account the socio-political articulation of the aspects of modernity that Lewis seeks to critique: a form of mass society in which a notionally democratic culture and its ever more aggressive inducements to free expression provide the ideological screen behind which the selves we are led obsessively to regard and express reveal themselves as a standardised product of society's mechanical rhythms, consumed and internalised via systems of education and mass media, and functioning to subsume the individual within a progressively more homogeneous social and political order. It is this aspect of modernity of which the popularised organicism that Lewis describes in *Time and Western Man* comes to be a symptom. The world of the time-mind seeks to replace a determining objective order with the primacy of subjective apprehension of a dynamic reality, always in a state of flux. In so doing, it offers the fantasy of almost absolute subjective autonomy, even as it disguises the extent to which it fundamentally troubles standard notions of the autonomous individual or self. In this context, the organicist transformation of matter into 'mind' or 'organism' – on the popular view, he argues, the two are roughly equivalent – has specific consequences for the understanding of the self. When matter comes to be understood as organism, he contends, 'something … happen[s] to you as well – the "you" that is the counterpart of what formerly has been for you a material object. You become no longer one, but many. What you pay for the pantheistic immanent oneness of "creative," "evolutionary" substance, into which you are invited to merge, is that you become a phalanstery of selves'.[17]

This image of an emergent phalanstery of selves replacing the stable and static ego and its relations to a world of equally stable material objects is, however, less straightforward than it might initially seem. As Joel Nickels has noted, there is a substantial body of critical opinion that views Lewis as 'one of modernism's most vocal advocates of the static, self-contained ego', for whom 'Individuality and stability … are the last lines of fortification against the sensationalism of crowd-life'.[18] With that in mind, Lewis's caustic description of the merging into an ostensibly creative immanent oneness might be taken as preparation for a rearguard action in defence of the unitary self. However, as Nickels argues, readings of Lewis as spokesman for the stable ego fail to account for important aspects of his work – increasingly prevalent in the 1920s and 1930s – that are actively anti-egoic, and seek in particular to situate 'the artist's sensibility … as an embodiment or analogue' of the 'collective realities' associated with the 'consciousness of the crowd'.[19] Indeed, Nickels draws our attention to an especially important development of this anti-egoism whereby Lewis 'does not represent the ego as *beset* by extra-egoic forces that overwhelm or subvert it. Instead, he often seems to recommend the active and deliberate suspension of egoic boundaries'.[20] In that context, the phalanstery of selves comes to seem something of an ambivalent figure. On the one hand, it evokes the lure of a solipsistic retreat into incommunicably idiosyncratic subjective experience that promises autonomy in the face of the determining rhythms of both physical and social realities understood in mechanical terms, but which ultimately functions to facilitate a deindividualising fusion within a collective state of immanent oneness. On the other, it bespeaks Lewis's recognition that a defence of stable, self-enclosed and self-determining individuality is neither feasible nor desirable – that the 'phalanstery of selves' inescapably describes individuality more adequately than notions of unitary selfhood, and that the pursuit of autonomy must, as Nickels puts it, 'venture … into territory in which "many individuals" coexist in an unpredictable play of qualities and forces'.[21]

Lewis's engagement with popular organicism thus functions to isolate the problem of individuality as a defining feature of late modernity. By this I do not mean to invoke standard modernist accounts of subjective fragmentation or the degradation of individual agency, but the more fundamental question of how to understand the individual as a category. His critique simultaneously warns of the dangers of the fusional subsumption of individuals within homogenising collective totalities, and expresses dissatisfaction with atomistic notions of individuality. He decries the erosion of individuality and the critical autonomy associated with it, and recognises the extent to which individuals are constituted by the 'unpredictable play of qualities and forces' that define the supra-individual fields of relation in which they are imbricated. In that context, the problem becomes not one of deciding whether or not it is possible or desirable to rehabilitate the individual contra the corrosive force of various collective realities, but of redefining individuality itself outside of the terms provided by this opposition. And in this, Lewis ironically finds himself squarely within the terms of one of the animating problems of the organicist biology and philosophy whose popular reception he derides. Ludwig von Bertalanffy, for example, emphasises the consequences of precisely this sort of redefinition in his account of organicist biology when he writes in *Problems of Life* that:

> the purport of biology for modern intellectual life is … deeply rooted. The world-concept of the nineteenth century was a physical one. Physical theory, as it was then understood – a play of atoms controlled by the laws of mechanics – seemed to indicate the ultimate reality underlying the worlds of matter, life, and mind, and it provided the ideational models also for the non-physical realms, the living organism, mind, and human society. Today, however, all

sciences are beset by problems which are indicated by notions such as 'wholeness,' 'organization,' or *'gestalt'* – concepts that have their root in the biological field.[22]

Bertalanffy thus describes a shift away from a scientific model that is essentially analytic: driven to decompose composite entities into ever-smaller constituent individuals whose ontological status is defined intrinsically, independent of the external forces and relations that govern their interactions. In noting the emergence of what he characterises as the properly biological problems associated with wholeness and organisation, however, he does not argue merely for a shift of focus from atomistic individuals to the composite wholes in which they participate, but a fundamental reconceptualisation of both individuals and wholes.

In this regard, the specific notion of organisation to which Bertalanffy refers is key. Organisation, here, denotes not extraneous relations connecting the constitutive parts of an organism, but a fundamental ontological feature of life – what he calls 'the essential characteristic of living things as such – the arrangement or organisation of materials and processes'.[23] As Joseph Needham claims in a similar vein:

> organizational patterns and relations in living things, integrative hierarchies never exhibited in non-living material collocations, are the proper subject-matter of biological enquiry, and ... the recognition of their existence is in no sense a disguised form of Vitalism ... [B]iological order and organization are not just axiomatic either, but constitute a fundamental challenge to scientific explanation, and ... meaning can only be brought into the natural world when we understand how the successive 'envelopes' or 'integrative levels' are connected together, not 'reducing' the coarser to the finer, the higher to the lower, nor resorting to unscientific quasi-philosophical concepts.[24]

The notion of integrative hierarchies that Needham draws from the theoretical biology of J.H. Woodger describes a situation in which biological individuals inhabit a spatial hierarchy that exists on different orders of magnitude – the example he uses is of 'a protein molecule in a colloidal particle in a nucleolus in a liver-cell in a liver in a mammal'.[25] The individuals populating a given level of a hierarchy can be understood in terms of their participation in the level to which they belong, their relationship to the individuals populating a lower level of the hierarchy into which they can be analytically decomposed, or in terms of the ways in which they participate, as constituent parts, in the individuals that populate a higher level of the hierarchy.[26] Regardless of the level of a spatial hierarchy on which one chooses to focus, the individuals that populate it are inconceivable as entities susceptible to intrinsic definition; they are, rather, constituted by, and fundamentally dependent upon, the processes of organisation through which they are related not just to other individuals belonging to their level, but also to those that inhabit greater or lesser orders of magnitude within the hierarchy. Indeed, Whitehead's theory of 'organic mechanism' provides a more general account of this position, positing that 'the molecules may blindly run in accordance with the general laws, but the molecules differ in their intrinsic characters according to the general organic plans of the situation in which they find themselves'.[27] Consequently, Whitehead seeks to replace the category of the individual with that of the organism. 'The concrete enduring entities are organisms', he suggests:

> so that the plan of the *whole* influences the very characters of the various subordinate organisms which enter into it. In the case of an animal, the mental states enter into the plan of the total organism and thus modify the plans of the successive subordinate organisms until the ultimate smallest organisms, such as electrons, are reached. Thus an electron within a living body is different from an electron outside of it, by reason of the plan of the body.[28]

In so doing, he crucially extends the organicist model beyond the narrowly biological notion of individuality developed by the likes of Bertalanffy and Needham, insisting that this 'principle of modification is perfectly general throughout nature, and represents no property peculiar to living bodies'[29] – equally applicable to electrons, cells, inanimate objects, living creatures, or the societies in which they participate.

This model of individuality responds productively to the concerns animating Lewis's account of the popular reception of organicism in a number of different ways. If Whitehead contends that, in the case of animals, mental states enter into the organisational processes of the organism, the flat ontology he will predicate upon this observation – 'the things experienced and the cognisant subject', he writes, 'enter into the common world on equal terms'[30] – does not entail the wholesale extension of mentality to the material world of which Lewis complains. Indeed, he echoes both Bertalanffy's and Needham's insistence that the form of organicism for which he advocates in no way seeks to diminish the importance of physico-chemical modes of explanation, or to replace the physical laws of nature with metaphysical principles. The goal, rather, is to replace linear, mechanical causal relations with a more complex form of causality that can account for both the maintenance of enduring individuals and processes of emergence or creation. While it challenges the traditional split between stable subjects and objects, Whitehead's theory of organism by no means promotes the sort of dissolution into an indifferent 'immanent oneness' that Lewis associates with popular organicism, and recognises the fundamental importance of accounting for enduring individuals. 'The mere fusion of all that there is', Whitehead insists, 'would be the nonentity of indefiniteness', an assertion that leads him to argue that the 'endurance of things has its significance in the self-retention of that which imposes itself as a definite attainment for its own sake. That which endures is limited, obstructive, intolerant, inflecting its environment with its own aspects'.[31] Whitehead thus marks the importance of recognisable individuals which are not merely a product of their environment, but whose processes of individuation grant them a certain agency, the power to inflect their environment. He is, however, quick to note that such an individual or organism is not self-sufficient: 'The aspects of all things enter into its very nature. It is only itself as drawing together into its own limitation the larger whole in which it finds itself. Conversely it is only itself by lending its aspects to this same environment in which it finds itself'.[32] This relationship of reciprocal constitution – not mere mutual determination – applies equally to animate and inanimate entities, and the model of distributed agency it entails fundamentally shifts the terrain upon which Lewis stages his critique of popular organicism.[33] Without dissolving the individual in an undifferentiated flux, or fusionally subsuming it within a totalised whole, the organicist tradition in which Bertalanffy, Needham and Whitehead participate dislodges the presiding conceptual oppositions that coordinate the form of modernity to which Lewis responds, and which underpin the dominant understanding of organicism within modernist culture. In this context, well-worn tensions between human and machine, animate and inanimate, or biological and physical no longer retain their familiar shape and function; by the same token, the question of agency can no longer be framed in terms of the comfortable opposition between the autonomous individual and constraining collective structures.

It is, perhaps, with regard to this problem of agency that Whitehead speaks most provocatively to the concerns that Lewis identifies. For if Whitehead's notion of organism accounts for the endurance, or self-retentiveness, of distinct individuals, it equally seeks to develop a complex non-linear causality that can accommodate processes of becoming which produce novel entities whose emergence cannot be derived from the properties of their constituent components. For example, Whitehead defines the organism as 'a unit of emergent value',[34] a claim he develops by distinguishing 'two sides to the machinery involved in the development of nature'. On the one hand, we have the aspect of the evolutionary picture emphasised by 'scientific materialism', whereby 'there is a given environment with organisms adapting themselves to it'.[35] On the other hand, he will emphasise not the determining givenness of an environment that 'dominates everything', but the

> neglected side of [the evolutionary machinery] … expressed by the word *creativeness*. The organisms can create their own environment. For this purpose, the single organism is almost helpless. The adequate forces require societies of cooperating organisms. But with such cooperation and in proportion to the effort put forward, the environment has a plasticity which alters the whole ethical aspect of evolution.[36]

If we remember that, for Whitehead, any given organism is by definition a society of organisms – produced by the organising relations to other organisms that constitute it – his emphasis on the social basis of creativeness suggests not an opposition between individual and collective action, but that the agency necessary to alter a given situation (at whatever order of magnitude) must be as understood as a product of the organising processes through which novelty can emerge, not the power of the isolated individual. Agency becomes, in other words, a distributed property of organic assemblages – assemblages composed of the animate and the inanimate, the human and the nonhuman, the mental and the physical.

To return to Lewis in this context, I have not sought to suggest that *Time and Western Man* actively pursues the form of organicism articulated by Whitehead and others, merely that it functions to disclose the limitations of a significant cultural dynamic for which its popular-isation provides the occasion. Indeed, Lewis's claim that he targets a popularised version of organicism, combined with his well-known rhetorical propensity to articulate or ventriloquise a variety of opposi-tional stances that bear no simple relationship to his own, make it difficult to discern the extent to which his characterisation of organicist thought reflects his own understanding of contemporary organicism, or to which it claims merely to offer a sort of intellectual reportage. In either case, it is safe to say that the description of organicism available in the pages of *Time and Western Man* is seriously inadequate to those articulated by the likes of Whitehead and Bergson, not to mention the biologists whose work falls outside the purview of Lewis's consideration. Ultimately, however, the reasons for Lewis's failure accurately to describe the organicist theories of his contemporaries remains less interesting than the extent to which his response dramatises the fate of organicist thought in dominant accounts of modernist culture. On the one hand, his text exemplifies the persistence with which the period's organicism is returned to the ambit of a post-Kantian theorisation against which it strains, and the ways in which it is consequently contained within a series of conceptual tensions (between organism and mechanism, subject and object, individual autonomy and collective constraint) that some of its variants would unsettle. On the other hand, however, the ways in which Lewis's encounter with organicism comes to frame the problem of individuality isolates a key aspect of late modernity to which Whiteheadian organicism offers a potentially transformative response – a response, moreover, that addresses the concerns raised by Lewis's ambivalence regarding models of autonomous individuality and the forms of agency they entail. If Lewis's role in this regard can largely be understood as a labour of the negative, let me conclude by turning to the roughly contemporaneous work of D.H. Lawrence in order to gesture very briefly towards a more affirmative contribution to this tradition of organicism. Lawrence presents an interesting figure in this context. He is generally viewed as representing a substantially different form of modernism from Lewis – to borrow Jessica Burstein's taxonomy, Lawrence undoubtedly embodies the 'hot modernism' against which Lewis's 'cold modernist' sensibility positions itself.[37] Moreover, the philosophical coordinates of Lawrence's and Lewis's work diverge substantially; the philosophical tradition extending from Whitehead and Bergson to Alexander and James, in relation to which Lewis situates himself, is relatively absent as an explicit influence from Lawrence's work, where Nietzsche and Schopenhauer emerge as much more significant interlocutors.[38] Despite these differences, Lewis and Lawrence converge in important and interesting ways during the 1920s, not least of all with regard to questions of mass democratic culture and individuality.

Lawrence articulates most forcefully the line of argument I seek to pursue in his work of the late teens and early twenties, prominently including his *Study of Thomas Hardy*, *Women in Love*, 'Democracy', and *Fantasia of the Unconscious*. His essay on 'Democracy', for example, is remarkably close to Lewis in its analysis of the homogenising effects of mass democracy as an important context within which to approach the problem of individuality. Lawrence will trace this process of homogenisation to two key sources. The first is a quantifying impulse that reduces individuals under the rubric of the 'Average Man', an abstraction necessary to equitably address the material needs of humanity, but fatal to what he will call the 'living self'. 'There are', he writes, 'two sorts of individual identity. Every factory-made pitcher has its own little identity, resulting from a certain mechanical combination of Matter with Forces. These are the material identities … . The other identity, however, is the identity of the living self'.[39] Lawrence's recourse here to an image of mass production is significant, as it situates this process of abstraction as the embodiment of a materialist impulse. 'Men and women', he argues, 'are thus turned into abstracted, functioning mechanical units. This is all the great ideal of Humanity amounts to: an aggregation of ideally-functioning units: never a man or woman possible'.[40] The second homogenising tendency that Lawrence identifies proceeds from a reaction to this materialist subsumption of the individual within a mechanical aggregate, which seeks to identify a countervailing organic principle of cosmic unity – what he will refer to as the 'One-Identity' or the 'En-Masse'. Here we arrive at something very close to the indifferent 'immanent oneness' against which Lewis's account of organicism struggles, and for Lawrence it is no less problematic; the One-Identity, he insists, is 'a horrible nullification of true identity and being'.[41]

Indeed, it is precisely this homogenising nullification that Lawrence contests in positing a 'new Democracy' that emerges with the proclamation: 'Not people smelted into a oneness … [but] released into their single, starry identity, each one distinct and incommutable'.[42] If this seems to return us to the familiar tension between collective constraint and individual autonomy, Lawrence is quick to forestall that conclusion. He does this, in part, by drawing a distinction between *personality* and *individuality* that provides the basis for a critique of egoistic individualism – personality designating the egoistic self as a pernicious inheritance, a 'horrible incubus' from beneath which the individual 'spends the rest of his life trying to drag his spontaneous self'.[43] More profoundly, though, Lawrence addresses this tension by seeking to reframe our understanding of individuality, a task he encapsulates in the paradoxical claim that 'the great development in collective expression in mankind has been a progress towards the possibility of purely individual expression. The highest Collectivity has for its true goal the purest individualism, pure individual spontaneity'.[44] No libertarian call for the withering away of collective constraint, this claim seeks, rather, to render the spontaneity or creativity that defines Lawrence's notion of the individual consequent upon participation in a certain form of collectivity. In this regard, the specific form of his argument is crucial: his definition of 'pure individuality' in terms of spontaneity suggests that we must understand the 'individual expression' for which this highest collectivity constitutes the condition of possibility not as the intentional act of a willing subject, but as the manifestation of the creativity embodied by the process of individuation. And in this, the singularity of the individual is constituted by its relations to its environment.

Lawrence gestures in this direction with his repeated recourse to the astronomical constellation as a means of figuring individuality, the 'starry identity' of which he writes: 'the myriad, mysterious identities, no one of which can *comprehend* another. They can only exist side by side, as stars do'.[45] If, however, this figure remains somewhat ambiguous, insisting on the irreducibility of the individual but potentially emphasising the separation of individuals as much as their relation, the constitutively relational notion of individuality towards which he works is developed in more detail three years later in *Fantasia of the Unconscious* (1922). In this context, the familiar Lawrencian notion of blood consciousness is refigured as the unconscious or 'dynamic consciousness' – a form of embodied consciousness that he locates in a series of nerve plexuses and ganglia located along the anterior and posterior surfaces of the human body.[46] Whereas the 'voluntary' system formed by the ganglia governs a consciousness of individuality based on the differentiation and separation of the subject from the world of objects, the 'sympathetic' system facilitated by the plexuses provides a consciousness of individuality as constituted by relation and conjunction. Lawrence envisions this system of nerve-centres as enmeshing the human individual within a complex network of vibratory affective relations connecting it to its environment: 'Between an individual and any external object with which he has an affective connection, there exists a definite vital flow … Whether this object be human, or animal, or plant, or quite inanimate, there is still a circuit'.[47] While this model of individuality certainly remains more anthropocentric and tied to consciousness than the theory of prehensive relations that Whitehead articulates, in positing a form of consciousness that participates in, rather than reflects upon or represents, its world, Lawrence moves towards the sort of flat ontology developed by Whitehead and others. In that context, he seeks to develop a notion of individuality whereby the singularity of the individual must be understood as a product not of its intrinsic properties, but of the ongoing processes of individuation that are tied to the affective relations of which it becomes capable – creativity indexed to the individual's capacity to affect and be affected by the myriad aspects of its world. As he puts it in his 1918 essay 'Life':

> We are not created of ourselves. But from the unknown, from the great darkness of the outside that which is strange and new arrives on our threshold, enters and takes place in us … This is the first and greatest truth of our being and of our existence. How do we come to pass? We do not come to pass of ourselves. Who can say, of myself I will bring forth newness? Not of myself, but of the unknown which has ingress into me … [B]ecause whilst I live, I am never sealed and set apart; I am but a flame conducting unknown to unknown, through the bright transition of creation.[48]

Notes

1   Alfred North Whitehead, *Science and the Modern World* (New York: Free Press, 1967), 103.

2   *The Concept of Nature*, delivered as the Tarner Lectures in 1919 and published the following year, is arguably his first major articulation of the problem, subsequently developed in *Science and the Modern World*, *Process and Reality*, and a variety of other late texts.

3   Raymond Williams, *Keywords: A Vocabulary of Culture and Society*, Revised edition (New York: Oxford UP, 1985), 227.

4   Morag Shiach, *Modernism, Labour and Selfhood in British Literature and Culture, 1890–1930* (Cambridge: Cambridge UP, 2004), 16.

5   Williams, *Keywords*, 277.

6   Tilottama Rajan, 'Organicism', *English Studies in Canada*, 30, no. 4 (2004): 46.

7   Scott Gilbert and Sahotra Sarkar summarise this orthodoxy as follows: 'Imagine a philosophy claiming that the entire physical universe operates solely according to the interactions of matter and energy. No "vital forces" exist, and all living phenomena consist only of chemical and physical processes. Such an ontologic position … is called materialism, and it provides the basis for contemporary natural science. Then imagine a materialistic philosophy that claims that all complex entities (including proteins, cells, organisms, ecosystems) can be completely explained by the properties of their component parts. Such an epistemological position is called reductionism, and it is the basis for most of physics and chemistry, and much of biology.' See Scott F. Gilbert and Sahotra Sarkar, 'Embracing Complexity: Organicism for the 21st Century', *Developmental Dynamics*, 219, no. 1 (September 2000): 1.

8   Joseph Needham, *Order and Life* (Cambridge Mass: MIT Press, 1968), 8–9.

9   Gilbert and Sarkar, 'Embracing Complexity', 4.

10 Anne Harrington, *Reenchanted Science: Holism in German Culture from Wilhelm II to Hitler* (Princeton, N.J: Princeton University Press, 1996), xxi.

11 Wyndham Lewis, *Time and Western Man*, ed. Paul Edwards (Santa Rosa: Black Sparrow Press, 1993), 421.

12 Lewis, *Time and Western Man*, 166.

13 Lewis, *Time and Western Man*, 421.

14 Lewis, *Time and Western Man*, 430.

15 Lewis, *Time and Western Man*, xix; Mary Ann Gillies's account of the widespread popularity of Bergson's work among a non-specialist readership in Britain provides suggestive context in this regard. She notes, for example, that 'the ideas that were … passed as the philosopher's own may well have born little resemblance to what Bergson had written. But those who expressed their admiration

of Bergson by unknowingly corrupting his ideas were responsible for the popularity of Bergsonism. Indeed, the major force behind Bergson's popularity in England before the war was not the philosophers who debated his works, nor the scientists who poured over his theories; it was the individuals who picked up some of his vocabulary and a piecemeal understanding of his concepts and who then imported this language into their own discussions of the world around them'. See *Henri Bergson and British Modernism* (Montreal, Que.: McGill-Queen's UP, 1996), 36.

16 Wyndham Lewis, *The Art of Being Ruled*, ed. Reed Way Dasenbrock (Santa Rosa: Black Sparrow Press, 1989), 148–49.

17 Lewis, *Time and Western Man*, 166.

18 Joel Nickels, 'Anti-Egoism and Collective Life: Allegories of Agency in Wyndham Lewis's *Enemy of the Stars*', *Criticism*, 48, no. 3 (2008): 347. For an account of Lewis's early critique of egoism in *Tarr*, see Paul Peppis, 'Anti-Individualism and the Fictions of National Character in Wyndham Lewis's *Tarr*', *Twentieth Century Literature*, 40, no. 2 (1994): 226–55.

19 Nickels, 'Anti-Egoism and Collective Life', 347.

20 Nickels, 'Anti-Egoism and Collective Life', 348.

21 Nickels, 'Anti-Egoism and Collective Life', 349; Paul Peppis makes a similar point when he claims that: 'To the Individualist view that persons are autonomous beings capable of independent action and personal liberation, Lewis opposed a picture of persons as overdetermined beings incapable of controlling the maelstrom of competing forces that constitute human identity'. See Peppis, 'Anti-Individualism and the Fictions of National Character in Wyndham Lewis's *Tarr*', 234.

22 Ludwig von Bertalanffy, *Problems of Life: An Evaluation of Modern Biological Thought* (London: Watts, 1952), ix; Joseph Needham makes an analogous case for the socio-cultural impact of changing scientific paradigms, and for the particular significance, in this regard, of organicism. See *Time: The Refreshing River (Essays and Addresses, 1932–1942)* (London: Allen & Unwin, 1943), 186.

23 Ludwig von Bertalanffy, *Modern Theories of Development: An Introduction to Theoretical Biology*, trans. J.H. Woodger (London: Oxford University Press, 1933), 35.

24 Needham, *Order and Life*, viii.

25 Needham, *Order and Life*, 111.

26 For Needham's extended description of spatial hierarchies, see Needham, *Order and Life*, 111–13.

27 Whitehead, *Science and the Modern World*, 80.

28 Whitehead, *Science and the Modern World*, 79.

29 Whitehead, *Science and the Modern World*.

30 Whitehead, *Science and the Modern World*, 89.

31 Whitehead, *Science and the Modern World*, 94.

32 Whitehead, *Science and the Modern World*.

33 In this regard, Whitehead anticipates the theory of 'thing power' articulated by Jane Bennett. Bennett invites us, for example to 'picture an ontological field without any unequivocal demarcations between human, animal, vegetable or mineral. *All* forces and flows (materialities) are or can become lively, affective, and signaling. And so an affective, speaking human body is not *radically* different from the affective, signaling nonhumans with which it coexists, hosts, enjoys, serves, consumes, produces, and competes. This field lacks primordial divisions, but it is not a uniform or flat topography. It is just that its differentiations are too protean and diverse to coincide exclusively with the philosophical categories of life, matter, mental, environmental. The consistency of the field is more uneven than that: portions congeal into bodies, but not in a way that makes any one type the privileged site of agency. The source of effects is, rather, always an ontologically diverse assemblage of energies and bodies, of simple and complex bodies, of the physical and the physiological'. See *Vibrant Matter: A Political Ecology of Things* (Durham: Duke University Press, 2010), 116–17.

34 Whitehead, *Science and the Modern World*, 107.

35 Whitehead, *Science and the Modern World*, 111.

36 Whitehead, *Science and the Modern World*, 111–12.

37 See Jessica Burstein *Cold Modernism: Literature, Fashion, Art* (University Park, PA: Pennsylvania State University Press, 2012). For Burstein, 'cold modernism' denotes a modernist sensibility focussed on a fundamentally ahuman world, whereas 'hot modernism' refers to a much more

widely recognised modernist attention to psychological and affective experience.

38 For an account of this philosophical influence on Lawrence's organicism, see Shiach, *Modernism, Labour and Selfhood*, 170–75.

39 D.H. Lawrence, 'Democracy', in *Reflections on the Death of a Porcupine and Other Essays*, ed. Michael Herbert (Cambridge: Cambridge UP, 1988), 73.

40 D.H. Lawrence, 'Democracy', 69.

41 D.H. Lawrence, 'Democracy', 73.

42 D.H. Lawrence, 'Democracy'.

43 D.H. Lawrence, 'Democracy', 75.

44 D.H. Lawrence, 'Democracy', 66.

45 D.H. Lawrence, 'Democracy', 72.

46 For a fuller account of Lawrence's system of dynamic consciousness, see Craig Gorden, *Literary Modernism, Bioscience & Community in Early 20th Century Britain* (New York: Palgrave Macmillan, 2007), 88–91; See also, Shiach, *Modernism, Labour and Selfhood*, 182–83.

47 D.H. Lawrence, *Psychoanalysis and the Unconscious and Fantasia of the Unconscious* (Cambridge: Cambridge UP, 2004), 153.

48 D.H. Lawrence, 'Life', in *Reflections on the Death of a Porcupine and Other Essays*, ed. Michael Herbert (Cambridge: Cambridge UP, 1988), 16–17.

16
Liquid crystal as chemical form and model of thinking in Alfred Döblin's modernist science

Esther Leslie

In 1933, Alfred Döblin, a novelist and doctor, published a long and complex book, titled *Unser Dasein* (*Our Existence*).[1] *Unser Dasein* is a difficult book and has enjoyed nothing of the success of *Berlin Alexanderplatz*, but it is also a book that presents in a variety of ways – scientific, fictionalised, philosophical amongst others – a mode of thought characteristic of Döblin, in which a kind of monism is at work, whereby science and art, scientific approaches and artistic responses, are presented as equally appropriate, equally evocative, equally generative of knowledge and understanding. If one were seeking the model of a crossover and combination of scientific and artistic work *Unser Dasein* would yield a curious but productive one. It utilises the montage form not just or even predominantly in terms of splicing scenes or genres and disciplines. It deploys it in the sense of yoking together that which is often kept apart. In so doing, it perhaps evinces a much deeper absorption of and commitment to a then recently discovered – but still marginalised – chemical and theoretical form, the liquid crystal, a form which is discussed in the course of the book. It is as if it combines this strangely contradictory and improbable form, which is liquid and crystal at once, into its mode of presentation and into its vision of the world and the human as part and counterpart of a world.

Unser Dasein was written more or less contemporaneously with Döblin's city-novel *Berlin Alexanderplatz*, which had appeared in 1929. *Berlin Alexanderplatz* was a montage novel, a compiling of documents and bawdy songs, of bus timetables and scientific pronouncements, tram routes, weather reports and stock exchange reports, radio broadcasts, mortality statistics, advertisements and melodramas from the press. In a 1930 review of Alfred Döblin's *Berlin Alexanderplatz*, titled 'Crisis of the Novel', Walter Benjamin argues that the novel can survive only if it adopts an epic, cinematic form.[2] Döblin's work gave an extension of life to the novel, set it on new grounds in the media age of technological reproduction. The traditionally isolated form of the novel – written and read alone – opens up to the technical imperative of the modern age and imports something of the mediatised collectivity into its pages. Alfred Döblin had called for a cinema style in 1913 already. For Döblin, this meant writing characterised by what he termed urgency and precision, three-dimensionality and liveliness. In practice it meant a development of montage methods through the inclusion of non-literary, reproducible materials into the writing. Literature imports such cinematic devices that play with space and time – scene shifting, close-up or flashback. A new kind of writing arises here and it is one that deals less with self-expression and more with the rendering of objective, social reality, though it never relinquishes moments of flight into subjectivism, or even expressionistic exaggeration. 'Authentic reality', the stuff of life, is incorporated into the writing, or, more, it is the story. The story is told through documents and urban ephemera. *Berlin Alexanderplatz* told the story of Franz Biberkopf, but it did so not as a communication from one individual about another.[3] Historically, the novel is a form, Benjamin maintains, in reflections on Döblin, that is written by a solitary and silent person, who cannot speak to the collective, but can only render individual experiences. What Döblin produces, in contrast, is a new form of epic, out of the fragmentations of montage: 'The montage explodes the framework of the novel, bursts its limits both stylistically and structurally, and clears the way for new, epic possibilities'. Benjamin understood *Berlin Alexanderplatz* to be specifically Berlinish, forged of its dialect and its streets, a 'low life Naturalism'. The novel absorbed the city, its technologies, bureaucracies, systems of governance and control into its language and its stories.

The contemporaneous book *Unser Dasein* similarly evaded generic definition. It reinvented form, presenting in its multiple modes of address, different linguistic registers and multiple disciplinary approaches, something unclassifiable. Like *Berlin Alexanderplatz*, it deployed montage, gleaning materials from widely differing sources. It used sudden cuts of scenes – cinematically again – in order to pursue not just the dissolution of the individual in the collective city experience that is Berlin, as *Berlin Alexanderplatz* had done, but rather something larger. *Our Existence* sought to pursue the origins of life in past times and the remnants of something characterisable as the whole in each individual part, each part of life, each organic and inorganic fragment. The book is a five-hundred page investigation of life, society, nature's forms, subjectivity, aesthetics, religion, morality, time and suffering. Its title – *Our Existence* – indicated something of its ambition, for it promised a panoramic theory of life, specifically human life, though that implied all other lifeforms and inorganic ones too. Its overall title echoed those of Ernst Haeckel's hugely popular turn-of-the-century study, *The Riddle of the Universe*, with its sections on 'Our Bodily Frame'; 'Our Life'; 'Our Embryonic Development'; 'Our Monistic Religion'; 'Our Monistic Ethics'.[4] Its section titles had the ring of Haeckel's authoritative voice: for example, 'The Self and the World of Things', 'The Counterpart of Nature: The Three Peculiarities of the Self', 'Transition to the Collective: Of Herds and Individuals'. *Our Existence* was as grand in scale and conception, and about as little read, as Döblin's earlier science fiction novel *Berge Meere und Giganten* (*Mountains Seas and Giants*), from 1924, which imagined several thousand years of history, from the First World War onwards far into the future.[5] As, in the course of the story, technology develops, the narrative describes large-scale wars and smaller-scale guerrilla wars, new machines and waves of neo-Luddism, the emergence of Shamanic societies and the return of oral storytelling. There is also the conquering of uninhabited lands. This last one demands the deliberate melting of Greenland's ice, through a harnessing of Iceland's volcanic energy, a process which reanimates prehistoric bones and plants. These fuse into monstrous and deadly hybrid forms. Like *Berlin Alexanderplatz*, *Mountains Seas and Giants* tries to render genre and language anew. Punctuation is missing and there is slipperiness between objects and subjects. The vast spans of time dealt with in nine books means that there are no characters who carry over from sub-book to sub-book and there is no unity of place. It all threatens to dissipate.

Unser Dasein shares this sense of discontinuity. The themes of life, wherever life finds itself, are explored in the eight 'books' and three interludes of *Our Existence*, conveyed through a mix of poetic and scientific registers, slipping at various points into bawdy or infantile rhyme and deviated quotations which have been played around with. What unites these books is the approach from the perspective of 'natural philosophy'. Döblin works with a natural-philosophical conception, meaning that his work's scientific basis has a Romantic inflection. In order to engage with the sciences of the day, Döblin feels compelled to reach back to the scientific methods and insights of the Romantics who engaged in Natural Philosophy.[6] Alongside his fascination with German

Idealist Romanticism, Döblin evinced an interest in Spinoza's ideas. These presented him with a version of philosophical Monism, in which God and nature are one and Nature is released from any transcendent force that brings it into movement.[7]

In December 1927, Döblin published an article in the *Vossische Zeitung* under the title 'Outsiders of Natural Science'. Some of these had been dismissed as mystics and Romantics, but Döblin perceived them as 'the intellects of tomorrow', because they 'reach far enough backwards'.[8] Amongst their number were included the doctor and metaphysician Oskar Goldberg, the palaeontologist Edgar Dacque and the philosopher of harmony Hans Kayser. Each worked on the margins of science, speculating on the whereabouts of Atlantis or the magical origins of language and number. Döblin reaches to those who worked the span between, on the one hand, a subjective and in a sense romantically accented study of nature and, on the other, the pursuit of significant technical and scientific discoveries. In contrast to the separation often declared between scientific thought and poetic expression, Döblin evokes in his various writings a number of investigators, present in particular in Germany, in whom the proximity or identity of scientific thought and poetic expression, or scientific expression and poetic thought, was to the fore. This is what he reaches back to. But there is another reaching back. This is the reaching back to the origins of humans, the beginnings of nature. For the Romantics, as for Döblin, the beginnings lodge still in the ends. There is a quest for origins, but furthermore there is the denial of linear time, or better a denial of a simple idea of progress and the refinement of forms and the separation out of realms. Humans were once mineral, are still mineral. Minerals share characteristics of humans. Vegetables are rooted humans. Humans are rootless vegetables that have grown nerves and muscles and so on. Such a perspective is one that undermines another widely held idea of conventional science, the opposition between humans and nature. When undermined by Romanticism, a vision of nature out there is proposed, in which the plants, the rocks, the stars, are, like humans, the possessors of subjectivity and agency. With another inflection, a scientific one, a similar empathetic sense argues that plants, rocks and stars are composed of the same matter as us. In either case, a world of life, and even of non-life and life, in some sort of unity is proposed. This book was preceded by a book of similar tenor titled *Das Ich über der Natur*, a paean to connectedness of all things in the world, and the presence of a primal spirit or intellect that moves through all that exists in the world.[9]

And so in *Unser Dasein* Döblin argues that the animal is a mineral and, at the same time, the vegetable-animal form of the mineral. Animals and minerals share forms, modes, processes and elements. Such forces are evident in the ways in which the universe rearticulates itself: the earth is warm, but covered by a hard stony crust. It is like the polar bear, covered with a layer of thick fur. 'The thought should be possible: The earth resembles those animals, or the earth is of the same kind as these animals'.[10] Döblin's Universalism considers the universe as a whole organism, each part affecting the others, just as the waxing and waning of the moon affects the sea and animals. Seeking the nature of humans, there is no definitive opposition between souled and unsouled beings, between matter and spirit, between the hard stony crust of the earth with its soft vegetation and the soft pliable skin that bounds the human, its bony skeleton within.

Döblin's mode of perceiving unity and connections has antecedents, specifically those who had affinities to Romantic natural philosophy and natural history. In an early English incarnation, it can be recognised in a book of natural history, begun in 1774, an eight-volume work by the poet Oliver Goldsmith. It is titled *A History of the Earth and Animated Nature*. A Romantic work of scholarship, Goldsmith decided that the best way to depict the wonders of the natural world was 'to write from our own feelings and to imitate nature'.[11] In his book, nature, already described by other naturalists, is re-described through an observing eye whose look is informed by identification and empathy. If the first object of natural history is the apprehension of nature and its knowing, then the second object destabilises this, penetrates further into nature's realm in order to realise how much of nature is not known, or not fully known or is known in new ways, within an ever-widened prospect. The second – and real – object of natural history, inflected by philosophy, is the natural object remade in thought and imagination. This is the utopian axis of a nature infused by concept and idea and word. The title of Goldsmith's book relays something about a Romantic stance towards nature. The nature he writes of is animate, which at its simplest means that he wishes only to write of nature that can be described as alive, properly alive – as, for example, are plants and animals. But beyond that he also indicates his approach: that nature is precisely something spirited, as lively and inter-connected and multiply related, sometimes through enmity, sometimes solidarity, across its chain of being. Animated nature is historical, changing over time, dynamic, and all the vital elements of the universe, from the highest, the human, to the lowest, the insects, are animated by spirit, which is to suggest – as Darwin makes clearer later that all of nature is unified by what Coleridge in 1796, in his poem 'The Eolian Harp', called 'the one Life within us and abroad'. Nature is us and we perceive parts of ourselves reflected in all its elements. It was a perspective echoed in many thinkers, as for example, in Johann Gottfried Herder's *Outline of a Philosophy of the History of Humanity* (1784/91):

> The more we learn of Nature, the more we observe these indwelling powers, even in the lowest orders of creatures, as mosses, funguses, and the like which almost inexhaustibly reproduces its own likeness, in the muscle, which moves briskly and variously by its own irritability, the existence of these powers cannot be denied: and thus all things are full of organically operating omnipotence. We know not where this begins, or where it ends; for, throughout the creation, wherever effect is, there is power, wherever life displays itself, there is internal vitality. Thus there prevails in the invisible realm of creation, not only a connected chain, but an ascending series of powers; as we perceive these acting before us, in organized forms, in its visible kingdom.[12]

For Döblin, there are echoes across all parts of the world and its various kingdoms. But his is also a philosophy of conflict. The self is in the world, part and not part of it. *Unser Dasein* was a reflection on the make-up of the I and the make-up of the world and the relations between the two. Is the world an illusion, invented by the self? Is the I formed only by its environment? Döblin's answer was a dialectical one, if poetic too:

> It is not an illusory world, but a real world, but it has its reality in us. In the sea of being the temporal world is a wave. Or a pearl.[13]

There is *Sein* ('being') and *Dasein* ('existence'). There is *Leben* ('life') and *Erleben* ('experience'). Each of these is co-constitutive of the other. Experience occurs through time, like the moment of a wave rising, specifically out of the sea of being. Or it is like a formation – a pearl – a nugget that is made through time and passes away in time. The I arises or the I is made, two different accounts of how being might exist. The I experiences the world, reflecting it like a mirror, as it experiences it directly. In this regard it is a part of the world. But it also remains separate from it, a counter-part, which can withdraw into itself, experiencing its own transitoriness. It is both part of and part against nature. But even as it does this, the world acts on and with it. The self is an object for the world.

The first book of *Unser Dasein* has the heading 'The thing world and me'. The writer sits at his writing bureau in a study and inquires after the 'I', in relation to animals, objects, sounds, people, the world within and outside. This I has a body comprising nails, bones, teeth, eyes, ears, muscles and sinews. It also possesses a mess of organs. Hair propagates as if one were a mountain with trees on top of the head or a verdant meadow that must be regularly mowed. The I appears as a factory, an incubator, a business. But this factory of the body is also a carnal agent, which experiences and suffers, thinks and desires. *Leben* – life – has to *erleben* – experience. The I is drenched by the world and the world is unlocked again and again by the feeling, acting person. We, the I, everything is a part and a counter-part of nature. There is natural form and there is autonomy, but all takes place in the context of the universal principle of 'resonance', the interconnections back and forth between humans and their environment, a *Kraftfeld* ('energy field') within which life and experience occur.

After this introduction, a next section, titled 'Summer Love', explores social atomisation through the story of some lovers in the city. Its protagonist is a man in his thirties, a civil court judge, who is 'an isolated frozen animal'. His self is exposed to an indication of its 'transcendental homelessness', as Georg Lukacs termed it in his work of 1916, *The Theory of the Novel*.[14] The self is aware of its loss of a place in the world and the loss of any greater purpose to existence, yet the desire to inhabit a home and to have meaning persists. The self is a forsaken being, alone in the world but his selfhood appears to be massified endlessly. He is not alone in his loneliness. In the course of his work, the protagonist falls in love with a happy-go-lucky woman. He melts, seems to himself to be rain, a stream, suffusing all things and people, sinking in an ocean of exhilaration and delight. It is as if he could kiss every mouth, gaze into all eyes. He feels 'more elastic'. He learns to identify himself by entering into this voyage of yearning. The self is realised in an other. But what is also formed is one great collective woman who is loved and a community of lovers. Once he has reached this recognition, the woman who kindled the passion is long gone, and he has already seen her with her new beau. But his love is not quashed. He falls in love with their love, partaking in their desire. Love is fluid, flows with the streams of city life, which are powered by drives, in whirling spaces of the urban masses, in transport systems, beneath and within dynamic advertising. It is a passage into life.

Döblin moves from here to a section titled 'The Unlocking of Nature', which returns, in another way, to the themes of what the I is and how it exists in the world. Here he considers such topics as the plant in relation to what he calls the nerve-muscle human and he reflects on the possibility of love in plants and animals. World and I are recognised as 'open systems' that are incomplete and are always seeking completion, each by the other. Completion is never reached. There is a drive to form and to completed form – this drive is utopian – but it is never finished. This is the dialectical tension of which life and experience are composed. It takes its driving force from a sense of process that Döblin draws from his biological understanding. Another tension is at work too, a scientific one. Across his various writings a certain tension as to where life comes from, or what it means. Perhaps it is less a tension than the indication of an openness to concepts. Just as he sets life in a tension between the crystallising and dissolving aspects of formation, Döblin himself is caught between the mechanical theories of life as a result of chemistry and number, and the Vitalist ones, which insist on an innate building agency in the organic. The Vitalists, or 'Neo-Vitalists' as they were known from the turn of the century, ascribed life to a somewhat mysterious self-directed and indivisible force of nature. It was something like gravity or electricity or the newly discovered radioactivity. The Neo-Vitalists' leading light was Hans Driesch who gave the universal life force a name: 'Entelechy', borrowed from Aristotle.[15] In opposition to this conception was the Mechanists' claim that life was to be comprehended as an isolated chemical function occurring within a physical substance.

Between these poles exists Döblin's natural-philosophical stance. In as much as it is dialectical, it might also be said to be evinced in his literary approach – exemplified in the varied modes of expression, of writing form, of disciplines and aspects, all elements of a whole that can never be made whole. The open system of the human pulls it always towards dissolution into the environment, and yet, against this, form continually asserts itself. The 'innate tragedy of all formed things' resides in this interplay of openness and integrity.[16] All form dissolves. But dissolution achieves new forms. Such is the book before us, much as it was, if more tentatively, the case with his other books.

In *Unser Dasein*, Döblin explores the relation between the I and the organic world. Original life was, he claims, vegetable, but the human, like most animals, eventually emerged as something unrooted. Such thinking is usefully contextualised within a wider scientific field of the time, a period in which intensive research was going on into vitamins and what researchers in Germany termed *biologische Wirkstoffe* ('biolog-ically active agents'), which were found in various substances, isolated, analysed and synthesised.[17] The unrootedness of the human-vegetable compelled the evolution of other organs to bolster the humans' I – nerves and muscles. Despite this evolution of organs, he emphasises similarities across forms. The plant may have no mouth, stomach, intestine, but, like the person, it takes in nourishment, stores it or expels it. The plant knows stimulation too, and movement – not enough to make it need those organs of nerves and limbs, but enough to make it close to us. Döblin considers inorganic nature too, observing that organic nature comprises inorganic elements. The human too is mineral and, like all things in nature, is involved in a dance with form, which is the province of the mineral, the crystal. What counts as an organ in the animal world takes, in the world of mineral, the shape of a space lattice. The principles of nature that generate form are, according to Döblin, rhythm, number and repetition. Such are the principles of crystal formation – but they extend into all organic living things, for example:

> The expression of number in phyllotaxy, in the laws pertaining to the ordering of leaves on the stem, in the position of scales on a spruce cone. The ornamental forms of leaves.[18]

Number explains elements of form, or what might be termed the crystal side of life. And yet there is something that number does not capture. Experience is not accounted for wholly by number. Experience is fluid and unstable. Knowledge of details might not open up the experience of nature, he suggests, when he notes, in relation to water that we know to be made of repetitions of two hydrogen and one oxygen molecule:

> From this formula one can never arrive chemically at the particular form or un-form of water.[19]

Döblin reflects on the relationships of cells and crystals through H_2O, a peculiar form that can be hard as in a snow or ice crystal, or a fluid in the form of water. Abundant in the world, it is found in the cell too. Water yokes organic and inorganic worlds. Döblin considers regular, geometric formation, which is a characteristic of crystals, and observes it in plants and animals, such as spruce cones and radiolarians, corals, anthropods, worms, feathers, the compound eyes of bees and the symmetrical segmented forms of the inner organs of vertebrates. Beauty is inorganic form, with its patterning and regularity. Not just a spatial element, Döblin extends the principle of number and geometry to rhythmic movement in time, animals and astronomy. How could it be otherwise, he asks, when organic forms are composed of inorganic elements? We humans, who over time experience existence as sometimes stable and sometimes labile, that is, in German, *stabil* and *labil*, are made mainly of water, being moveable seas that must be constantly refilled. Everything is in formation. Even a squalid blur of smut has a crystalline structure. But nothing in the world is finished. Everything is a fragment of something else.

Döblin elaborates a history of the world and the I. Light brought new life forms. Once the surface of the earth is heated enough by the sun, water is joined by protein jelly (*Eiweissgallerte*). From this amalgamation come animal and plant forms. The crystal forms that had dominated, with their harsh edges, are softened by water and made rounder. Once water and jelly exist, metabolism, the exchange of stuff, is essential. Osmosis, light: forms of life proliferate in water, and then on land. Protoplasm, a labile thing, liquid-like or jelly-like, is what emerges after a hard crystal epoch. Döblin describes jellies as miscarried crystals with an enormous will to become a crystal. But other circumstances in the swirly oceans reform the crystal and it becomes otherwise. Earth's crystal epoch is over as softer forms develop, though these forms that emerge can easily step back into a crystalline existence, can easily harden or mineralise once more. What is the crystal, Döblin asks rhetorically. His answer is that it is the fighting form of matter against fire and water, those fluid, consuming forms. Life forms form themselves against water – non-dissolvable, waterproof – by forming skins. They become, in so doing, in developing cells, crystalline, to some degree. And so Döblin outlines a vision of a universe that is reflected in all its parts in us: we are composed of animal, vegetable, mineral, planetary forms. It is not sequential – we were not once planet, then crystal, cell, plant, animal and, finally, human. Everything is a part of us still. Time is shattered.

In Döblin, the purpose of existence is the 'continuation of forming'. This is a simple procedure for the crystal, but a difficult effort for the 'plasma being', with its organs and its movements. But as a final point the whole lot inclines towards stiffening – *Erstarrung*. The watery states of the plasma beings account for that, as they have the capacity to freeze. Forms harden over time, and such hardening is a movement towards death, but it is also part of the process of new lives forming. The crystal achieved this state very early in its existence, hardening out of the melting flow. Petrified existence is also a part of life, in its mineral expression.

In Döblin, the liquid and the crystal are brought together, exchange properties, exist in a dance. Such proximity of liquid and crystal forces had been made available scientifically in Germany since 1888, when a momentary phase of liquid crystallinity was caught in view for the first time. Liquid crystals were closely observed, if unnamed, in 1888 when an Austrian chemist and botanist in Prague, at the Institute of Plant Physiology, named Friedrich Reinitzer, experimented with cholesterol of carrots in the form of cholesteryl benzoate and cholesteryl acetate.

It appeared that this substance could melt into a liquid as the temperature increased, but also he discovered that in its cloudy phase it polarised light in the way that a crystal does, that is, it refracts it twice, in different directions. Colours flashed up during phase transitions. Reinitzer had to conclude that the substance seemed to be crystalline and liquid at one and the same time. Furthermore, these crystal-like things 'melted' during cooling. Reinitzer was perplexed and passed his findings to a physicist in Germany, Otto Lehmann. Reinitzer and Lehmann exchanged letters and samples by post.[20] They pondered on what might be at work as the bright colours briefly flashed up. Lehmann was adept with a microscope and the first task was to examine the substances closely using his home-built one, a crystallisation microscope, with polarisers and a moveable object table. Later, he integrated micro-photographic and screening mechanisms into the microscope, making it possible to reproduce and objectify that which the eye could perceive and so make it available to others and to research and classification. Lehmann continued studying this form his life long. Lehmann wrote under the influence of the biologist Ernst Haeckel, who, reciprocally, on hearing about liquid crystals as an old man, deemed them to be the missing link between inorganic and living systems and devoted his final work, *Crystal Souls: Studies of Inorganic Life*, to them in 1917. Other researchers perceived in liquid crystals something crucial, indeed, something that was set at the heart of life and conceptions of life. A paper published in German and English in 1930, by Friedrich Rinne, provided a pithy image of such thinking. Titled 'Sperm as Living Liquid Crystal', its abstract asserted:

> It is customary to draw the boundary between living organic and inorganic matter so that crystals represent the highest form of inorganic material and low organisms form the beginning of the organic world, with a definite and deep physiological gap between the two categories. In my opinion, this gap does not exist, since the sperms, which are undoubtedly living, are at the same time liquid crystals.[21]

Döblin knew, like many other scientists of the time, of Ernst Haeckel's pinning of the liquid crystal as the intermediate form between organic and inorganic life in his book 'Crystal Souls', but he absorbed similar ideas more directly through the work of his contemporary Hans Kayser, a theorist of harmony. Harmony, Kayser conceived, as the connecting element between the material and the mental world, the physical and the metaphysical. Döblin owned a copy of Kayser's *Orpheus: Morphologische*

Fragmente einer allgemeinen Harmonik, which was printed in a handful of copies in Berlin in 1924.[22] Notions of resonance were undoubtedly provoked by Kayser's assertion of harmonics and his interest in waves of sound and the matter waves of the new quantum physics. The centrality of the crystal was also affirmed by him, as basis of life and as harmonic form. A rhythmic principle expresses itself in developmental stages from crystal to plant to animal. Kayser asserted the existence of mathematical laws, in these various forms. Nature creates and humans can observe these creations in relation to harmonic intervals and numerical patterns. The discovery of liquid crystal makes clear to him the threads of connection between inanimate and animate worlds, though he is keen also to emphasise the dividing lines. Kayser was a reader of the Romantic experimenter J.W. Ritter, who was in Jena around 1800, in contact with Romantic natural philosophers and poets. Ritter, using his own body as a testing ground, eventually fatally in 1810, worked on notions of polarity in relation to light and electricity. Kayser also knew of Otto Lehmann, the man who had taken up the discovery of liquid crystals in 1888, and held to them as other scientists demurred.

Döblin's book goes on to explore ethics and art, questions of free will, the flow of time and causality. Art is a manifestation of the desire for recognition of resonance, even as the I struggles not to be melted into the earth. Through art, the human discovers connections to the basis of the world and the self. Art speaks to the inorganic composition of humans and never more so, apparently, than in Döblin's time, the time of the Modern Art movements, as he draws for his argument on Cubism, Constructivism and abstraction.[23] These mobilise number, rhythm, the formed, all that which is inorganic. Art, for Döblin, is primarily a re-remembering of the non-human forces of formation in the human. Artworks strengthen our vegetable, crystalline, inorganic substance. Art is something that aims at the completion of our incomplete selves. It does not achieve it.

The book continues with reflections on human types and behaviour, with reflections on modes of love affairs, suicide, and questions of collectives, herds and individuals, including the 'Jewish Volk-Nichtvolk, which he interpreted from the perspective of suffering'.[24] Here in these analyses of peoples he draws in the phenomenon of paragenesis, the mode whereby contact between mineral deposits affects each other's formation. The touching points between entities produce form. Döblin applies it to the people, arguing that the environment marks itself on physiognomy, gesture and stance. Bodies are formed by social and historical forces, as tracts are in minerals.[25]

The final book examines the peoples of the occident and their mode of rule by powerful states, which become like monsters with uncontrollable organs. Poised before an economic catastrophe, it is set against the free development of the I, which is always a collective subject, a we, and which as such makes the world with nature. But what exists is capitalism, which needs profits and melds needs to their ends, and thereby the entrepreneur reforms the whole of humanity biologically. Against this is set hundreds of years of class struggle, culminating, in Germany, in 1918–19 and the attempted social revolution. From the writer at his desk to the fertile anarchy of revolution,[26] Döblin has answered the question of what is the I, variously, biologically, socially, psychologically, chemically, emotionally, politically and cosmically.

The liquid crystal view of things and self – a dialectical, historical materialist view – might be counterposed to an undialectical, ahistorical idealist view, which arises in the same period. This is embodied in the Fascist insistence, in various registers, on the crystalline. In those ideas that burgeoned happily in the environment of the Third Reich, there was a tendency to donate high value to the crystal. The imagined crystal of the Third Reich is solid and unchanging. It takes the place of life, instead of being a form of life. This may be found variously in the mountain film genre, many of which were directed by Arnold Fanck and starred the actress Leni Riefenstahl.[27] The crystalline view is present in Arctic myths of Aryan origin, and in the Nazi approved Cosmic Ice Theory of Hanns Hoerbiger.[28] This glacial cosmology became the popularly accepted cosmology of the Third Reich, after Hörbiger's death in 1931. It gained political backing from Himmler, who thought that Aryans had reached Earth from the skies in sperms preserved in the Cosmic Ice. Hitler intended to commemorate the theory in a building designed by Albert Speer, in Linz. The building was to showcase three world pictures: Ptolemy's, Copernicus's and the Cosmic Ice Theory. The theory received institutional backing in Himmler's research organisation *Ahnenerbe* ('ancestral heritage'), which sought to elaborate pre-Christian practices and beliefs. Towards the end of the 1920s, Cosmic Ice Theory began to be framed as the 'German antithesis' to the 'Jewish' theory of relativity.[29] The polar theories evoked in the Third Reich do not include polarity within their own epistemological field, for they set their resources against melting and mixing. Metaphorically, Fascism favours the moment of ice, of freezing, of the pure white driven snow, the North, the cold.

A contemporary of Döblin, Ernst Jünger, a warrior Fascist, wrote a treatise titled *Der Arbeiter: Herrschaft und Gestalt* (*The Worker: Domination and Gestalt*), in 1932. It set terms that the coming Nazi regime of 1933 came closer to fulfilling than the democratic one that preceded it. Modern existence, Jünger argues, has produced change and unrest, but it will be succeeded by a settling of all into a landscape that 'is more constructive and dangerous, colder and more luminous'.[30] The crystalline wins out. All change is solidified into order, a steely, cool order, in which humanism is evacuated. All cosiness has disappeared. It is possible already, in this world of the early 1930s, he notes, to traverse areas that are like 'dead moonscapes, governed by a vigilance that is as invisible as it is omnipresent'.[31] Here is the proper home of the mobilised worker. This is a worker who is something like a combination of soldier and machine, militarised and industrialised, and it is one who will develop a sensibility for the 'icy geometry of light'. This worker-warrior had survived the trenches of the First World War and knew how to operate under chaotic conditions. This worker-warrior is a 'type', and, as one type crystallises, another should be condemned to melt into history:

> The mass is in essence amorphous, and therefore the purely theoretical equality of individuals, which are its building stones, is sufficient. In contrast, the organic construction of the twentieth century is a formation of a crystalline kind, and therefore it demands of the type that occurs in it, quite a different measure of structure.[32]

In the nineteenth century, Jünger argues, the mass was constant and the individual was variable. In the twentieth century, the mass, the formations of life, the situations that demand energy and participation, are variable, while the individual is constant. In this context, he claims, mathematics, or number, comes to play a greater role in life, as a socially organising force.

> The crystal is an ideal object. The crystal is the form that evinces picture-perfect symmetry, against modern chaos. The crystal is a symbol of a well-formed, regular, transparent world. Complete symmetry signals death. The mobilised life is a life that stares death constantly in the face.

In any case, where Jünger thrived, if critically, Döblin's universalism finds no resonance at all in Nazi Germany. When *Our Existence* appeared in April 1933, Döblin was already in Zurich, in exile. His writings were not banned until 1935, so the book escaped the flames, but it died the death of being unread. It met a world where hardness is a virtue and where the human is made into matter that can be formed into ornaments in rallies and on the battlefield, and does not form itself, liquidly *and* crystallinely, out of enlivened matter, out of the conscious and exuberant intermingling of humans and environment.

Notes

1   Alfred Döblin, *Unser Dasein* (Munich: Deutscher Taschenbuch Verlag, 1988).
2   See Walter Benjamin's 'Krisis des Romans; Zu Döblins *Berlin Alexanderplatz*', *Gesammelte Schriften*, vol. III (Frankfurt: Suhrkamp, 1992 [1930]), 230–236.
3   Alfred Döblin, *Berlin Alexanderplatz: Die Geschichte vom Franz Biberkopf* (Munich: Deutscher Taschenbuch Verlag, 2002).
4   Ernst Haeckel, *Die Welträthsel: Germeinverständliche Studien über monistische Philosophie* (Bonn: Strauß,, 1899). In English: *The Riddle of the Universe*.
5   Alfred Döblin, *Berge Meere Giganten; Ein Roman* (Berlin: Fischer, 2013).
6   For one study of Döblin as 'natural philosopher', see the contribution by Roland Dolinger, 'Alfred Döblins Naturphilosophie in den Zwanziger Jahren', in *Philosophia naturalis: Beiträge zu einer zeitgemässen Naturphilosophie*, eds. Thomas Arzt, Roland Albert Dollinger, and Maria Hippius-Gräfen Dürckheim (Würzburg: Königshausen & Neumann, 1996), 135–150.
7   See Dolinger, 'Alfred Döblins Naturphilosophie', 137.
8   See Thomas Keil, *Alfred Döblins 'Unser Dasein': Quellenphilologische Untersuchungen Würzburger Beiträge zur Deutschen Philologie*, vol. xxix (Würzburg: Königshausen & Neumann, 2005), 118. For another setting of Döblin's ideas within their context and in relation to influence of 'Lebensphilosophie', in particular Bergson, Nietzsche, Driesch and Plessner, see Ursula Elm, *Literatur als Lebensanschauung: Zum ideengeschichtlichen Hintergrund von Alfred Döblins 'Berlin Alexanderplatz'* (Bielefeld: Aisthesis, 1991).
9   For discussion of this text, and its efforts to integrate scientific discoveries into metaphysical speculations, see David Midgely, in 'Metaphysical Speculation and the Fascination of the Real: On the Connections between Döblins Philosophical Writings and his Major Fiction before *Berlin Alexanderplatz*', in *Alfred Döblin: Paradigms of Modernism*, eds. Steffan Davies and Ernest Schonfield (Berlin: Walter de Gruyter, 2009), 7–27.
10 Döblin, *Unser Dasein*, p. 139.
11 Oliver Goldsmith, 'Preface', *A History of the Earth and Animated Nature, Vol. 1.* (London: Wingrave et al., 1805), p. xiii.
12 Johann Gottfried Herder, *Outlines of a Philosophy of the History of Man*, trans. T. Churchill, vol. 1 (London: J. Johnson), 191.
13 Döblin, *Unser Dasein*, 132.
14 Georg Lukacs, *The Theory of the Novel* (London: Merlin Press, 1963), 70.
15 See H. Driesch, *Geschichte des Vitalismus* (Leipzig: Verlag von Hans Ambrosius Barth, 1922).
16 Döblin, *Unser Dasein*, 97.
17 For discussion of this, see the introduction to Alexander von Schwerin, Heiko Stoff, and Bettina Wahrig, eds., *Biologics: A History of Agents Made From Living Organisms in the Twentieth Century* (London: Pickering & Chatto, 2013).
18 Von Schwerin, Stoff, and Wahrig, *Biologics*, 123.
19 Von Schwerin, Stoff, and Wahrig, *Biologics*, 121.
20 Peter M. Knoll and Hans Kelker, *Otto Lehmann. Erforscher der flüssigen Kristalle. Eine Biographie mit Briefen an Otto Lehmann* (Frankfurt: Ettlingen, 1988), 50.
21 Letters to Editor, *Nature* 126 (23 August 1930): 27.
22 Keil, *Alfred Döblins 'Unser Dasein'*, 117.
23 Döblin, *Unser Dasein*, 245.
24 For a discussion of its status, see Keil, *Alfred Döblins 'Unser Dasein'*, 52ff. See also Hans Otto Horch's introduction to Alfred Doeblin's *Schriften zu jüdischen Fragen*, (Munich: Deutscher Taschenbuch Verlag, 1997), 7–78 and Klaus Mueller Salget, 'Döblin and Judaism' in Roland Albert Dollinger, Wulf Köpke, Heidi Thomann

Tewarson, eds., *A Companion to the Works of Alfred Döblin* (Camden House, NY, 2004), 233–246.

25 Salget, 'Döblin and Judaism', 353.

26 For a survey of Döblin's political stances during the years of the Weimar Republic, see Wulf Koepke's 'Döblin's Political Writings during the Weimar Republic', in *A Companion to the Works of Alfred Döblin*, eds. Roland Albert Dollinger, Wulf Köpke, Heidi Thomann Tewarson (New York: Camden House, 2004),183–214. For Döblin's retrospective sense of the German Revolution of 1918, see Michael W. Jennings, 'Of Weimar's First and Last Things: Montage, Revolution, and Fascism in Alfred Döblin's *November 1918* and *Berlin Alexanderplatz*', in *Politics in German Literature*, eds. Beth Bjorklund and Mark E. Corey (New York: Camden House, 1998), 132–152.

27 See a recounting of some of this in Leni Riefenstahl, *Kampf in Schnee und Eis* (Leipzig: Hesse & Becker, 1933).

28 For further explorations of the rise of snowy visions in the Third Reich, see Friedrich Paul Heller and Anton Maegerle, *Thule: Vom völkischen Okkultismus bis zur Neuen Rechten* (Stuttgart: Schmetterling, 1998), 72ff.

29 Heller and Maegerle, *Thule: Vom völkischen Okkultismus bis zur Neuen Rechten*, 72ff. See also Milena Wazeck, *Einstein's Opponents: The Public Controversy about the Theory of Relativity in the 1920s*, trans. Geoffrey S. Coby (Cambridge: Cambridge University Press, 2014).

30 Ernst Jünger, *Der Arbeiter: Herrschaft und Gestalt* (Stuttgart: Metzler, 2014), 175.

31 Jünger, *Der Arbeiter*.

32 Jünger, *Der Arbeiter*, 145.

17

'I am attracted to the natural order of things': Le Corbusier's rejection of the machine

Tim Benton

Between the completion of the Villa Savoye in 1931 and that of the Villa Le Sextant aux Mathes, in 1935, something strange happened.[1] Le Corbusier threw out the 'Five Points of a New Architecture', formulated in 1927.[2] Where are the pilotis, the roof garden, the long window, the free façade and free plan?

Two projects, designed while the Villa Savoye was still on the drawing board, help us understand this transition. In October 1929, in Buenos Aires, Le Corbusier met Matias Errázuriz and they signed a contract to build a summer villa in Chile. Between December 1929 and April 1930, Le Corbusier's office worked up the project. Although some of the syntax of the Villa Savoye remains – the ramp and the picture window – the rest of the Five Points have gone. The free plan and free façade have gone. The concrete pilotis have become pine trunks, rough stone has replaced reinforced concrete and a butterfly roof the roof garden.

The case of the vacation house he designed for Hélène de Mandrot at Le Pradet near Toulon in the Vars is more curious.[3] We have to begin with the mass housing prototype he designed to meet the requirements of the Loi Loucheur, passed in August 1928. This law contained within it a contradiction, born of the political compromise required to get the law through the Assembly. Louis Loucheur, an industrialist whose construction company had made a fortune during the war, wanted industrial means – standardisation and prefabrication – to be used to bring down unit costs. But the Socialist and Communist vote had to be assuaged by ensuring that builders, craftsmen and local materials should be employed.

The Maisons loucheur project, first sketched out in December 1928, developed in 1929 to include a range of building types, thereby juxtaposing two contradictory philosophies: a hand-built stone wall 45cm thick constructed by local masons, and a prefabricated steel structure clipped to this wall and faced with zinc sheets like a Parisian roof. The Villa de Mandrot project began with a version of the Maisons Loucheur, between December 1929 and January 1930.[4] Instead of accommodating two dwellings of 45m² each, the de Mandrot house would have used both halves. In January 1930 the house was adapted to site, with a driveway which encircled the house to deliver cars to a garage under the house.

In the spring of 1930, however, when Pierre Jeanneret found it difficult to obtain a good price for the steel components, the project went through a number of variants, including a version based closely on the Errázuriz project. In the course of corresponding with companies advertising steel construction, Pierre came across a company called Constructions Métalliques Fillod. They produced prefabricated steel panels, incorporating an air gap, which could be assembled into little bungalows using a standard square module of 4.93m².

A sketch labelled '7 metallic cells', became the basis of the finished house, retaining the cellular composition of square cells dimensioned to the Constructions Métalliques Fillod plan, but articulated now in rough stone walls. The *mur diplomatique* and local craftsmen of the Maisons Loucheur had fought back. The local Italian mason, Aimonetti (who 'has everything in his pocket', according to Mme de Mandrot), was eventually fully incorporated into the scheme. The brutal contrast between hand and machine remained, however, with the insertion of massive steel prefabricated windows. For a detailed account of this project and its design implications, see the publications by myself and Bruno Reichlin.[5]

What does this sudden shift towards vernacular construction and the use of natural materials represent? What does it mean and how did it come about? The answer comes in five parts, five counterpoints to the new architecture:

> the belief that harmonious form eventually derives from nature
> the conviction that well-being can only come from contact with nature
> the admiration for 'honest' people attuned to working with nature
> the idea, derived from Adolf Loos, that only peasants have culture; only peasants can design a home.
> the understanding that the role of the creator is to impose order on nature without spoiling it (the male–female analogy).

To give this a political flavour, you could summarise this in two points: Nature (or harmony) is the dictator and the 'Real man' is the client.

These ideas sit alongside the traditional modernist position of the need to respond to the machinist age. This is a book-sized topic, so my responses for the purposes of this chapter are selective.

The idea that all form derives from nature comes in two parts. In the first part, the proposition that nature could be prompted to produce geometry and order. In the second, that nature could stimulate the senses and the imagination. The young Charles-Édouard Jeanneret was trained at the arts and crafts school at La Chaux-de-Fonds, to believe that all beauty derived from nature. Charles Blanc and John Ruskin taught that immersion in nature – close observation followed by rigorous analysis – could lead to decorative art and therefore to architecture, since architecture, for Ruskin, was decorated structure.[6]

Analysis consisted chiefly of seeking geometric structures that could bring order to the confusion of nature and make it serve the purposes of repetitive pattern. Nature submitted to geometric control could also become structure.

This touches the heart of the theme I want to explore. Nature, in its infinite variety and mysterious laws of growth and decay had to be understood. But geometry belongs to another Nature – let us give this one a capital 'N', from which the so-called Laws of Nature are derived. Absolute, unchanging, universal, the Laws of Nature are the product of human reason but are thought to derive from some higher order. Modern architecture derived from geometry (= 'Nature') but nature (the wet, messy, sensual and gratifying kind) was also essential for human satisfaction. Le Corbusier could never reconcile nature and Nature.

That nature could produce geometry was of course a major theme of nineteenth- and twentieth-century architectural discourse. I don't know whether Le Corbusier read D'Arcy Thompson, but he certainly knew Matila Ghyka's (1931) *Le Nombre d'or: rites et rythmes pythagoriciens dans le développement de la civilisation occidentale.*[7]

One of the classic arguments used to support the 'Nature = geometry' argument was to look for natural forms incorporating interesting proportional series. The Nautilus shell, with its Pythagorean spiral, had fascinated Le Corbusier at least since 1923 when he ordered a set of photos of sea shells from Hendrik Wijdefeld, the editor of the Dutch *Wendingen* magazine. He claimed to have long been fascinated by sea shells. He illustrated these shells in a number of his books including *L'Art decoratif d'aujourd'hui.*[8] Pythagoras, whose life in Edouard Shuré's *Les Grands Initiés* received Le Corbusier's closest attention, appeared to demonstrate that scientific enquiry and the spiritual awakening of the Initiate (and the artist and Prophet which Le Corbusier thought of himself as) could work together.[9] After the war, all this would come together in the work towards the Modulor, published in two volumes in 1950 and 1954.[10]

Even in the high Purist phase of the 'White villas', Le Corbusier habitually used nature as a metaphor for architectural quality. Seducing Madame Pierre Meyer in a 'letter' of sketches of April 1926 showing the villa he has designed for her, he turns from describing the building to evoking the wonders of nature: 'This is not at all like a garden in the French style but a wild wood where, thanks to the tall trees of the Park St James, you could imagine yourself far from Paris.'[11] Nature here works as a proof of the quality of life provided by the concrete architecture. Nature is both the product of architecture and proof of its excellence. In an earlier drawing of October 1925 to Madame Meyer, Le Corbusier had gone even further.[12]

> This project was not born out of the hasty pencil of a studio draughtsman in between telephone calls. It was matured slowly, caressed, during days of perfect calm in the presence of a landscape of high classicism.

Here nature has given birth to geometry.

The idea that nature could grow geometry is well represented in his sketch of a group of Villa Savoyes, hanging like apples from a tree in the remote pampas in Argentina.[13] In his fifth lecture in Buenos Aires in October 1929, Le Corbusier finished his description of his newly designed Villa Savoye by saying:

> This same house, I should set it down in a corner of the beautiful Argentinian countryside; we shall have twenty houses rising from the high grass of an orchard where cows continue to graze.[14]

The drawing seems to reflect a yearning on Le Corbusier's part to make his prismatic Purist architecture look natural in the countryside. It also reflects his view of the countryside as completely untamed and un-landscaped.

But these ideas of the relationship between Nature and architecture began to take on a deeper resonance in the 1930s. Nature was to take on the role of dictator. This is a painting called *The Hand and the Flint Stone* of 1932.[15] The artist (and architect) is represented by a pen and by a schematic box of matches which stands for smoking which, for Le Corbusier, represented creativity. The artist's hand blends into a natural form – the flint – which simultaneously takes on a fleshy and erotic quality. This painting served as the frontispiece for his important book *La Ville radieuse* (1935).[16] Among the preparatory papers for Le Corbusier's *La Ville radieuse* is a sketch of this painting with the legend 'the despot'.[17] As published, the word 'despot' was replaced by a caption that includes the statement, 'Objective and subjective are the two poles between which springs human creation born of matter and the spiritual'.[18] On the typescript draft of the title page, the book is still called: '"La Ville Radieuse", ou le despote'.[19] The 'despot' is the law which will rule the modern world once it has penetrated every heart, and the law is 'harmony'. The despotism of 'harmony' implies subservience to universal laws of nature – the objective – while Le Corbusier also understood nature as subjective, in its physical and sensual sense, with all its informality and uncontrollable confusion.

This tension, some might call it a contradiction, between the local and the universal, between universal laws of harmony and the close observation of natural forms, of physicality and spirituality, is the theme of this chapter. Another apparent paradox in Le Corbusier's political approach in the 1930s is his belief in the possibility of absolute personal freedom within the context of an authoritarian political state. I will follow Robert Fishman, Mary McLeod and Mark Antliff in reminding us of the peculiar political circles in which Le Corbusier circulated in the 1930s.[20]

So to my second 'counterpoint'. Let's go back to La Chaux-de-Fonds. The heart of the teaching of Le Corbusier's master Charles L'Eplattenier was that you had to get out and about, into the fresh air, the snow, the high Alpine passes, before you could understand anything of nature. Experience was the key. His father was the president of the local Alpine Club and the family took every occasion to penetrate up the valleys of the Jura to the alpine slopes. The young Charles-Edouard Jeanneret (Le Corbusier's real name) took photographs of these family expeditions, and there are photographs of Jeanneret and his student colleagues out sketching in the forest. This is the axis of direct experience of nature.

Later on, Le Corbusier discovered the seaside. Le Corbusier liked nothing better than leaving the city and immersing himself in nature – and contact with 'natural people' – typically on his travels in Brittany, Spain or North Africa or at Le Piquey on the Bassin d'Arcachon, where he took his vacations every summer from 1926 until 1936 or among the disturbing red rocks on the north shore of Brittany.[21]

It was here that his painting changed, from the strict rule-based compositions of his Purist compositions to a freer, more sensuous style. In these paintings, the sensual, the tactile and the metaphoric dominate.[22] Touch and feel are as important as visual analysis and everything seems to refer to something else. The still unspoiled landscape of the Bassin d'Arcachon evoked for Le Corbusier a simpler life dictated by the sun and moon, the passage of the tides and the fundamental occupations of working men and women.

It was here that he accumulated much of his personal collection of shells, bones and pieces of driftwood, which he referred to as *objets à reaction poétique* ('objects stimulating creation'). The butcher's bone appeared in many of his paintings and even in a short piece of film which he shot in September 1936. In 1939 he even painted it onto the wall of the guest bedroom of the house designed by Eileen Gray for Jean Badovici (1926–9).[23]

In September 1936 he took hundreds of photographs, revisiting the kinds of object that had inspired his earlier paintings. In particular, he took hundreds of photographs of small sections of sand or mud, tracing the mysterious ways the water moulded the surface.[24] These photographs look like the drawings he made of the desert in the valley of the M'zab, from an airplane in 1933. He also wrote about the 'cosmic forces' that governed the waters of the lagoon at Arcachon, where the passage of the tides created a continual evolution of the shoreline. Le Corbusier went on to spell out his cosmic metaphors, the male and female (sun and moon), the dictatorship of the sun and the law of water (the law of the meander), illustrated by sketches recalling his flights in South America and North Africa. The great cosmic forces of sun and water, wind, earth and fire governed human existence. All this was challenged and negated by man-made cities. In all of this Le Corbusier was following his friend Dr Pierre Winter, who had co-written an article about preventative medicine entitled 'Les grilles cosmiques de l'homme' in April 1931. This article was based on the assumption that, 'the sick man is a healthy man whose internal rhythms are no longer in accord with the cosmic rhythms which surround him.'[25] Winter described these cosmic forces:

> The earth – that is to say the centrifugal force, weight. The moon, a tangential lateral force. The earth force weighs man down and compresses him from top to bottom. The sun force lightens him and draws him up from the ground upwards. The moon force makes him supple and draws him out sideways, working on the weak points, the joints.[26]

Winter believed that sickness could only be addressed by confronting man's spiritual problems. It is never easy to know to what extent Le Corbusier shared the views of his friends, but it is clear that he was influenced by Winter's view of the cosmic forces at work in the world.

The opening words of *La Ville radieuse* are 'I am attracted to all natural organisations ...' but what needs figuring out is what Le Corbusier meant by nature. In a literal sense, Le Corbusier was moving from an obsession with machines to a close bond with natural forms and materials. The link between physical nature and the ineffable sphere of creation is described by Le Corbusier in *La Ville radieuse* not in terms of things but of people. In the first few lines he explains that he flees the city to find 'des hommes en instance d'organisation'.[27]

> I look for savages, not to find them barbarous but to measure their wisdom. In the Americas or Europe, peasants or fishermen. I understand: I go where men do the work necessary to feed themselves and reduce their sufferings. They also do what is necessary to obtain, free of charge, the joys of sociability: their trade, family, collectivity. I reckon that, as architect and urbanist, I come to learn the basics of my trade from man, or from men.[28]

He was referring principally here to the people he met and drank with at Le Picquey or in Brittany. A few lines further down, he notes the paradox of the modern world (the radio, the newspapers) sharing the world of these 'savages'. He accepts too that these men were not without faults, simple-minded and pusillanimous, the weaknesses of those who are disempowered by the modern world.

The manuscript draft of the introduction – 'Je suis attiré ...' – was labelled 'Piquey, Trégastel 1934'.[29] There is no doubt that the holidays spent at Le Picquey on the Bassin d'Arcachon and Trégastel in Brittany were fundamental for reordering Le Corbusier's attitudes to nature. Le Corbusier's moving account in *Une Maison, un palais* of the fishermen's houses in the pine forests of the Bassin d'Arcachon is his most heart-felt piece of writing on domestic architecture.[30] Le Corbusier noted 'the isolation, the separation from the rest of the world' of the tongue of sand dunes, between the lagoon of Arcachon and the Atlantic. The fishermen, who worked there in the summer, 'only came there with the idea of living "from day to day"'. This precariousness puts them into the paradigmatic situation of the house builder; they are making a shelter for themselves, somewhere to live, no more, in all simplicity and honesty. They are carrying out a pure programme unencumbered with claims to history, to culture, to the taste of the day. They're building a shelter, somewhere to live, with the materials that come to hand'. And suddenly, says Le Corbusier, he realised that these houses were Architecture (or as he puts it, they were 'palaces').[31] What is remarkable is that the book in which he wrote this passage was essentially about two of his designs, the luxurious Villa Stein de Monzie and the competition entry for the League of Nations Headquarters in Geneva which was eventually built to a classical design. The cover of the book illustrates the latter project, but above it Le Corbusier places not his magnificent villa but one of the 'palaces' from the Bassin d'Arcachon.[32]

There are other signs of Le Corbusier's preference for the most simple and primitive conditions for relaxing and living in the summer. On 10 July 1932 he sketched a wooden shack, raised on stilts, rented by friends of his for a vacation. He noted: 'life unfolds entirely under the pilotis …' and he clearly considered this casual existence idyllic.[33] He noted, 'The lesson is the extension of the 'inside' to the 'outside'. It is the economy of the cell and the generosity of its extension outside. It is the participation of the fundamental elements: sun, sky and greenery.'[34]

As we have seen in the case of the Villa Savoye prototypes 'planted in the virgin Argentinian pampas',[35] for Le Corbusier, landscape was ideally wild and untouched: the 'prairie' rather than 'paysage'. This is what Le Corbusier is referring to in his sketch confusingly labelled 'le dehors est toujours un dedans'.[36] Pure architecture in a pure landscape.

The extent to which Le Corbusier was prepared to push the idea of untouched nature emerged after the war, on observation of the roof garden of his apartment in Boulogne sur Seine. Le Corbusier had designed a block of flats in Boulogne-sur-Seine in 1933–4, overlooking the Jean Bouin and Parc des Princes stadia. Until the outbreak of war, when Le Corbusier and Yvonne left Paris, the garden was well looked after. Untended for five years, a new nature installed itself, brought by the wind and the birds. Writing to a Dutch horticulturalist John Voges of Hillegom on 17 September 1956, Le Corbusier explained:

> I decided to leave it without any compost, without any pruning, in fact any upkeep of any kind, even the removal of dead leaves. During the war, the wind or the birds brought seeds. False sycamore, now 5 metres high, laburnum 4 metres high; the roses have regressed to 3 metre high sweetbriars. The boxwood has flourished particularly well, the yucca to an extraordinary degree. The lilac blooms in spring, but very briefly. In general, the flowers have been killed by the weeds; I never allowed any weeding. The ivy is

rampant, etc. Could you suggest some flowers which might survive in these conditions?[37]

Similarly, since Le Corbusier had built himself a guest room overlooking the roof of his mother's house at Corseaux, he had given it close attention.

> Two or three years ago almost the whole roof was covered with forget-me-nots (albeit very poor). In the autumn little wild geraniums appear, replacing the clover and dandelions etc. In fact, it's a real meadow and what is remarkable is that it has never been watered artificially. In the summer the grasses dry up and you think all is lost, but from September it springs to life. This is my question: do you think that in these conditions of absolute penury some attractive flowers could survive? Crocuses, tulips, perhaps even hyacinths, snow-drops, ranunculus, daffodils (false narcissus).[38]

In this way, Le Corbusier believed, pure architecture and untrammelled nature could cohabit.

The model of course was the Parthenon. The Purist prism or Doric temple is transformed and humanised by its natural context.[39] It suggests that the radical principles of the Laws of Nature – geometry, order, reason – could blend with nature in its primal state.

> They erected on the Acropolis temples determined by a single idea and which gathered around them a desolate landscape which they subjected to their composition.[40]

This sketch on the processional route at Delphi crystallises this thought. Pure, geometrical form contrasted with wild, untamed landscape. Nature and nature. Illustrating this sketch in *Une Maison, un palais*, he wrote: 'Dominating the canyons and the valleys, at Delphi, these three stone slabs, pure and violent witnesses, evoke the sublime.'[41]

At its most basic, the form of juxtaposition of 'Nature' and 'nature' that Le Corbusier envisaged was the sudden confrontation of vertical and horizontal. This is the axis of contrast. He described this in a lecture reproduced in *Une Maison, un palais*, where the sight of a vertical rock set against the horizontal backdrop of the shore made him stop and wonder: 'Suddenly we stop, astonished, measuring and appreciating: a geometric phenomenon has appeared before our eyes: rocks erect like menhirs against the relentless horizontal of the sea and the meander of the beaches. By the magic of relationships here we are in the land of dreams.'[42] In 1937, photographing the rocky shoreline around Plougrescant, he repeatedly picked up incidents like this: the place of the right angle. In his lectures in 1929, he called this 'the place of all measures' (le lieu de toutes les mesures).[43] This evolved into an elaborate mythology of cosmic dimensions: male and female, sun and moon, active and passive, all crystallised in the sign of the right angle, which became his trade mark.

> you engage in a pact of solidarity with nature: this is the place of the right angle. Upright before the sea, there you are, vertical on your own two feet.[44]

In one of his lectures he let slip that the word *mer* could be spelled 'mer' (sea) or 'mère' (mother). Sea, mother, woman, nature, all are one. I'm not going to explore the erotic or alchemical aspects of all this, although Le Corbusier does so in great detail in the *Poème de l'angle droit*.[45]

It became a given for Le Corbusier that creation occurs when order (Nature, reason, the machine) and natural chaos (nature, sensuality, the vernacular) are fused in a perfect balance. To achieve this required a new kind of poet and prophet, the harmoniser. The outcome, the plans for la Ville Radieuse, can be criticised on a wide range of social, architectural and urban grounds, but it cannot be understood without comprehending the impossible juxtapositions it was meant to resolve: nature and the machine, free form and geometry, male and female and the violent conflicts of the cosmic forces of nature.

Notes

1   Benton, Tim. 'Le Corbusier E Il Vernacolare: Le Sextant a Les Mathes 1935'. In *Le case per artisti sull'isola Comacina*, ed. Andrea Canziani (Como: NodoLibri, 2010), 22–43. For Villa Savoye see www.fondationlecorbusier.fr/ corbuweb/morpheus.aspx?sysName= redirect64&sysLanguage=en-en&Iris ObjectId=7380&sysParentId=64, and for the Villa Le Sextant see www.fondationlecorbusier.fr/corbuweb/ morpheus.aspx?sysId=13&IrisObjectId= 5449&sysLanguage=en-en&itemPos= 63&itemSort=en-en_sort_string1%20 &itemCount=79&sysParentName= &sysParentId=64, accessed June 2018.

2   First presented in a lecture at the Weissenhof Siedlung in Stuttgart in July 1927 but best known in their publication in the first volume of the *Oeuvre complète*, eds. Le Corbusier, Pierre Jeanneret, Willy Boesiger, Oscar Stonorov, Max Bill, Le Corbusier, Le Corbusier, et al. (Zurich: Éditions Girsberger, 1930), 128).

3   See Tim Benton, 'La villa Mandrot i el lloc de la imaginacio', *Quaderns d'arquitectura i urbanisme* 163 (October, November, December, 1984): 36–47 (reprinted as Tim Benton, 'The Villa de Mandrot and the Place of the Imagination', in *Massilia*, ed. Michel Richard (Marseilles: Editions Imbernon, 2011), 92–105;

and C.C. Collins, Le Corbusier, et al., *La casa Errázuriz de Le Corbusier: encuentro entre dos culturas ficticias* (Santiago de Chile, Ediciones de la Escuela de Arquitectura Pontificia Universidad Católica de Chile, 1988).

4   See www.fondationlecorbusier.fr/corbuweb/morpheus.aspx?sysId=13&IrisObjectId=5415&sysLanguage=en-en&itemPos=66&itemCount=78&sysParentId=64&sysParentName=, accessed June 2018.

5   See note 3.

6   The earliest and still fundamental work on Le Corbusier's training in La Chaux-de-Fonds is Mary Patricia May Sekler, *The Early Drawings of Charles-Edouard Jeanneret (Le Corbusier) 1902–1908*, Outstanding Dissertations in the Fine Arts (New York: Garland Pub., 1977). See also Stanislaus von Moos and Arthur Rüegg, *Le Corbusier Before Le Corbusier: Applied Arts, Architecture, Painting, Photography, 1907–1922* (Baden; New Haven: Yale University Press, 2002).

7   D'Arcy Wentworth Thompson, *On Growth and Form* (Cambridge: Cambridge University Press, 1917); Matila Costiescu Ghyka and Paul Valéry, *Le Nombre d'or: rites et rythmes pythagoriciens dans le développement de la civilisation occidentale précédé d'une lettre de Paul Valéry*, second edition (Paris: Gallimard, 1931).

8   Le Corbusier, *L'Art décoratif d'aujourd'hui*, collection de 'l'esprit nouveau' (Paris: G. Crès, 1926).

9   Édouard Schuré, *Les Grands Initiés, esquisse de l'histoire secrète des religions*, (Paris: Perrin, 1889).

10 Le Corbusier, *Le Modulor: essai sur une mesure harmonique à l'échelle humaine applicable universellement à l'architecture et à la mécanique, Collection Ascoral* (Boulogne, Seine: Éditions de l'architecture d'aujourd'hui, 1950) and Le Corbusier, Modulor 2 1954 (La parole est aux usagers), suite de 'Le modulor' '1948', Collection Ascoral (Boulogne, Seine: Éditions de l'architecture d'aujourd'hui, 1955).

11 Comment on a set of sketches sent to Madame Meyer in October 1925 (FLC 31525). FLC = Fondation Le Corbusier, set up by Le Corbusier before he died and now the legal owner of his intellectual property rights.

12 For the original of the letter to Madam Meyer see www.fondationlecorbusier.fr/corbuweb/morpheus.aspx?sysId=13&IrisObjectId=6402&sysLanguage=en-en&itemPos=5&itemSort=en-en_sort_string1&itemCount=5&sysParentName=Home&sysParentId=11, accessed June 2018.

13 Le Corbusier, *Precisions on the Present State of Architecture and City Planning*, new edition of the 1991 English translation of the first edition, *Précisions* (1930) (Zurich: Park Books, 2015), 138.

14 Le Corbusier, *Precisions*, 139. The drawing was not one of those sketched in front of his audience, but created especially for the book in which he published the lectures.

15 Le Corbusier, *La main et le silex*, 1930 (FLC 101).

16 Le Corbusier, *La Ville radieuse: éléments d'une doctrine d'urbanisme pour l'équipement de la civilisation machiniste* (Paris, Genève, Rio de Janeiro, Sao Paolo, Montevideo, Buenos-Aires, Alger, Moscou, Anvers, Barcelone, Stockholm, Nemours, Piacé (Boulogne (Seine): Éditions de l'architecture d'aujourd'hui, 1935), facing title page. See www.fondationlecorbusier.fr/corbuweb/morpheus.aspx?sysId=13&IrisObjectId=6473&sysLanguage=en-en&itemPos=14&itemSort=en-en_sort_string1%20&itemCount=47&sysParentName=&sysParentId=25, accessed June 2018.

17 FLC B2(4)4 and B2(10)320.

18 'L'objectif et le subjectif sont les deux pôles entre lesquels surgit l'œuvre humain faite de matière et d'esprit'.

19 FLC B2(4)3.

20 Robert Fishman, *Urban Utopias in the Twentieth Century: Ebenezer Howard, Frank Lloyd Wright, and Le Corbusier* (New York: Basic Books, 1977); Mary Caroline McLeod, 'Urbanism and Utopia: Le Corbusier from Regional Syndicalism to Vichy' (Unpublished thesis, Princeton University, 1986) and M. Antliff, 'La Cité Française: Georges Valois, Le Corbusier, and Fascist Theories of Urbanism', in *Fascist Visions: Art and Ideology in France and Italy*, eds. Matthew Affron and Mark Antliff (Princeton: Princeton University Press, 1997), 134–70.

21 See www.fondationlecorbusier.fr/corbuweb/morpheus.aspx?sysName=redirect42&sysLanguage=fr-fr&IrisObjectId=8891&sysParentId=42, accessed June 2018.

22 See www.wikiart.org/en/le-corbusier/nature-morte-la-racine-et-au-cordage-jaune-1930, accessed June 2018.

23 See Tim Benton, *Le Corbusier, peintre à Cap Martin* (Paris: Éditions du Patrimoine, 2015), p. 71. See www.the guardian.com/artanddesign/2015/ may/02/eileen-gray-e1027-villa-cote-dazur-reopens-lost-legend-le-corbusier accessed June 2018.

24 See www.architectural-review.com/ rethink/reviews/the-secret-photography-of-le-corbusier/8654765.article, accessed June 2018.

25 'Le malade est un être sain dont le rhythme intérieur n'est plus en accord avec les rhythmes cosmiques qui l'entourent.'

26 'La terre, c'est-à-dire: influence centripète, la pesanteur; le soleil: l'influence centrifuge ascensionelle; la lune: influence tangentielle latérale. La force Terre alourdit et tasse l'homme de haut en bas. La force Soleil l'allège et l'étire de bas en haut. La force Lune l'assouplit et le disloque transversalement, agissant au point faible où l'os manque, aux articles.' P. Winter and M. Martiny, 'Les Grilles cosmiques de l'homme' *Plans*, 4 (April 1931), 43–44.

27 'Men engaged in organizing life.'

28 'Je cherche les sauvages, non pour y trouver la barbarie, mais pour y mesurer la sagesse. Amérique ou Europe, paysans ou pêcheurs. Je comprends: je vais là où des hommes pratiquent des travaux servant à les nourrir et prennent des initiatives dont l'effet est d'alléger leur peine. Ils font aussi ce qu'il faut pour obtenir sans frais ni dépense, les joies de la sociabilité: métier, famille, collectivité. Je mesure donc qu'étant architecte et urbaniste, je viens apprendre les choses de mon métier chez l'homme ou chez les hommes.' Le Corbusier, *La Ville radieuse,* 6.

29 FLC B2(4)7.

30 Le Corbusier, *Une Maison, un palais*, Collection de 'L'Esprit Nouveau' (Paris: Editions Connivances, 1989 [G. Cres, Paris, 1928]), 48–52.

31 For a moving description of these fishermen's huts on the shores of the Bassin d'Arcachon, near Bordeaux, see Le Corbusier, *Une Maison, un palais* (Turin: Bottega d'Erasmo, 1975 [G. Cres, Paris, 1928], 46.

32 See www.fondationlecorbusier.fr/ corbuweb/morpheus.aspx?sysId=13&Iris ObjectId=6462&sysLanguage=en-en& itemPos=4&itemSort=en-en_sort_string 1&itemCount=4&sysParentName= Home&sysParentId=11, accessed June 2018.

33 The original sketches are in Le Corbusier's Sketchbook, I, B6, 389 and 390 (redrawn later and included in Le Corbusier, *La Ville radieuse*, 29).

34 'La leçon, c'est l'extension du "dedans" vers le "dehors". C'est l'économie de la cellule et l'ampleur des dégagements. C'est la participation des éléments fondamentaux: soleil, ciel et verdure.' *La Ville radieuse*, 29.

35 In the English edition of *Precisions*, the phrase reads: 'set down in a corner of the beautiful Argentinian countryside': Le Corbusier, *Precisions*, 139.

36 *Précisions*, 78. The original drawing for this illustration to his lecture 'Architecture in everything, city planning in everything' in Buenos Aires in October 1929 survives in the Fondation Le Corbusier (FLC 33519).

37 'A la déclaration de la guerre, en 1939, ce jardin s'est trouvé privé de tous soins et j'ai décidé a cette époque de le laisser ainsi sans aucun engrais, sans aucune coupe de sécateur, sans aucun entretien de quoique ce soit, sans enlevage de feuilles mortes, etc … Pendant cette période de 1939–1956, le vent ou les oiseaux ont apporte des graines d'arbres: faux sycomore actuellement de 5 mètres de haut; cytise, 4 mètres de haut, les rosiers se sont transformés en églantines de 3 mètres de haut. Les buis ont particulièrement prospéré, la lavande aussi, le yucca extraordinairement. Les lilas fleurissent régulièrement au printemps mais avec une floraison très brève. En général les fleurs ont disparu sous l'effet de la mauvaise herbe (je n'ai jamais permis qu'on remue la terre). Le lierre a pousse énormément, etc. Etc. Il s'agit d'une terre devenue d'une pauvreté extrême (privée d'engrais) … Pouvez-vous me proposer des fleurs capables de tenir dans ces conditions? C'est une expérience que je voudrais faire …'. FLC H2(5)584.

38 FLC H2(5)584.

39 See www.architectural-review.com/ buildings/le-corbusier/the-classical-ideals-of-le-corbusier/8619974.article, accessed June 2018.

40 'On a dressé sur l'Acropole des temples qui sont d'une seule pensée et qui ont été ramassé autour d'eux le paysage désolé et l'ont assujetti à la composition': Le Corbusier, *Vers une architecture* (Paris: G. Crès et Cie., 1923), 166.

41 'Dominant les golfes et les vallées,
à Delphes, ces trois dés de pierre,
témoignages violents et purs, parlent
du sublime': *Une Maison, un palais*,
14.

42 'D'un coup nous nous arrêtons,
saisis, mesurant, appréciant: un
phénomène géométrique se développe
sous nos yeux: roches debout comme
des menhirs, horizontale indubitable
de la mer, méandre des plages. Et par
la magie des rapports, nous voici
au pays des songes': *Une Maison,
un palais*, 22.

43 *Précisions*, 76 and 77. The original
drawing for the sketch which he made to

illustrate this anecdote is in the Fondation
Le Corbusier (FLC 33505).

44 '… tu contractes avec la nature un pacte
de solidarité: c'est l'angle droit. Debout
devant la mer, vertical te voilà sur tes
jambes': Le Corbusier, *Poème de l'angle
droit: Lithographies originales [de l'auteur]*
(Paris, Tériade (impr. de Mourlot frères)
1955), 30–31.

45 See www.fondationlecorbusier.fr/
corbuweb/morpheus.aspx?sysId=
13&IrisObjectId=6474&sysLanguage=
fr-fr&itemPos=19&itemSort=fr-fr_
sort_string1+&itemCount=47&sys
ParentName=&sysParentId=25,
accessed June 2018.

Epilogue: Science after modernity

Frank A.J.L. James and Robert Bud

In the essays published in this volume, we have seen the centrality of science as knowledge, as practice and as a strong symbol of modernity. In many areas of human endeavour in the period from roughly 1890 to 1950, the interpretation of science was at the core. As the contributors to this book have demonstrated, it underpinned many aspects of contemporary literature, the visual arts, music, dance and at least parts of the humanities and the interpretation of religion. This conversation with science was undertaken, amongst other ways, through a new rich resource of scientific metaphor, through enhanced consciousness of engineering in architecture and design, for example in transport whether in the London Underground, aircraft, automobiles or the bicycle, and through mass communication, wireless and cinema.

In its association with modernity, or with the notion of 'being modern', science commanded enormous and widespread authority, buttressed by an understanding of its history. Whereas many of the references to innovations and to developments in engineering would today be identified as 'technology', that term was less used at the time, and science was the referent. At the same time, of course, as this volume has also shown, there was no agreement on the precise meanings either of 'modernity' or of 'science'.

Across the early years of the twentieth century, the degree of lay engagement with science was criticised as insufficient. Although in its most bitter and divisive form, the public debate about this presumed lack was a phenomenon that began in the mid-1950s, the attitudes expressed then had already been nurtured in the 1920s and 1930s. For instance, in a book about science education that was submitted as one of the first doctoral dissertations in the history of science in Britain, Dorothy Turner in 1927 referred to

That hypothetical individual, the average man, [who] pretends at least to know a little Latin and Greek and to be qualified to give an opinion on modern poetry, on music, and on painting. He would blush if he pronounced incorrectly a name in classical mythology or was ignorant of the scores of the latest test match. Yet this individual would not mind expressing a complete ignorance of the simplest facts of science.[1]

Turner was however writing at a moment when it seemed that 'progress' would, if all-too slowly, lead to the passing of such attitudes. She certainly looked ahead to another time. Such criticism was itself an indication of common assumptions. In 1929, Julian Huxley and a diverse group of friends founded a new magazine which it was hoped would change culture in Britain. The launch issue of their new vehicle, *The Realist*, announced, 'We stand for making the specialist understood, for introducing the laboratorist [*sic*], who has lived too long with symbols, to letters, and so giving him that important public which has no time to listen to a man who cannot express himself'.[2] The very first article chosen to express this ambition addressed 'the progress of the novel' by one of Britain's most popular writers, Arnold Bennett. Held over lugubrious meals in private rooms at distinguished restaurants, the meetings of the editorial board were themselves significant cultural events. The board included such protagonists for science as *Nature* editor Richard Gregory, physiologist J.B.S. Haldane and Francis Freeth, head of research at the giant chemical cartel, ICI. Other members were writers such as Bennett, Aldous Huxley, Rebecca West and H.G. Wells and cultural critics such as Herbert Read (who would be a co-founder of London's Institute of Contemporary Arts immediately after the Second World War) and the architect Frank Baines responsible for the magnificent modernist building housing ICI as well as academics such as political scientist George Catlin (husband to Vera Brittain), anthropologist Bronislaw Malinowski and historian Eileen Power.[3]

Although the funding, from ICI chairman Sir Alfred Mond, ended in the aftermath of the Wall Street Crash and the magazine went through just three volumes, its significance was considerable with an important legacy in the science series that coincidentally emerged just at the time *The Realist* came to its early death. Thus, the BBC's lead science broadcaster, Gerald Heard, had been *The Realist*'s literary editor and Huxley had important roles in both initiatives. The magazine's intellectual liveliness and the reputation of its supporters testified to the recognition of the issues it addressed. The articles in this book show their more general purchase on intellectual imagination.

The modernities described in this book were circumscribed by space. The references applied to Britain and most of Europe and America as was shown by the world fairs of the time.[4] They applied even to the newly communist Soviet Union. But there were boundaries. When the biologist J.B.S. Haldane denounced the policy of British governments in an article in *The Realist*, he opened with the sentence: 'Western civilisation rests on applied science.'[5] Faith in 'Western' modernity defined by science was also circumscribed by time. Despite the fearful experience of wartime, science continued to retain its authority in Britain after the Second World War. The retention of scientists in positions of high government authority after 1945 and science's key place in the 1951 Festival of Britain both testify to the widespread high regard in which science and its practitioners had continued to be held.[6] By the end of the 1960s, however, that authority had begun to be challenged from many directions, for instance by the environmental movement and the counter culture, by the association of 'science' with the devastating effects of weapons of mass destruction, questioning of such scientific *grands projets* as civil nuclear power, and a general suspicion of authority in any guise.[7] This transition has often been identified as the transition from 'modernity' to 'post-modernity'.

The process of transformation out of the era described in this book was however more complex than the emergence of 'post-modernism'. Reflecting on the United States, Berman in his classic *All that is Solid Melts into Air* identified a post-Second World War 'splitting-off of modernism from modernisation'.[8] Replacing the dialectic of pre-war years, he argued that modernist writers of the 1950s, unlike their pre-war predecessors, ignored their current modern environment. Correspondingly he found that such an apostle of 'modernity' as Robert Moses came to be divorced from the concerns of cultural leaders. There was therefore in Berman's view a division between the elites whose mutual engagement before the war had been so characteristic. In Britain, one can find close parallels in the anxious discussion that seemed to offer both significant agency and a vocabulary to express the split of 'science' and 'culture', the so-called 'Two Cultures' debate.[9]

This followed from the May 1959 Cambridge University Rede lecture of that title delivered by the former chemist turned civil servant and novelist, Charles Snow.[10] Much has been written about this, doubtless to be continued, but we want to conclude this book by briefly reflecting on the role of science and modernity in the debate, its historical roots and how it helped shape the current place where science finds itself in contemporary culture.[11] This has been portrayed as part of an ahistorical continuum; however, in recent years Guy Ortolano has emphasised its historicity and specific location at the end of the 1950s. He suggests that Snow can be compared to slightly younger American contemporaries who founded the journal *Commentary* and would later be described as 'Neoconservatives'. Ortolano's work provides a basis for understanding the ending in Britain of the intellectual engagement described in this book.

Culturally, Snow had been formed between the wars. Trained as a chemist, he had turned to writing and sought to span Culture as he saw it. His biographer, Lawrence de la Mothe, has portrayed him as an 'heroic' figure of modernity.[12] We shall focus upon his particular concerns and qualities, yet the debate he provoked was so influential and widely discussed that its significance transcended him. In August 1959 *Science* in the USA chose for its week's editorial an extract from the work of Snow headed 'Two Cultures'.[13] For the years between the giving of the lecture in 1959 and 1970 alone, 'Google Scholar' reports 1560 works referring to the usage (as indicated by the appearance jointly of 'Snow' and 'Two Cultures'). By 2017 that number had risen to 17,400![14]

Writing thirty years after Turner, Snow sounded less optimistic than his predecessor in her 1920s denunciation of popular culture. Though with little hope, he argued that a properly educated person should know both about science and literature and proposed what would be, notoriously, a literacy test of knowing both a play of Shakespeare and the second law of thermodynamics (though he later changed this to knowing about DNA). Possessing such wide knowledge was important for the state since, Snow believed, the higher echelons of government and the civil service were commanded overwhelmingly by literary intellectuals who were running Britain without the necessary scientific and technical knowledge so to do.

Observing Snow's career, we see both that the expression of his personal views changed over time and that the intellectual environment had evolved. Whereas, through the earlier years of Snow's life, science had been important in cultural debate, in his Rede lecture he passed over that period in silence. His early novel *The Search* published in 1932 had been reprinted in 1958 with a new preface.[15] At the heart of the novel was a character, Constantine, bearing a close relationship to the polymath X-ray crystallographer J.D. Bernal, in real life a close friend of such artists as Picasso and who also introduced a catalogue of the work of sculptor Barbara Hepworth. In the new preface however, Snow was more pessimistic about the chance of real integration of culture. Instead, he argued that the mathematics at the heart of science was impenetrable to a large number of people whose intelligence was not suited to it and whose science education should therefore not require it. In his lecture, he downplayed the integration of science within government, for instance ignoring the significant roles of civil service scientists, of which, as a former Civil Service Commissioner with special responsibility for scientific recruitment, he would have been aware. By the late 1950s Snow was an idealistic spokesman of a group whose most famous expression was Harold Wilson's 1963 speech promising that the Labour Party would make Britain ready to be forged in the 'white-heat' of the scientific revolution.

It might have been expected that a technocrat such as Snow, the self-appointed voice of the scientific community, would have believed that science and modernism would go well together. Yet in a recent paper Ortolano draws attention to Snow's attack on modernist authors active in the period 1914 to 1950.[16] Specifically referring to Yeats, Pound and Wyndham Lewis as examples of 'nine out of ten of those who have dominated literary sensibility in our time', Snow asked whether they were 'not only politically silly, but politically wicked? Didn't the influence of all they represent bring Auschwitz that much nearer?'[17]

Curiously, in his 1954 novel *The New Men* set during the recent world war, Snow's main protagonist (and alter ego) was Lewis Eliot, a legally trained civil servant working on the atomic bomb programme. When considering during 1943 a document calculating the casualties that such a weapon would cause, Eliot reflected that 'there must exist *memoranda* about concentration camps: people must be writing their views on the effects of a reduction in rations, comparing the death rate this year with last'.[18] Snow was thus entirely familiar with the connections between Fascism and technocratic, amoral, approaches to management.

By 1959 Snow had decided to omit any reference linking Fascism to science and engineering, but to concentrate instead on what he took to be the reactionary nature of modernist literature. Where Berman cited Robert Moses's rejection of modernist aesthetics in his post-war town planning, Snow, who would be one of Britain's first ministers for technology, denounced modern writers for their scorn for industry. Since many occupied positions of power both in government and civil service, they were, Snow claimed, unlikely to implement the ideas that he considered necessary for a better world. One of these ideas was that since industrialisation was essential to end world poverty, science and a large supply of trained scientists were thus crucial to economic development and prosperity. By 1959, he had concluded that industry, with significant scientific input, was the only hope for the world's poor. Thus, those literary intellectuals who raised problems about industrialisation needed to be criticised in turn. Referring to the social effects of the so-called 'industrial revolution', Snow commented:

> Plenty of them [writers] shuddered away, as though the right course for a man of feeling was to contract out; some, like Ruskin and William Morris and Thoreau and Emerson and Lawrence, tried various kinds of fancies which were not in effect more than screams of horror.[19]

He then proceeded to outline the benefits that industrialisation had already brought to the poor.

Ortolano suggests that Snow's expression of the two cultures articulated a division of liberalism. Snow's opponent Leavis had respect for science too, as an intellectual endeavour, but was furiously opposed to a technocratic 'modern civilisation' unchallenged by humanistic criticism.[20] Curiously, then, the opposition between Snow and Leavis was based more on what they stood for in the late 1950s, than their own historic views.

In his Rede lecture, Snow omitted his own past dismissal of the role of scientists in industry.[21] Like Leavis, historically his respect had been for the 'high science' of the Cavendish Laboratory in Cambridge which produced basic physical knowledge.[22] He was not interested in, for example, the large number of scientists who worked in the numerous agricultural and horticultural research stations spread throughout the country.[23] Those men and women were far more typical of the scientific community than Cavendish physicists and doubtless contributed to how science was generally perceived as a modern practice. One suspects that Snow would not have found them or their workplaces sufficiently glamorous, since those laboratories were not, like the Cavendish, 'stiff with Nobel Prize winners'.[24]

Snow's intellectual descendants, still to be found in areas of science journalism, have kept up his line of attack as Simon Lock has pointed out. These journalists sustain Snow's hostility to literary intellectuals referring, presumably as some sort of joke, to publications such as Vita Sackville West's posthumous letters or Lawrence's annotated laundry lists.[25] With equal lack of humour, the former is an impossibility and no one, apart from those journalists, seem ever to have seen the latter.

Part of Snow's problem (and one that he seems to have shared with Leavis) was the belief that literature and culture were the same thing. Snow hardly ever expressed any interest in the visual arts, music or architecture. His personal antipathy towards literary modernism and his indifference to other arts goes a long way to explain why he was not interested in the multiple interconnections that he could have observed during his lifetime. Snow's simplistic dialectic, highly memorable and easily deployed in argument, has contributed significant damage to dividing culture into unnecessarily fractious camps.

Neither science nor culture, however, need be rendered as impenetrable, unitary, black boxes. The essays in this volume have provided strong, and we hope, compelling evidence, that such a regrettable position is not innate or essential to the nature of culture. Diverse papers presented here have shown how engagement with the contested boundaries within and around science proved productive. Calls are now heard once again for the better integration of the arts with science, technology, engineering and medicine, indeed for 'STEAM' to join 'STEM' in the litany of acronyms. This book has shown how in the still-recent past, diverse visions of modernity were creatively interwoven.

Notes

1   Dorothy M. Turner, *History of Science Teaching in England* (London: Chapman and Hall, 1927),190.
2   *The Realist: A Journal of Scientific Humanism* 1, no. 1 (1929): 179.
3   Robert Bud, '"The Spark Gap is Mightier than the Pen": The Promotion of an Ideology of Science in the Early 1930s', *Journal of Political Ideologies* 22, no. 2 (2017): 169–181.
4   Robert H. Kargon, Karen Fiss, Morris Low and Arthur P. Molella, *World's Fairs on the Eve of War: Science Technology & Modernity, 1937–1942* (Pittsburgh, PA: University of Pittsburgh Press, 2015).
5   J.B.S. Haldane, 'The Place of Science in Western Civilisation', *The Realist* 2 (November 1929), 49–164. For more on attitudes to 'Western' civilisation and anxieties about it in Britain, see R.J. Overy, *The Morbid Age: Britain between the Wars* (London: Allen Lane, 2009).
6   On science at the Festival of Britain, see S. Forgan, 'Festivals of Science and the Two Cultures: Science, Design and Display in the Festival of Britain, 1951', *British Journal for the History of Science* 31, no. 109 (1998): 217–240. Patterns in the overall visibility and evaluation of science in the press have been studied by

Martin Bauer and his colleagues. See for example, 'Public Attention to Science 1820–2010 – a 'Longue Durée' Picture', in *The Sciences' Media Connection – Communication to the Public and its Repercussions*, eds. S. Rödder, M. Franzen and Peter Weingart, *Sociology of the Sciences Yearbook* (Dordrecht: Springer, 2011), 1; Martin W. Bauer et al., 'Long-Term Trends in the Public Representation of Science across the "Iron Curtain": 1946–1995', *Social Studies of Science* 36, no. 1 (1 February 2006): 99–131.
7   Rupert Cole, 'The Common Culture: Promoting Science at the Royal Institution in post-war Britain' (PhD thesis, University College London, 2017). For the more recent period, see the broadcast Reith Lectures, Onora O'Neill, *A Question of Trust* (Cambridge: Cambridge University Press, 2002).
8   Marshall Berman, *All That is Solid Melts into Air: The Experience of Modernity* (New York: Simon and Schuster, 1982), 309.
9   For the origin of the term, see Rupert Cole, 'A Tale of Two Train Journeys: Lawrence Bragg, C.P. Snow and the "Two Cultures", *Interdisciplinary Science Reviews*

41 (2016): 133–147. Snow initially used it as a title of a 1956 *New Statesman* article.

10 C.P. Snow, *The Two Cultures and the Scientific Revolution* (Cambridge: Cambridge University Press, 1959). There are many later printings and editions, the best being the 1993 Canto edition with an introduction by Stefan Collini.

11 On the rich existing literature, see, for example, Guy Ortolano, *The Two Cultures Controversy: Science, Literature and Cultural Politics in Postwar Britain* (Cambridge: Cambridge University Press, 2009).

12 John de la Mothe, *C.P. Snow and the Struggle of Modernity* (Austin: University of Texas Press, 1982).

13 C.P. Snow, 'Two Cultures', *Science* 130 (21 August 1959): 419.

14 'Google Scholar', consulted September 2017. For the immediate response, see 'The Two Cultures. A Discussion of C. P. Snow's Views', *Encounter* 13 (1959): 67–73. Also see the discussion by Collini in his 'Introduction' to the 1993 Canto edition, 27–28.

15 C.P. Snow, 'A Note', *The Search* (New York: Charles Scribner's Sons, 1959), np.

16 Guy Ortolano, 'Breaking Ranks: C.P. Snow and the Crisis of Mid-Century Liberalism, 1930–1980', *Interdisciplinary Science Reviews* 41 (2016): 118–132.

17 Snow, *Two Cultures*, 8.

18 C.P. Snow, *The New Men* (London, MacMillan, 1954), Chapter 19.

19 Snow, *Two Cultures*, 26.

20 See Collini, 'Introduction' to Leavis, *Two Cultures?*, 27–28.

21 Frank A.J.L. James, 'Introduction: Some Significances of the Two Cultures Debate', *Interdisciplinary Science Reviews* 41 (2016): 107–117, 111–112.

22 See C.P. Snow, *The Physicists* (London: Macmillan, 1981).

23 Paul Smith, *The Development of Horticultural Science in England, 1910–1930* (Unpublished PhD thesis, University College London, 2016).

24 John Halperin, *C.P. Snow: An Oral Biography* (Brighton: Harvester, 1983), 21.

25 Simon J. Lock, 'Cultures of Incomprehension? The Legacy of the Two Cultures Debate at the End of the Twentieth Century', *Interdisciplinary Science Reviews* 41 (2016): 148–166.

Select bibliography

Adams, Mary Grace Agnes. *Science in the Changing World*. London: George Allen & Unwin, 1933.

Adrian, E.D. *The Basis of Sensation: The Action of the Sense Organs*. London: Christophers, 1928.

Akin, William E. *Technocracy and the American Dream: The Technocrat Movement, 1900–1941*. Berkeley: University of California Press, 1977.

Albright, Daniel. *Quantum Poetics: Yeats, Pound Eliot, and the Science of Modernism*. Cambridge: Cambridge University Press, 1997.

Anduaga, A. *Wireless & Empire: Geopolitics, Radio Industry & Ionosphere in the British Empire, 1918–1939*. Oxford: Oxford University Press, 2009.

Antliff, Mark, and Patricia Leighten, eds. *A Cubism Reader: Documents and Criticism, 1906–1914*, translated by Jane Marie Todd, Jason Gaiger, Lydia Cochrane, Lois Parkinson Zamora and Ivana Horacek. Chicago: University of Chicago Press, 2008.

Arceneaux, N. 'The Wireless in the Window: Department Stores and Radio Retailing in the 1920s', *Journalism & Mass Communication Quarterly* 83 (2006): 581–595.

Ardis, Ann L. *Modernism and Cultural Conflict, 1880–1922*. Cambridge: Cambridge University Press, 2002.

Armstrong, Tim. *Modernism: A Cultural History*. Cambridge: Polity, 2005.

Arpad, Oliver, Istvan Botar and Isabel Wünsche. *Biocentrism and Modernism*. Aldershot: Ashgate Publishing, Ltd., 2011.

Ash, Mitchell G. 'Multiple Modernisms? Episodes from the Sciences as Cultures, 1900–1945'. In *Jewish Musical Modernism, Old and New*, edited by Philip V. Bohlman, 31–54. Chicago: University of Chicago Press, 2008.

Avery, T. *Radio Modernism: Literature, Ethics and the BBC*. Aldershot: Ashgate, 2006.

Baade, Christina. *Victory Through Harmony: The BBC and Popular Music in World War II*. New York: Oxford University Press, 2012.

Baader, Gerhard, Veronika Hofer and Thomas Mayer, eds. *Eugenik in Österreich. Biopolitische Strukturen von 1900 bis 1945*. Vienna: Czernin-Verlag, 2007.

Baker, Malcolm, and Brenda Richardson, eds. *A Grand Design: The Art of the Victoria and Albert Museum*. New York: Harry N. Abrams, 1997.

Baker, Robert S. 'Aldous Huxley: History and Science between the Wars', *Clio* 25 (1996): 293–300.

Bauer, Martin W., et al. 'Public Attention to Science 1820–2010 – a 'Longue Durée' Picture'. In *The Sciences' Media Connection – Communication to the Public and its Repercussions*, edited by S. Rödder, M. Franzen and Peter Weingart, *Sociology of the Sciences Yearbook*. Dordrecht: Springer, 2011.

Bauman, Zygmunt. *Modernity and the Holocaust*. Cambridge: Polity Press, 1989.

Beard, Mary. 'Cast: Between Art and Science'. In *Les moulages de sculptures antiques et l'histoire de l'archéologie*, edited by Henri Lavagne and François Queyrel, 157–166. Geneva: Droz, 2000.

Beer, Gillian. *Open Fields: Science in Cultural Encounter*. Oxford: Oxford University Press, 1996.

Beer, Gillian. '"Wireless": Popular Physics, Radio and Modernism'. In *Cultural Babbage: Technology, Time and Invention*, edited by F. Spufford and J. Uglow, 149–166. London: Faber and Faber, 1996.

Beers, L. and G. Thomas. *Brave New World: Imperial and Democratic Nation-Building in Britain between the War*. London: Institute of Historical Studies, 2012.

Bell, Ian. F.A. *Critic as Scientist: The Modernist Poetics of Ezra Pound*. London; New York: Methuen, 1981.

Beller, Steven, ed. *Rethinking Vienna 1900*. Oxford and New York: Berghahn, 2001.

Bennett, Jane. *Vibrant Matter: A Political Ecology of Things*. Durham: Duke University Press, 2010.

Bentley, Michael. *The Life and Thought of Herbert Butterfield: History, Science and God*. Cambridge: Cambridge University Press, 2011.

Benton, Tim. *The Villas of Le Corbusier and Pierre Jeanneret, 1920–1930*, rev. edn. Basel: Birkhauser, 2007.

Benton, Tim. *Le Corbusier, peintre à Cap Martin*. Paris: Éditions du Patrimoine, 2015.

Bergson, Henri. *Creative Evolution*, translated by Arthur Mitchell. London: University Press of America, 1983 [1911].

Berman, Marshall. *All That is Solid Melts into Air: The Experience of Modernity*. New York: Simon and Schuster, 1982.

Bijsterveld, K. *Mechanical Sound: Technology, Culture, and Public Problems of Noise in the Twentieth Century*. Cambridge: MIT Press, 2008.

Birkhoff, Garrett, and M.K. Bennett. 'Felix Klein and his "Erlanger Programm"', in *History and Philosophy of Modern Mathematics*, edited by William Aspray and Philip Kitcher, 145–176. Minneapolis: University of Minnesota Press, Minneapolis, 1988.

Birksted, Jan. *Le Corbusier and the Occult*. Cambridge, MA.: MIT Press, 2009.

Bland, Lucy, and Lesley A. Hall. 'Eugenics in Britain: The View from the Metropole'. In *The Oxford Handbook of the History of Eugenics*, edited by Alison Bashford and Philippa Levine. Oxford: Oxford University Press, 2010.

Blom, Philip. *Der taumelnde Kontinent: Europa 1900–1914*, translated by MGA, seventh edition. Munich: deutsche taschenbuchverlag, 2015 [2011].

Blumenberg, Hans. *Paradigms for a Metaphorology*, translated by Robert Ian Savage. Ithaca: Cornell University Press, 2010.

Boon, T.M. '"The Shell of a Prosperous Age": History, Landscape and the Modern in Paul Rotha's *The Face of Britain* (1935)'. In *Regenerating England: Science, Medicine and Culture in the Interwar Years*, edited by C.J. Lawrence and Anna Mayer, *Clio Medica* vol. 60, 107–48. Amsterdam: Rodopi, 2000.

Botar, Oliver A. J. and Isabel Wunsche, eds. *Biocentrism and Modernism*. Farnham: Ashgate, 2011.

Bowler, Peter J. *Reconciling Science and Religion: The Debate in Early Twentieth Century Britain*. Chicago: University of Chicago Press, 2010, first edition 2001.

Bowler, Peter J. *Science for All: The Popularization of Science in Early Twentieth-Century Britain*. Chicago: University of Chicago Press, 2009.

Bradshaw, D. 'The Best of Companions: J.W.N. Sullivan, Aldous Huxley, and the New Physics', *Review of English Studies* 47 (1996): 188–206 and 352–368.

Bramwell, Anna. *Ecology in the 20th Century: A History*. New Haven and London: Yale University Press, 1989.

Brennan, E.J.T., ed. *Education for National Efficiency: The Contribution of Sidney and Beatrice Webb*. London: Athlone Press, 1975.

Brey, Philip. 'Theorizing Modernity and Technology'. In *Modernity and Technology*, edited by Thomas J. Misa, Philip Brey and Andrew Feenberg, 33–72. Cambridge: MIT Press, 2003.

Briggs, A. *The History of Broadcasting in the United Kingdom*. Three volumes. Oxford: Oxford University Press, 1961–1965.

Brown, K.D. *Factory of Dreams: A History of Meccano Ltd, 1901–1979*. Lancaster: Crucible, 2007.

Brückweh, Kerstin, Dirk Schumann, Richard F. Wetzall, and Benjamin Ziemann, eds. *Engineering Society: The Role of the Human and Social Sciences in Modern Societies*. Basingstoke: Palgrave Macmillan, 2012.

Bruyninckx, Joeri. 'Sound Science. Recording and Listening in the Biology of Bird Song, 1880–1980'. Unpublished PhD thesis, University of Maastricht, 2013.

Bud, Robert. '"The Spark Gap is Mightier than the Pen": The Promotion of an Ideology of Science in the Early 1930s', *Journal of Political Ideologies* 22 (2017): 169–181.

Bunker, J. *From Rattle to Radio*. Studley: K.A.F. Brewin Books, 1988.

Burnham, Jack. *Beyond Modern Sculpture: The Effects of Science and Technology on the Sculpture of This Century*. New York: George Braziller, 1968.

Burstein, Jessica. *Cold Modernism: Literature, Fashion, Art*. University Park, PA: Pennsylvania State University Press, 2012.

Bussey, G. *Wireless: The Crucial Decade: History of the British Wireless Industry 1924–34*. London: Peter Peregrinus, 1990.

Caddy, Davinia. *The Ballets Russes and Beyond: Music and Dance in Belle-Epoque Paris*. Cambridge, Cambridge University Press, 2012.

Campbell, T.C. *Wireless Writing in the Age of Marconi*. Minneapolis: University of Minnesota Press, 2006.

Carey, James W. *Communication as Culture, Revised Edition, Essays on Media and Society*. London: Routledge, 2009.

Carpenter, Humphrey. *The Inklings: C.S. Lewis, J.R.R. Tolkien, Charles Williams and Their Friends*. London: HarperCollins, 2006.

Cheng, John. *Astounding Wonder: Imagining Science and Science Fiction in Interwar America*. Philadelphia: University of Pennsylvania Press, 2012.

Chessa, Luciano, and Luigi Russolo. *Futurist: Noise, Visual Arts, and the Occult*. Berkeley: University of California Press, 2012.

Clair, René. 'The Art of Sound'. In *Film Sound: Theory and Practice*, edited by Elisabeth Weis and John Belton. New York: Columbia University Press, 1985.

Clarke, A. *Childhood Ends: The Earliest Writing of Arthur C. Clarke*. Rochester, Michigan: Portentous Press, 1996.

Clarke, I. 'Negotiating Progress: Promoting "Modern" Physics in Britain, 1900–1940'. Unpublished PhD thesis, University of Manchester, 2012.

Clarke, I. 'How to Manage a Revolution: Isaac Newton in the Early Twentieth Century', *Notes and Records of the Royal Society* 68 (2014): 323–337.

Coen, Deborah R. *Vienna in the Age of Uncertainty: Science, Liberalism and Private Life*. Chicago: University of Chicago Press, 2007.

Cohen, Jean-Louis. 'Le Corbusier's Modulor and the debate on proportion in France', *Architectural Histories*, 2, no. 1: 23, 1–14 [DOI : http://dx.doi.org/10.5334/ah.by].

Cole, Rupert. 'A Tale of Two Train Journeys: Lawrence Bragg, C.P. Snow and the "Two Cultures"'. *Interdisciplinary Science Reviews* 41 (2016): 133–147.

Cole, Rupert. 'The Common Culture: Promoting Science at the Royal Institution in Post-War Britain'. Unpublished PhD thesis, University College London, 2017.

Cooke, Leighton Brett. 'Ancient and Modern Mathematics in Zamyatin's *We*'. *Zamyatin's We; A Collection of Critical Essays*, edited by Gary Kern, 149–167. Ann Arbor, MI.: Ardis, 1988.

Corry, Leo. 'How Useful is the Term "Modernism" for Understanding the History of Early Twentieth-Century Mathematics?' In *Science as Cultural Practice: Modernism in the Sciences, ca. 1900–1940*, edited by Moritz Epple and Falk Müller. Berlin: Akademie Verlag, (forthcoming).

Crawford, Robert. *Young Eliot: From St Louis to The Waste Land*. London: Jonathan Cape, 2015.

Crawford, T. Hugh. *Modernism, Medicine, and William Carlos Williams*. Norman; London: University of Oklahoma Press, 1995.

Crimp, Douglas. *On the Museum's Ruins*. Cambridge: MIT Press, 1993.

Crossley, Robert. 'The First Wellsians: A Modern Utopia and its Early Disciples', *English Literature in Transition, 1880–1920* 54 (2011): 444–469.

Cuddy-Keane, M. 'Virginia Woolf, Sound Technologies, and the New Aurality'. In *Virginia Woolf in the Age of Mechanical Reproduction*, edited by P.L. Caughie, 69–96. New York and London: Garland, 2000.

Daston, Lorraine, and Peter Galison, *Objectivity*. Cambridge: Zone Books, 2007.

Davison, Annette, and Julie Brown, eds. *The Sounds of the Silents in Britain*. New York: Oxford University Press, 2012.

de la Mothe, John. *C.P. Snow and the Struggle of Modernity*. Austin: University of Texas Press, 1982.

Denton, Peter H. *The ABC of Armageddon: Bertrand Russell on Science, Religion, and the Next War, 1919–1938*. Albany, NY: State University of New York Press, 2001.

Desmarais, Ralph John. '"Promoting Science": The BBC, Scientists, and the British Public, 1930–1945'. Unpublished MA dissertation IHR University of London, 2004.

Desmarais, Ralph John. 'Jacob Bronowski: A Humanist Intellectual for an Atomic Age, 1946–1956', *British Journal for the History of Science* 45 (2012): 573–589.

Dilthey, Wilhelm. *Wilhelm Dilthey: Selected Works, Volume I: Introduction to the Human Sciences*, edited by Rudolf A. Makkreel and Frithjof Rodi. Princeton: Princeton University Press, 1991.

Dolinger, Roland. 'Alfred Döblins Naturphilosophie in den Zwanziger Jahren'. In *Philosophia naturalis: Beiträge zu einer zeitgemässen Naturphilosophie*, edited by Thomas Arzt, Roland Albert Dollinger and Maria Hippius-Gräfen Dürckheim, 135–150. Würzburg: Königshausen & Neumann, 1996.

Donald, James, Anne Friedberg, and Laura Marcus. *Close Up 1927–1933: Cinema and Modernism*. Princeton: Princeton University Press, 1999.

Doxiadis, Apostolos, Christos H. Papadimitriou, Alecos Papadatos and Annie Di Donna. *Logicomix: An Epic Search for Truth*. London: Bloomsbury, 2009.

Drouin, Jeffrey S. *James Joyce, Science and Modernist Print Culture: 'The Einstein of English Fiction'*. New York: Routledge, 2015.

Eccles, J.C. 'Alexander Forbes and his Achievement in Electrophysiology', *Perspectives in Biology and Medicine* 13 (1970): 388–404.

Eliot, T.S. *The Complete Prose of T.S. Eliot: The Critical Edition: Apprentice Years, 1905–1918*, edited by Jewel Spears Brooker and Ronald Schuchard. Baltimore: The Johns Hopkins University Press, 2014.

Eliot, T.S. *The Complete Prose of T.S. Eliot: The Critical Edition: The Perfect Critic 1919–1926*, edited by Anthony Cuda. Baltimore: The Johns Hopkins University Press, 2014.

Emmitt, Robert Joseph. 'Scientific Humanism and Liberal Education: The Philosophy of Jacob Bronowski'. Unpublished PhD thesis, University of Southern California, 1982.

Epple, Moritz. 'Styles of Argumentation in Late 19th Century Geometry and the Structure of Mathematical Modernity'. *In Analysis and Synthesis in Mathematics: History and Philosophy*, edited by Michael Otte and Marco Panza, 177–198. Dordrecht: Kluwer, 1997.

Esposito, Mauritzio. *Romantic Biology, 1890–1945*. London: Routledge, 2015.

Falby, A. *Between the Pigeonholes: Gerald Heard, 1889–1971*. Newcastle: Cambridge Scholars Publishing, 2009.

Fawcett, R. *Nuclear Pursuits. The Scientific Biography of Wilfrid Bennett Lewis*. Montreal: McGill-Queen's University Press, 1994.

Feldman, M., E. Tonning and H. Mead, eds. *Broadcasting in the Modernist Era*. London: Bloomsbury, 2014.

Fischer, Gerhard. *Mathematische Modelle* (Plates) and *Mathematical Models: From the Collections of Universities and Museums*, 2 vols. Braunschweig: Vieweg, 1986.

Fishman, Robert. *Urban Utopias in the Twentieth Century: Ebenezer Howard, Frank Lloyd Wright, and Le Corbusier*. New York: Basic Books, 1977.

Forgan, S. 'Festivals of Science and the Two Cultures: Science, Design and Display in the Festival of Britain, 1951'. *British Journal for the History of Science* 31 (1998): 217–240.

Forgan, Sophie. 'From Modern Babylon to White City: Science, Technology and Urban Change in London, 1870–1914'. In *Urban Modernity: Cultural Innovation in the Second Industrial Revolution*, 75–132. Cambridge: MIT Press, 2010.

Forman, Paul. 'The Primacy of Science in Modernity, of Technology in Postmodernity, and of Ideology in the History of Technology,' *History and Technology* 23 (2007): 1–152.

Foucault, Michel. *Power/Knowledge: Selected Interviews and Other Writings 1972–1977*, edited by Colin Gordon. Hemel Hempstead: Harvester Wheatsheaf, 1980.

Galison, Peter. 'Aufbau/Bauhaus: Logical Positivism and Architectural Modernism'. *Critical Inquiry* 16 (1990): 709–752.

Galison, Peter L. *Einstein's Clocks and Poincaré's Maps: Empires of Time*. New York: W. W. Norton, 2003.

Ghyka, Matila Costiescu, and Paul Valéry. *Le Nombre d'or: rites et rythmes pythagoriciens dans le développement de la civilisation occidentale précédé d'une lettre de Paul Valéry*, second edition. Paris: Gallimard, 1931.

Giedion, Siegfried. *Space, Time and Architecture: The Growth of a New Tradition*. Cambridge, MA: Harvard University Press, 1941.

Giedeon, Siegfried. *Mechanization Takes Command: A Contribution to Anonymous History*. New York: Oxford University Press, 1948.

Gilbert, Scott F., and Sahotra Sarkar, 'Embracing Complexity: Organicism for the 21st Century'. *Developmental Dynamics* 219 (2000): 1–9.

Gillies, Mary Ann. *Henri Bergson and British Modernism*. Montreal, Que.: McGill-Queen's University Press, 1996.

Goble, M. *Beautiful Circuits: Modernism and the Mediated Life*. New York: Columbia University Press, 2010.

Gordon, Craig A. *Literary Modernism, Bioscience, and Community in Early 20th Century Britain*. New York: Palgrave Macmillan, 2007.

Gould, Stephen Jay. *Ontogeny and Phylogeny*. Cambridge: Harvard University Press, 1977.

Graham, Loren R. 'The Socio-Political Roots of Boris Hessen: Soviet Marxism and the History of Science', *Social Studies of Science* 15 (1985): 705–722.

Gray, Jeremy J. 'Modern Mathematics as a Cultural Phenomenon'. *The Architecture of Mathematics*, edited by José Ferreirós and Jeremy Gray, 371–396. Oxford: Oxford University Press, 2006.

Gray, Jeremy J. *Plato's Ghost; The Modernist Transformation of Mathematics*. Princeton and Oxford: Princeton University Press, 2008.

Greene, John C. 'The Interaction of Science and World View in Sir Julian Huxley's Evolutionary Biology', *Journal of the History of Biology* 23 (1990): 39–55.

Greenhalgh, Paul. *Ephemeral Vistas: The Expositions Universelles, Great Exhibitions and World's Fairs, 1851–1939*. Manchester University Press, Manchester, 1988.

Greenhalgh, Paul, ed. *Modernism in Design*. London: Reaktion Books, 1990.

Greenhalgh, Paul. *The Modern Ideal: The Rise and Collapse of Idealism in the Visual Arts from the Enlightenment to Postmodernism*. London: V&A Books, 2005.

Hahn, Hans. *Empiricism, Logic, and Mathematics: Philosophical Papers*, edited by Brian McGinness. Dordrecht: Reidel, 1980.

Hall, A. Rupert. *The Scientific Revolution 1500–1800: The Formation of the Modern Scientific Attitude*. London: Longmans, Green and Company, 1954.

Hall, A. Rupert. 'Historical Relations of Science and Technology', *Nature* 200 (1963): 1141–1145.

Halperin, John. *C.P. Snow: An Oral Biography*. Brighton: Harvester, 1983.

Haraway, Donna Jeanne. *Crystals, Fabrics and Fields: Metaphors of Organicism in Twentieth-Century Developmental Biology*. New Haven: Yale University Press, 1976.

Haring, K. *Ham Radio's Technical Culture*. Cambridge: MIT Press, 2007.

Harrington, Anne. *Reenchanted Science: Holism in German Culture from Wilhelm II to Hitler*. Princeton: Princeton University Press, 1996.

Harrod, Tanya. 'Introduction Craft, Modernism and Modernity', *Journal of Design History* 11 (1998): 1–4.

Hawkins, Thomas. 'The *Erlanger Programm* of Felix Klein: Reflections on its Place in the History of Mathematics', *Historia Mathematica* 11 (1984): 442–470.

Hennessy, B. *The Emergence of Broadcasting in Britain*. Lympstone: Southerleigh, 2005.

Henry, Holly. *Virginia Woolf and the Discourse of Science: The Aesthetics of Astronomy*. Cambridge: Cambridge University Press, 2003.

Hofer, Veronica. 'Rudolph Goldscheid, Paul Kammerer und die Biologen des Prater-Vivariums in der liberalen Volksbildung der Wiener Moderne'. In *Wissenschaft, Politik und Öffentlichkeit: von der Wiener Moderne bis zur Gegenwart*, edited by Mitchell G. Ash and Christian Stifter, 149–184. Vienna: Wiener Universitätsverlag, 2002.

Hogg, Jonathan. *Nuclear Culture: Official and Unofficial Narratives in the Long 20th Century*. London: Bloomsbury, 2016.

Hughes, J. '"Modernists with a Vengeance": Changing Cultures of Theory in Nuclear Science, 1920–1930', *Studies in History and Philosophy of Modern Physics* 29 (1998): 339–367.

Hughes, J. 'Craftsmanship and Social Service: W.H. Bragg and the Modern Royal Institution'. In *'The Common Purposes of Life': Essays on the History of the Royal Institution of Great Britain, 1799–1999*, edited by F.A.J.L. James, 225–247. Aldershot: Ashgate, 2002.

Hughes, J. 'Insects or Neutrons? Science News Values in Inter-War Britain'. In *Journalism, Science and Society: Science Communication between News and Public Relations*, edited by M.W. Bauer and M. Bucchi, 11–12. New York and Abingdon: Routledge, 2007.

Hughes, Robert. *The Shock of the New: Art and the Century of Change*. London: British Broadcasting Corporation, 1980.

Hughes, Thomas P. 'The Evolution of Large Technological Systems'. In *The Social Construction of Technical Systems: New Directions in the Sociology and History of Technology*, edited by Wiebe E. Beijker, Thomas P. Hughes and Trevor Pinch, 51–82. Cambridge: MIT Press, 1987.

Hull, A. 'War of Words: The Public Science of the British Scientific Community and the Origins of the Department of Scientific and Industrial Research, 1914–16', *British Journal for the History of Science* 32 (1999): 461–481.

Ihmig, Karl-Norbert. 'Ernst Cassirer and the Structural Conception of Objects in Modern Science: The Importance of the "Erlanger Programm"', *Science in Context* 12 (1999): 513–529.

Jacob, M.C. *Scientific Culture and the Making of the Industrial West*. Oxford: Oxford University Press, 1997.

James, F.A.J.L. 'Presidential Address: The Janus Face of Modernity: Michael Faraday in the Twentieth Century', *British Journal for the History of Science* 41 (2008): 477–516.

James, Frank A.J.L. 'Alfred Rupert Hall 1920–2009 Marie Boas Hall 1919–2009', *Biographical Memoirs of Fellows of the British Academy* 9 (2012): 353–408.

James, Frank A.J.L. 'Introduction: Some Significances of the Two Cultures Debate'. *Interdisciplinary Science Reviews* 41 (2016): 107–117.

Jenkin, John. *William and Lawrence Bragg, Father and Son: The Most Extraordinary Collaboration in Science.* Oxford: Oxford University Press, 2008.

Jodock, Darrell, ed. *Catholicism Contending with Modernity: Roman Catholic Modernism and Anti-Modernism in Historical Context.* Cambridge: Cambridge University Press, 2000.

Jolivette, Catherine, ed. *British Art in the Nuclear Age.* Farnham: Ashgate, 2014.

Jones, A.C. 'Mary Adams and the Producer's Role in Early BBC Science Broadcasts', *Public Understanding of Science* 21 (2012): 968–963.

Jones, A.C. 'Speaking of Science: BBC Science Broadcasting and its Critics, 1923–64'. Unpublished PhD thesis University College, London, 2010.

Juler, Edward. *Grown but Not Made: Biological Metaphor and Experimental Art in England, 1930–39.* Manchester University, 2008.

Kagan, Jerome. *The Three Cultures: Natural Sciences, Social Sciences, and the Humanities in the 21st Century.* Cambridge: Cambridge University Press, 2009.

Kahn, Joel S. *Modernity and Exclusion.* London: Sage, 2001.

Kargon, Robert H., Karen Fiss, Moriss Low and Arthur P. Molella. *World's Fairs on the Eve of War: Science Technology & Modernity, 1937–1942.* Pittsburgh, PA: University of Pittsburgh Press, 2015.

Kern, Stephen. *The Culture of Time and Space, 1880–1918: With a New Preface.* Harvard University Press, 1983.

Kidwell, Peggy. 'American Mathematics Viewed Objectively: The Case of Geometric Models'. In *Vita Mathematica: Historical Research and Integration with Teaching,* edited by Ronald Callinger, 197–208. Mathematical Association of America, 1996 [MAA Notes No. 40].

Kittler, Friedrich A. *Gramophone, Film, Typewriter,* translated by Geoffrey Winthrop-Young and Michael Wutz. Stanford: Stanford University Press, 1999 [1986].

Knoll, Peter M. and Hans Kelker. *Otto Lehmann. Erforscher der flüssigen Kristalle. Eine Biographie mit Briefen an Otto Lehmann.* Frankfurt: Ettlingen, 1988.

Koch, Ludwig. *Memoirs of a Birdman.* London: Phoenix House, 1955.

Lacey, Kate. *Listening Publics. The Politics and Experience of Listening in the Media Age.* Cambridge: Polity Press, 2013.

Lawrence, D.H. 'Life'. In *Reflections on the Death of a Porcupine and Other Essays,* edited by Michael Herbert. Cambridge: Cambridge University Press, 1988.

Lawrence, D.H. *Psychoanalysis and the Unconscious and Fantasia of the Unconscious.* Cambridge: Cambridge University Press, 2004 [orig. pub. 1921].

Le Corbusier. *Towards a New Architecture,* translated by Frederick Etchells. London: Architectural Press, 1927.

Le Corbusier. *Precisions on the Present State of Architecture and City Planning,* new edition of the 1991 English translation of the first edition, *Précisions.* Zurich: Park Books, 2015 [1930].

Ledebur, Sophie. *Das Wissen der Anstaltspsychiatrie in der Moderne: Zur Geschichte der Heil- und Pflegeanstalten Am Steinhof in Wien.* Vienna: Böhlau Verlag, 2015.

LeMahieu, D.L. *A Culture for Democracy: Mass Communication and the Cultivated Mind in Britain Between the Wars.* Oxford: Clarendon Press, 1988.

Lenoir, Timothy. *Instituting Science: The Cultural Production of Scientific Disciplines.* Stanford: Stanford University Press, 1997.

Lewis, Wyndham. *Time and Western Man,* edited by Paul Edwards. Santa Rosa: Black Sparrow Press, 1993.

Loach, Judi. 'Le Corbusier and the Creative Use of Mathematics', *British Journal for the History of Science* 31 (1998): 185–216.

Loach, Judi. 'Entre Valéry et Le Corbusier: la coquille et le Nombre d'or'. In *La Ville et la coquille: Huit essais d'emblématique,* edited by Paulette Choné, 109–133. Paris: Beauchesne, 2016.

Lock, Simon J. 'Cultures of Incomprehension? The Legacy of the Two Cultures Debate at the End of the Twentieth Century'. *Interdisciplinary Science Reviews* 41 (2016): 148–166.

Lofthouse, Richard A. *Vitalism in Modern Art, C. 1900–1950: Otto Dix, Stanley Spencer, Max Beckmann and Jacob Epstein.* Lewiston: Edwin Mellen Press, 2005.

Logan, Cheryl A. *Hormones, Heredity and Race: Spectacular Failure in Interwar Vienna.* New Brunswick, NJ: Rutgers University Press, 2013.

Lovell, B., and D.G. Hurst. 'Wilfrid Bennett Lewis', *Biographical Memoirs of Fellows* 34 (1988): 451–509.

Mabey, Richard. *Whistling in the Dark: In Pursuit of the Nightingale.* London: Sinclair Stevenson, 1993.

Macdonald, Helen. 'What Makes You a Scientist is the Way You Look at Things': Ornithology and the Observer 1930–1955', *Studies in History and Philosophy of Biological and Biomedical Sciences* 33 (2002): 55–56.

Mach, Ernst. *The Analysis of Sensations and the Relation of the Physical and the Psychical*, translated by C.M. Williams, revised and supplemented by Sydney Waterlow. New York: Dover Publications, 1959 [1886].

Mach, Ernst. *The Science of Mechanics: A Critical and Historical Account of Its Development*, sixth edition, translated by Thomas J. McCormack, 612–613. LaSalle, Ill.: Open Court Press, 1960.

MacLachlan, James. 'Experimenting in the History of Science', *ISIS* 89 (1998): 90–92.

Mansell, James. 'Rhythm, Modernity and the Politics of Sound'. In *The Projection of Britain: A History of the GPO Film Unit*, edited by S. Anthony and J. Mansell, 161–167. Basingstoke: BFI/Palgrave, 2011.

Mansell, James. *Hearing Modernity: Sound and Selfhood in Early Twentieth-Century Britain*. Urbana, IL: University of Illinois Press, 2016.

Mayer, Anna-K. 'Moralizing Science: The Uses of Science's Past in the 1920s', *British Journal for the History of Science*, 30 (1997): 51–70.

Mayer, Anna-K. '"A Combative Sense of Duty": Englishness and the Scientists'. In *Regenerating England: Science, Medicine and Culture in Inter-War Britain*, edited by Christopher Lawrence and Anna-K. Mayer, vol. 60, 67–106. Amsterdam: Rodopi, 2000.

Mayer, Anna-K. 'Setting Up a Discipline: Conflicting Agendas of the Cambridge History of Science Committee, 1936–1950', *Studies in the History and Philosophy of Science* 31 (2000): 665–689.

Mayer, Anna-K. 'Fatal Mutilations: Educationism and the British Background to the 1931 International Congress for the History of Science and Technology', *History of Science* 40 (2002): 445–472.

Mayer, Anna-K. 'Setting Up a Discipline, II: British History of Science and "The End of Ideology", 1931–1948', *Studies in the History and Philosophy of Science* 35 (2004): 41–72, 55–56.

Mayerhöfer, Josef. 'Ernst Mach in Wien', *Mitteilungen der Österreichischen Gesellschaft für Wissenschaftsgeschichte* 9 (1989): 19–42.

McAleer, Joseph. *Popular Reading and Publishing in Britain 1914–1950*. Oxford, 1992.

McDonald, Gail. *Learning to Be Modern: Pound, Eliot, and the American University*. Oxford: Clarendon Press, 1993.

Mehrtens, Herbert. *Moderne Sprache Mathematik: Eine Geschichte des Streits um die Grundlagen der Disziplin und des Subjekts formaler Systeme*. Frankfurt am Main: Suhrkamp, 1990.

Mehrtens, Herbert. 'Mathematical Models'. In *Models: The Third Dimension of Science*, edited by Soraya de Chadarevian and Nick Hopwood, 276–306. Stanford: Stanford University Press, 2004.

Meyer, Steven. *Irresistible Dictation: Gertrude Stein and the Correlations of Writing and Science*. Stanford: Stanford University Press, 2003.

Michaud, Eric. 'Esthétique scientifique de Charles Henry ou l'avènement du normal'. In *Vers la science de l'art: l'esthétique scientifique en France, 1857–1937*, edited by Jacqueline Lichtenstein et al., 133–144. Paris: Presses de l'Université Paris-Sorbonne, 2013.

Migirdicyan, Eva. 'Les modèles mathématiques dans l'art du XXe siècle'. Unpublished PhD thesis, Université de Paris Ouest Nanterre La Défence, 2011.

Misa, T., P. Brey and A. Feenberg, *Technology and Modernity*. Cambridge, MA.: MIT Press, 2003.

Moore, Jerry D. *Visions of Culture*. Lanham: Altamira, 2004.

Musil, Robert. 'The Mathematical Man'. In *Precision and Soul: Essays and Addresses*, edited and translated by Burton Pike and David S. Luft, 39–43. Chicago: University of Chicago Press, 1990 [1913].

O'Brien, Charles. *Cinema's Conversion to Sound: Technology and Film Style in France and the U.S.* Bloomington: Indiana University Press, 2004.

O'Neill, Onora. *A Question of Trust*. Cambridge: Cambridge University Press, 2002.

Ortolano, Guy. *The Two Cultures Controversy: Science, Literature and Cultural Politics in Postwar Britain*. Cambridge: Cambridge University Press, 2009.

Ortolano, Guy. 'Breaking Ranks: C.P. Snow and the Crisis of Mid-Century Liberalism, 1930–1980'. *Interdisciplinary Science Reviews* 41 (2016): 118–32.

Overy, Richard. *The Morbid Age: Britain and the Crisis of Civilization, 1919–1939*. London: Penguin, 2009.

Papapetros, Spyros. *On the Animation of the Inorganic: Art, Architecture, and the Extension of Life.* Chicago: University of Chicago Press, 2012.

Parkinson, G. *Surrealism, Art and Modern Science.* New Haven and London: Yale University Press, 2008.

Pegg, M. *Broadcasting and Society, 1918–1939.* London: Croom Helm, 1983.

Peppis, Paul. *Sciences of Modernism: Ethnography, Sexology, and Psychology.* Cambridge: Cambridge University Press, 2014.

Perloff, Marjorie. *The Futurist Moment: Avant-Garde, Avant Guerre, and the Language of Rupture.* Chicago: University of Chicago Press, 1986.

Poggi, Christine. *Inventing Futurism: The Art and Politics of Artificial Optimism.* Princeton: Princeton University Press, 2009.

Pollen, Annebella. *The Kindred of the Kibbo Kift: Intellectual Barbarians.* London: Donlon Books, 2015.

Pollen, Annebella. 'Utopian Futures and Imagined Pasts in the Ambivalent Modernism of the Kibbo Kift Kindred'. In *Utopia: The Avant-Garde, Modernism and (Im)Possible Life,* edited by David Ayers, Benedikt Hjartarson, Tomi Huttunen and Harri Veivo. Berlin: De Gruyter, 2015.

Price, Katie. *Loving Faster than Light: Romance and Readers in Einstein's Universe.* Chicago: University of Chicago Press, 2012.

Price, Katie. 'Gender and the Domestication of Wireless Technology in 1920s Pulp Fiction'. In *Domesticity in the Making of Modern Science,* edited by D. L. Opitz et al., 129–150. Basingstoke: Palgrave Macmillan, 2016.

Putnam, Tim. 'The Theory of Machine Design in the Second Industrial Age'. *Journal of Design History* 1 (1988): 25–34.

Pyenson, Lewis. *Neohumanism and the Persistence of Pure Mathematics in Wilhelmian Germany.* Philadelphia: American Philosophical Society, 1983.

Pyenson, Lewis. *The Young Einstein: The Advent of Relativity.* Bristol and Boston: Adam Hilger, 1985.

Pyenson, Lewis. 'Western Wind: The Atomic Bomb in American Memory', *Historia Scientiarum* 6 (1997): 231–241.

Pyenson, Lewis, and Susan Sheets-Pyenson. *Servants of Nature: A History of Scientific Institutions, Enterprises, and Sensibilities.* London: HarperCollins; New York: W.W. Norton, 1999.

Rajan, Tilottama. 'Organicism'. *English Studies in Canada* 30, no. 4 (2004): 46.

Richards, I.A. *Poetries and Sciences: A Reissue with a Commentary of Science and Poetry (1926, 1935).* New York: W.W. Norton and Company, 1970.

Rieger, B. *Technology and the Culture of Modernity in Britain and Germany, 1890–1945.* Cambridge: Cambridge University Press, 2005.

Roach, Rebecca, Laura Marcus, and David Bradshaw, eds. *Moving Modernisms: Motion, Technology and Modernity.* Oxford: Oxford University Press, 2016.

Ross, William T. *H.G. Wells's World Reborn: The Outline of History and its Companions.* Selinsgrove, PA: Susquehanna University Press, 2002.

Rotha, Paul. *Documentry Diary: An Informal History of the British Documentary Film, 1928–1939.* London: Secker and Warburg, 1973.

Rowe, David E. 'Mathematical Models as Artefacts for Research: Felix Klein and the Case of Kummer Surfaces', *Mathematische Semesterberichte* 60 (2013): 1–24.

Russell, Bertrand. *My Philosophical Development.* London and New York: Routledge, 1993.

Scales, R.P. *Radio and the Politics of Sound in Interwar France, 1921–1939.* Cambridge: Cambridge University Press, 2016.

Scannell, P. and D. Cardiff. *A Social History of British Broadcasting. Volume One 1922–1939. Serving the Nation.* Oxford: Basil Blackwell, 1991.

Schivelbusch, Wolfgang. *Disenchanted Night: The Industrialization of Light in the Nineteenth Century,* translated by Angela Davis. Berkeley: University of California Press, 1995.

Schoenberg, Arnold. 'Composition in Twelve Tones'. In *Style and Idea: Selected Writings of Arnold Schoenberg,* edited by Leonard Stein, 217–218. Berkeley: University of California Press, 1984.

Scott, David A., and Donna Stevens. 'Electrotypes in Science and Art', *Studies in Conservation* 58 (2013): 189–198.

Schaffer, Simon. 'How Disciplines Look'. In *Interdisciplinarity: Reconfigurations of the Natural and Social Sciences,* edited by Andrew Barry and Georgina Born. London: Routledge, 2013.

Shiach, Morag. *Modernism, Labour and Selfhood in British Literature and Culture, 1890–1930.* Cambridge: Cambridge University Press, 2004.

Shuttleworth, Sally. *George Eliot and Nineteenth-Century Science: The Make-Believe of a Beginning*. Cambridge: Cambridge University Press, 1984.

Sleigh, Charlotte. *Six Legs Better: A Cultural History of Myrmecology*. Baltimore, Johns Hopkins University Press, 2007.

Sleigh, Charlotte. 'Science as Heterotopia: The British Interplanetary Society before World War II'. In *Scientific Governance in Britain, 1914–1979*, edited by Don Leggett and Charlotte Sleigh. Manchester University Press, 2016.

Smith, Roger. 'Biology and Values in Interwar Britain: C.S. Sherrington, Julian Huxley and the Vision of Progress', *Past & Present* 178, no. 1 (1 February 2003): 210–242.

Snapper, Ernst. 'The Three Crises in Mathematics: Logicism, Intuitionism and Formalism'. *Mathematics Magazine* 52 (1979): 207–216.

Snow, C.P. *The Two Cultures and the Scientific Revolution*. Cambridge: Cambridge University Press, 1959.

Snow, C.P. *Variety of Men*. London: Macmillan, 1967.

Stadler, Friedrich. *The Vienna Circle: Studies in the Origins, Development and Influence of Logical Empiricism*, translated by S. Schmidt, J. Golb, T. Ernst and C. Nielsen. Vienna: Springer Verlag, 2001.

Staley, R. *Einstein's Generation: Origins of the Relativity Revolution*. Chicago: University of Chicago Press, 2009.

Stepan, Nancy. *The Idea of Race in Science: Great Britain, 1800–1960*. London: Macmillan, 1982.

Stichweh, Rudolph. 'The Sociology of Scientific Disciplines: On the Genesis and Stability of the Disciplinary Structure of Modern Science', *Science in Context* 5 (1992): 3–15.

Stöltzner, Michael. 'Vienna indeterminism: Mach, Boltzmann, Exner', *Synthèse* 119 (1999): 85–111.

Street, Sean. *The Poetry of Radio. The Colour of Sound*. Oxford: Routledge, 2012.

Terranova, Charissa N. and Tromble, Meredith. *The Routledge Companion to Biology in Art and Architecture*. London: Routledge, 2016.

Thompson, D'Arcy Wentworth. *On Growth and Form*. Cambridge: Cambridge University Press, 1917.

Thompson, Emily. *The Soundscape of Modernity: Architectural Acoustics and the Culture of Listening in America, 1900–1933*. Cambridge: MIT Press, 2004.

Topp, Leslie. *Architecture and Truth in Fin-de-Siècle Vienna*. Cambridge: Cambridge University Press, 2004.

Trotter, D. *Literature in the First Media Age: Britain Between the Wars*. Cambridge, MA; Harvard University Press, 2013.

Trotter, David. *Paranoid Modernism: Literary Experiment, Psychosis, and the Professionalization of English Society*. Oxford: Oxford University Press, 2001.

Tworek, H.J.S. 'The Savior of the Nation? Regulating Radio in the Interwar Period', *Journal of Policy History* 27 (2015): 465–491.

Volkert, Klaus Thomas. *Die Krise der Anschauung: Eine Studie zur formalen und heuristischen Verfahren in der Mathematik seit 1850*. Göttingen: Vandenhoeck & Ruprecht, 1986.

Walter, Christina. *Optical Impersonality: Science, Images, and Literary Modernism*. Baltimore: Johns Hopkins University Press, 2014.

Ward, F.A.B. 'Physics in Cambridge in the Late 1920s'. In *The Making of Physicists*, edited by R. Williamson. Bristol: Adam Hilger, 1987.

Washburn, Dorothy K., and Donald W. Crowe. *Symmetries of Culture: Theory and Practice of Plane Pattern Analysis*. Seattle: University of Washington Press, 1988.

Weart, Spencer. *Nuclear Fear*. Cambridge, MA: Harvard University Press, 1988.

Weingart, Peter. 'A Short History of Knowledge Formations'. In *The Oxford Handbook of Interdisciplinarity*, edited by Robert Frodeman, Julie Thompson Klein and Carl Mitcham. Oxford: Oxford University Press, 2010.

Weis, Elisabeth, and John Belton, eds. *Film Sound: Theory and Practice*. New York: Columbia University Press, 1985.

Whitehead, Alfred North. *Science and the Modern World*. New York: Free Press, 1967.

Whitworth, Michael. *Einstein's Wake: Relativity, Metaphor, and Modernist Literature*. Oxford: Oxford University Press, 2001.

Whitworth, Michael. 'Science in the Age of Modernism'. In *The Oxford Handbook of Modernisms*, edited by Peter Brooker, Andrzej Gasiorek, Deborah Longworth and Andrew Thacker, 445–460. Oxford: Oxford University Press, 2010.

Whitworth, Michael. 'Natural Science'. In *T.S. Eliot in Context*, edited by Jason Harding, 336–345. Cambridge: Cambridge University Press, 2011.

Wilson, Daniel. 'Machine Past, Machine Future: Technology in British Thought, C. 1870–1914'. Unpublished PhD thesis, Birkbeck University of London, 2010.

Zweiniger-Bargielowska, Ina. *Managing the Body: Beauty, Health and Fitness in Britain, 1880–1939*. Oxford: Oxford University Press, 2010.

Index

Badovici, Jean 378
Bahr, Hermann, 'The unsalvageable ego' 29
Baines, Frank 387
Balfour, Arthur 107
Ballets Russes 211
Barbanera, Marcello 190
Barfield, R.H. 257
Barnes, Ernest 132
Barney, Natalie Clifford 241n113
Barron, D.H. 266
Basch, Victor 216, 238n75, 238n76
  'Aesthetics and the Science of Art' 216
Basic English 159
Bassin d'Acachon 377, 378, 379–80
Bavarian National Museum 201n1
BBC (British Broadcasting Company) 387
  and development of the wireless 247, 248,
    251, 256, 260, 264, 267
  Koch's wartime broadcasts 294, 297–302
  radio as new medium 118–20
BBC Natural History Unit 293, 301–2,
  308n33
BBC Written Archive 294
Beaver, Jack 44, 55n23
Beer, Gillian 16, 187n57, 311, 325
  Darwin's Plots 4
Beers, Laura 247
Behrens, Peter 214
Belle Epoque 188, 196, 199
Beller, Steven, *Rethinking Vienna* 23
Belloc, Hilaire 118
Belton, John 41
Benjamin, Walter 358
Bennett, Arnold 387
Bénoni-Auran, Benoît 197
benzene 131
Bergius, Alfred 109
Bergriffsgeschichte ('conceptual history') 99
Bergson, Henri 318, 321, 341, 342, 343, 344,
  351, 354–5n15
  Creative Evolution 318
Berman, Marshall 390
  All that is Solid Melts into Air 388
Bernal, Desmond 139, 145n50
Bernal, J.D. 389
Bernard, Claude 317–18
Bertalanffy, Ludwig von 16, 341, 348, 349
  Problems of Life 346–7
Bicycle Thieves (film, 1948) 281
bicycles 15, 122n6, 275
  categorised as 'slow traffic' 282
  in cinema 281–2
  and commuting 279, 280–1, 287–9
  and gender mobility/freedom 277–8
  as green machine 285
  in literature 277, 278, 287n4
  and middle-class modernism 277, 278–9
  military connection 284–5, 289n26
Bijsterveld, Karin 41, 42
biocentrism 10, 318, 321–4
biogenetic law 327–9, 332–3n2
biology 16, 23, 346–7
Bird Watching and Bird Behaviour (radio
  programme) 297
birdsong 9, 15–16
  background 293–4

and citizenship through knowing 302–4,
  309n68
employed as a 'silence' 305–7
fighting for British nature 304–5
listening carefully to recordings 294–7
meanings/uses of Koch's broadcasts 302–7
nightingale broadcast 306, 310n86
wartime broadcasts for 'all classes of
  society' 297–302
Blackett, P.M.S. 119, 264
Blanc, Charles 375
Bland, Lucy 314
Bletchley Park 268
Blood of a Poet (film, 1930) 49
Bloomsbury group 66
Bloy, Léon 230
Blumenberg, Hans 99
Boas, Franz 330
  Primitive Art 330
Boccioni, Umberto 8, 13, 197–8
  Antigrazioso 199
  Development of a Bottle in Space 197–201
  Dynamism of a Cyclist 278
Bodleian Library (Oxford) 114
Boer war 102
Boeuf sur le Toit restaurant (Paris) 49
Bohr, Niels 63
Boltanski, Luc 81
Boltzmann, Ludwig 27, 31–2, 36
Bon Marché 279
Bonifas, Charles 231
Borchard, Klaus 191
Botar, Oliver A.I. 10, 16, 311, 318, 321
Bowler, Peter J. 63, 101, 132, 138
  Science For All 62
Boy Scouts 311, 312
Boyle, Robert 140, 141
Bradford Girls Grammar School 300
Bragg, Robert 136
Bragg, Sir William 118, 119, 136
Bramwell, Anna 322
Brancusi, Constantin 220
Braque, Georges 220
Brecht, Bertolt, *Kuhle Wampe, or: To Whom
  Belongs the World?* 281
Briggs, Asa 252
Brill, Alexander 188, 191, 201n1
Bristol Naturalist Society 300
British Association for the Advancement of
  Science (BAAS) 107, 113, 118
  Liverpool meeting (1923) 245–7
British Empire Exhibition (Wembley, 1924)
  110–11, 112
British Expeditionary Force (BEF) 254
British Medical Journal (BMJ) 107
British Museum 192
British Society for Literature and Science 4
British Trust for Ornithology 303
Brittain, Vera 387
Bronk, Detlev 266, 267
Bronowski, Jacob 19n29
Brooklands (Surrey) 253
The Brothers Karamazov (film, 1931) 53
Brunn, Heinrich 191, 192
Bryan, George 253
Buchelin, Léo 231

Hall, G. Stanley 328
    Youth: Its Education, Regimen and Hygiene
        328
Hall, Lesley A. 314
Hall, Marie Boas 131, 139–40, 143
Hall, Rupert 131, 139–40, 143
    The Scientific Revolution 1500–1800: The
        Formation of the Modern Scientific
        Attitude 10–11, 141
Haraway, Donna 10
Hargrave, John 312, 330, 334n26, 334n28,
        334n36
    being modern battle-cry 327–8
    and evolution 325
    founds Kindred of the Kibbo Kift 312
    interest in the occult 331, 335–6n75
    and knowledge 317–18
    natural eugenics 313–14
    The Confession of Kibbo Kift 317–18
    The Great War Brings it Home: The Natural
        Reconstruction of an Unnatural Existence
        313
    'A Short Exposition of the Philosophic Basis
        of the Kibbo Kift' 321–3
Harrington, Anne 340–1
Harrison, Beatrice 306
Harrisson, Tom 297, 301
Harrod, Tanya 6
Harvard University 82, 114, 226
Hawkins, Desmond 301, 308n33
Heard, Gerald 263–4, 387
Hecht, N., 'Aerials' 255
Heinroth, Oscar, *Feathered Mastersingers*
        295
Heisenberg, Werner 252
Hellerau 210, 214, 228
Helmholtz, Hermann 33
Henry, Charles 216, 217, 219
    'Lumière, couleur et forme' 217
    Quelques aperçus sur l'esthétique des formes
        217
Hepworth, Barbara 389
Herbart, Johann Friedrich 192
Herder, Johann Gottfried, *Outline of a*
        Philosophy of the History of Humanity
        362
Hessen, Boris 136, 137, 138, 139
    'The Social and Economic Roots of
        Newton's "Principia"' 137
Hewlett, Maurice 333n19
Hill, A.V. 266–7
Himmler, Heinrich 369
history of science 130, 131–2
    avoidance of immediate past 140–2
    Cambridge University committee
        139–40
    and culture, society and general polity
        141–2
    debates on approaches to 140–1
    and development of modern science 141
    historians vs scientists 139
    International Congress (1931) 136–9
    modern practice of 143
    Royal Institution treatment of 135–6
Hitler, Adolf 369
HMS *Vernon* 255

Hodler, Ferdinand 229, 230, 232
    Night 230
Hoerbiger, Hanns 369
Hofer, Veronica 23
Hogben, Lancelot 139
Holländer, Friedrich 49, 50
Hollingworth, J. 256, 257
Holmyard, Eric 139
Honegger, Arthur 49
Horace 194
Horder, Lord Tommy 42
Hornsey (London) 108
Hornsey town hall 12
Horton, Frank 253
Hoxie, M.C. 257
Hughes, Robert 1
Huntley, John, *British Film Music* 54
Huxley, Aldous 12, 118, 128, 131, 138, 387
    Brave New World 119, 120
Huxley, Julian 10, 15, 118, 120, 139, 264,
        293, 295, 297, 298, 302–3, 305, 306,
        308n28, 314, 315, 387
    The Modern Churchman 132–3
Huxley, Thomas 120

ICI 387
Ideal Home Exhibition 102
Idealism 231
Ilford Science Literary Circle 148, 159
Imperial Chemical Industries (ICI) 110
Imperial College for Science and Technology
        (London) 103, 119, 142, 253
Imperial General Staff 105
Imperial Wireless Committee 253
Impressionists 197
industrialisation 390–1
Inge, William, Dean of St Pauls 12, 116, 119,
        128n90, 131, 132–5, 138, 141
    'The Future of the Human Race' Discourse
        133–4
    Plotinus lecture course 133
    'Theocracy' lecture series 133
Inklings (literary discussion group) 117
Innes, Hammond 110
The Insect Men 158
Institut für Radiumforschung (Vienna) 251
Institut Henri Poincaré (Paris) 188, 196
Institut Jacques-Dalcroze 222
Institute of Contemporary Arts (ICA) 387
Institute of Inventors 122n6
Institute of Mechanical Engineers (IME) 265
Institute of Plant Physiology (Prague) 366
Instituto Da Guarda (A Coruña) 202n13
International Congress for the History of
        Science (1931) 131, 136–8
The International Journal of Ethics 88
Isle of Man 298
Italy 101–2, 124n22

Jablan, Slavik Vlado 193
Jackson, Admiral Sir Henry 255
    'The Educational Value of Wireless' 255
Jacques, Dennis A. 154
Jacques-Dalcroze, Émile 209–10, 210, 224,
        227–9, 232, 233, 243n147
James, Frank A.J.L. 121

Lightning Source UK Ltd.
Milton Keynes UK
UKHW020636301018

331411UK00002B/53/P